THE AGE OF
GLOBAL
WARMING

A HISTORY

THE AGE OF
GLOBAL
WARMING

A HISTORY

RUPERT DARWALL

LONDON   BOSTON

This paperback edition published in the United Kingdom and North America in 2014

Quartet Books Ltd
A member of the Namara Group
27 Goodge Street, London W1T 2LD

Interlink Publishing Group, Inc
46 Crosby Street, Northampton, MA 01060
www.interlinkbooks.com

A record for this book is available from the
British Library and US National Library of Congress

ISBN 978 0 7043 7339 6

Typeset by Josh Bryson

Printed and bound in Great Britain by
T J International Ltd, Padstow, Cornwall

CONTENTS

1
THE IDEA

It is ideas that make history ... Human society is an issue of the mind.

Ludwig von Mises

The orthodoxy produced by intellectual fashions, specialization, and the appeal to authorities is the death of knowledge, and that the growth of knowledge depends entirely upon disagreement.

Karl Popper

Polar bears or swallows?

The two symbolize the dilemma posed by global warming.

A male polar bear's desperate struggle for food was captured in the 2006 BBC television series *Planet Earth*. Eventually the exhausted animal finds a pod of walruses. Three times he attacks. Each time he fails, retreating to lie down and die. The burgeoning number of polar bears has replaced the giant panda as the icon of a species on the verge of extinction and of fragile nature endangered by man.

Swallows crossing the Sahara can lose up to half their bodyweight as they fly from Northern Europe to Southern Africa burning hydrocarbons, one of the highest density sources of energy found in nature. For the same reason, mankind has come to depend on hydrocarbons.

Global warming's entrance into politics can be dated with precision – 1988; the year of the Toronto conference on climate change, Margaret Thatcher's address to the Royal Society, NASA scientist James Hansen's appearance at a congressional committee and the establishment of the Intergovernmental Panel on Climate Change (IPCC).

By then, the world was ready.

Rounds of international climate change conferences and treaties followed – the Rio Earth summit and the United Nations Framework Convention on Climate Change in 1992, the 1997 Kyoto Protocol – culminating in the 2009 Copenhagen climate conference, interspersed with periodic pronouncements from the IPCC with four assessment reports (1990, 1995, 2001 and 2007) sounding a crescendo of alarm.

The global warming idea is composed of three propositions which proceed from two facts. The first fact is that carbon dioxide is one of a number of so-called greenhouse gases (the most abundant being water vapour). The second is that the proportion of carbon dioxide in the atmosphere has risen by around thirty-five per cent from pre-industrial times.[1]

The first proposition is about the past. Increased carbon dioxide, together with other greenhouse gases emitted by human activities, has caused global temperatures to rise.* In the words of the IPCC's Fourth Assessment Report, 'It is likely that there has been significant anthropogenic warming over the past fifty years over each continent (except Antarctica).'[2]

The second is about the future. Left unchecked, rising global temperatures will cause immense damage to the environment and humanity.

The third proposition is political. Developed countries should lead the world in making deep cuts in carbon dioxide emissions, preferably by substituting fossil fuels with renewable energy sources such as wind and solar power.

Why was the world ready for this new idea?

Global warming did not make itself evident like a solar eclipse. People had to be told it was happening. An event of such historical magnitude does not come out of nothing. It can be understood in terms of ideas, which in turn reflect particular values and visions.

The British mathematician and philosopher A.N. Whitehead said that the spiritual precedes the material; philosophy works slowly before mankind suddenly finds it embodied in the world. It builds cathedrals before the workmen have moved a stone.

So it is with global warming.

During the course of the twentieth century, mankind's relationship with nature underwent a revolution. At the beginning of the last century, human intervention in nature was regarded as beneficent and a sign of the progress of civilization. By its end, such interventions were presumed harmful unless it could be demonstrated they were not.

The entry of nature into politics was of very different character on either side of the Atlantic. In the US, it was healthy. Unlike the European versions, it was not defined by opposition to democracy or capitalism.

In Germany and Britain, to the degree that the British variant was influenced by German thinking, it emerged from the same swamp in which Nazi doctrines festered. Accordingly, only the American variant was viable after the Second World War.

The pivotal year was 1962 with the publication of Rachel Carson's *Silent Spring*. Within a decade, the notion that mankind's pillaging of the planet was leading to an environmental catastrophe – risking man's extinction – had taken hold.

> If all man can offer to the decades ahead is the same combination of scientific drive, economic cupidity and national arrogance, then we cannot rate very highly the chances of reaching the year 2000 with our planet still functioning and our humanity securely preserved

---

* According to the IPCC's Fourth Assessment Report, increased carbon dioxide emissions accounted for just over one hundred per cent of the net man-made global warming effect. IPCC, *Climate Change 2007: Synthesis Report Summary for Policymakers* (2007), p. 39.

– words written by Barbara Ward, the other seminal environmentalist of the era, in *Only One Earth*.[3] It had been commissioned for the first major UN conference on the environment at Stockholm in 1972.

Malthusian ideas had enjoyed previous bouts of popularity. The first was at the end of the eighteenth century, when Thomas Malthus predicted disaster as population growth would outstrip food production. In 1865, the foremost economist of the day, William Stanley Jevons, shifted the focus from food to resource depletion. Britain's economic demise was inevitable with the exhaustion of cheap coal. Similar scares afflicted the US before and after the First World War.

What was new in the late 1960s and early 1970s was the pervasiveness of doomsday predictions. 1972 also saw publication of the Club of Rome's *The Limits to Growth* and *A Blueprint for Survival*, signed by leading British scientists of the day, predicting the end of civilization.

At Stockholm, the West's environmental awakening had to contend with another force of the post-war era – the ambition of newly independent Third World nations to industrialize. A Third World boycott was averted by Maurice Strong, the UN conference organizer, and Ward formulating a political compact between First World environmentalism and Third World development aspirations. Under it, economic growth was deemed double-edged. When rich countries got richer, it harmed the environment; when poor countries grew, the environment benefited.

Sustainable development – the link forged by Strong and Ward between environmentalism and the Third World development agenda – was pushed onto the international agenda by the 1980 Brandt Report and the 1987 Brundtland Report. Whatever its accuracy as a description of reality, sustainable development was the political fiction environmentalism needed to buy developing nations' neutrality, a fiction that broke down when the increase in developing countries' greenhouse gas emissions overtook those of the developed world.

Then, as Sherlock Holmes explained to the Scotland Yard detective, there is the curious incident of the dog in the night-time. But the dog did nothing. 'That,' Holmes replied, 'was the curious incident.'

Marx and Engels condemned Malthus and his population theory. In turn, their labor theory of value put no value on pristine nature. The timing of the demise of Marxism as a living ideology meant that global warming never had to contend with opposition from the Left of the political spectrum. It is hard to conceive of the pre-Gorbachev Soviet Union being a party to global environmental treaties on ideological grounds, let alone during a strategic race to bury the West.

Viewed as an ideology, environmentalism took the Marxist concept of the alienation of the working class from the means of production and applied it to the rich man's alienation from nature. In doing so, environmentalism triumphed in developed societies, dominating the mainstream politics of the West where Marxism had entirely failed. 'By losing sight of our relationship with Nature, and its interdependent and holistic characteristics, we have engendered a profoundly dangerous alienation,' the Prince of Wales stated in 2009. Poorer societies were, in many ways, 'infinitely richer in the ways in which they live and organize themselves as communities.'[4]

That's not how those societies viewed their level of economic development. The corollary of environmentalism's success in the West is its limited appeal in poorer countries. Communism, based on a doctrine of scientific materialism, had much greater success for most of the second half of the twentieth century.

Thus the West and the East (or North and South) took different approaches to the climate change negotiations. The West, particularly the Europeans, viewed global warming diplomacy as about persuading other countries to be virtuous, like themselves. The stance of countries such as China and India was driven by their economic interests.

The comparison with Marxism also illuminates an important dimension of environmentalism – its relationship with science. Environmentalism exists in a similar relationship to its scientific base as communism did to the economics of *Das Kapital*. Science and ideology become so deeply entwined that in practice it is difficult to separate the two, the scientist and the environmentalist becoming one and the same person.

In his 2002 presidential address to the American Association for the Advancement of Science, Professor Peter Raven (a botanist) argued that human society had outstripped the limits of global sustainability (a key tenet of environmentalism). 'Simply appropriating as much as possible of the world's goods and processing them as efficiently as possible can never be a recipe for long-term success,' Raven asserted. 'Success' was about finding 'new ways of thinking about our place in the world and the ways in which we relate to natural systems' – a clearly ideological construct.[5]

Similarly the response to adverse evidence is often an ideologue's denial rather than a scientist's questioning of their hypothesis. In 1968, the American zoologist Paul Ehrlich's *The Population Bomb* declared that the battle to feed humanity was over. It was inevitable that hundreds of millions of people would starve to death. Ehrlich specifically predicted that India was doomed.

Yet the falsification of Ehrlich's prediction led him to make only trivial changes to his thesis. Forty years later, Ehrlich was comparing humanity's ability to feed itself to fruit flies living off rotting bananas. 'Our problem is we only have one pile of bananas.'[6] The remarkable technological accomplishments of modern human beings had merely delayed the timetable of doom.

When a Danish statistician (and an environmentalist to boot) decided to debunk the arguments of American economist Julian Simon that received views on the worsening state of the environment were mistaken, he found that, by and large, the evidence supported Simon's view. Would the evidence change minds? As Bjorn Lomborg recounts in *The Skeptical Environmentalist*, to begin with, I was surprised that the only reaction from many environmental groups was the gut reaction of complete denial … but I would have thought as the debate progressed that refusal would give place to reflection on the massive amounts of data I had presented.[7]

Lomborg's optimism was misplaced. The hostility he encountered wasn't because he had changed his mind. It was because he was an apostate.*

Unlike the discovery of the link between tobacco smoking and lung cancer, with carbon dioxide and global warming there was a pre-existing structure of ideas about the incompatibility of industrial society and nature to co-exist in harmony. Environmentalists were predisposed to anticipate harmful consequences from industrial societies' reliance on hydrocarbons. Because global warming is the most powerful argument in the environmentalist armory and because science provides its source code, global warming occasioned a profound and subtle shift in the nature of science.

According to Karl Popper, the twentieth century's leading thinker on the theory of science, the essence of the scientific method is critical argument and genuine attempts to refute theories with empirical tests yielding reproducible data. We cannot be sure of what is truthful. We can know what is false. Truth is therefore approached by discarding what has been proved false.

What hardened into the scientific orthodoxy of global warming does not meet the threshold of being a scientific theory because its predictions are not capable of being tested against nature and therefore refuted by evidence. Thus it would be more accurate to describe global warming as a speculation or conjecture. Instead, global warming must depend on the preponderance of scientific opinion in order to maintain its potency.

Popper's view of scientific theories as being provisional, only valid until they've been refuted, conflicts with the political need to characterize the science of global warming as settled. Science advances through challenge but maintaining the consensus subverts the needs of genuine scientific enquiry. Politicization of climate science and the de-legitimization of debate – risks inherent in the idea of global warming – led to a retreat from the standards that emerged during the Scientific Revolution of the seventeenth century to pre-scientific norms; principally, reliance on consensus, peer review and appeals to authority.

In contrast to the overstatement with respect to the science is the downplaying of the likely costs and the adverse consequences of adopting policies designed to delay the rise in global temperatures. This is the opposite of what can be reliably said about the science of global warming on the one hand and the economics on the other. The state of the climate in one hundred years' time and its impact on

---

* Raven described Lomborg as the latest in a line of 'false prophets and charlatans', concluding his American Association for the Advancement of Science presidential homily with the plea that scientists should learn to respect one another.

the environment are not just uncertain. They are unknowable. The impact on food prices of policies that divert corn into car fuel is not. We can be fairly certain that if more wind farms are built, more birdlife will be destroyed.

Yet the political imperative of global warming means that it is more important for governments to be seen to be doing something than assessing the likely consequences of their policies.

This license leads to the logical outcome of the global warming idea: policies ostensibly designed to deal with global warming bring about the very outcomes as if global warming was left unchecked – rising food prices, instability caused by food shortages, reduced biodiversity, and, with their threat to the international trading system, a poorer and more dangerous world. These are all examples of the Global Warming Policy Paradox – the therapy causing the malady it was designed to avert. The logic of collective action leads to collectively harmful outcomes but, for individuals and interest groups advocating such action, potentially advantageous ones.

Global warming now affects almost every aspect of our lives. Where and how we travel; the amount we pay for power; the way houses are built; the lights we read by; the food we eat (no lamb please, they burp methane), and how much it costs; the destruction of local habitats to make way for wind farms, palm oil plantations and tidal barrages; what children are taught in schools.

It has re-defined what constitutes ethical behaviour. Virtue is being seen to tread with a small carbon footprint – offsets being offered for sale to obtain one – and changes how people behave towards each other. Being kind to the planet appears to involve being less kind to other people. A Canadian study found that people act less altruistically and are more likely to cheat and steal after purchasing green products.[8]

To explore global warming is to journey through the mind of contemporary Western man. Cultivating the habit of thinking what we are doing was much over-rated, Whitehead observed. 'Civilization advances by extending the number of important operations which we can perform without thinking about them,' Whitehead argued. Could it be that in an age when so much mental activity is automated and so much information is available instantaneously that it is harder to reason from first principles, making it more susceptible to global warming?

'Operations of thought,' Whitehead went on, 'are like cavalry charges in a battle – they are strictly limited in number, they require fresh horses, and must only be made at decisive moments.'

It is time to mount the cavalry.

2
PROMETHEAN REVOLUTION

Knowing the power and the action of fire, water, air, stars, heavens and all the other bodies which environ us as distinctly as we know the various trades and crafts of our artisans, we might in the same way be able to put them to all the uses to which they are proper, and thus make ourselves, as it were, masters and possessors of nature.

René Descartes

Accounts of global warming usually start with the observation that carbon dioxide is one of a number of greenhouse gases and that human activities have increased the amount of carbon dioxide in the atmosphere. This isn't the real beginning for it leaves out why modern civilizations evolved to exhale carbon dioxide. In the words of the Nobel economist Ronald Coase, we need to know the value of what is obtained, as well as what is sacrificed to attain it – the benefits as well as the costs.[1]

A better place to start is at the beginning, or rather before the beginning, with the man who foresaw a new beginning when he turned his back on the Middle Ages.

In 1577 or thereabouts, the record isn't precise, the sixteen-year-old Francis Bacon was studying at Trinity College, Cambridge. He had been a favorite of Elizabeth I, answering her questions with a gravity beyond his years. Bacon grew up in an age of vitality and exuberance. According to his Italian biographer Paolo Rossi, Bacon was a medieval philosopher haunted by a modern dream. It was at Cambridge that the young Bacon had the decisive thought of his life, one that pointed to a new era in the history of mankind. In the words of his future private secretary, Bacon fell into the dislike of the philosophy of Aristotle, a pursuit strong only 'for disputations and contentions, but barren of the production of works, for the benefit of the life of man'.[2] That the pursuit of knowledge should be directed at deriving practical benefits for mankind and not the abstract hair-splitting of the medieval monastery became the guiding idea of Bacon's life.

Bacon spat at Aristotle, the pope of medieval philosophers, and Greek philosophy in general. Aristotle was a wretched sophist, his logic a manual of madness. Bacon denounced the 'degenerate learning' of the followers of Aristotle, who idled their lives in pointless disputes over logic and abstract speculation to prove points of no practical benefit. He also accused Plato of doing mankind a mortal injury by turning men's minds away from the observation of nature to 'grovel before our own blind and confused idols under the name of contemplative philosophy'.[3]

Examining the long history of human thought from the ancient Greeks, Bacon asked: Why had mankind's condition improved so little? Men's minds should turn to the observation of nature and to experimentation. By discovering what was written in nature's book, it could be harnessed to bettering man's lot and transforming his existence. 'We cannot command nature except by obeying her,' he wrote.

Bacon saw himself as an experimenter, scientist and inventor. But his scientific contributions were minimal. Even his famous discourses on scientific method had little influence on the development of science. His mind had fastened on to something larger.

> If a man could succeed, not in striking out some particular invention but in kindling a light in nature – a light which should in its very rising touch and illuminate all the border-regions that continue to circle our present knowledge; and so spreading further and further should presently disclose and bring into sight all that is most hidden and secret in the world – that man (I thought) would be benefactor indeed of the human race – the propagator of man's empire over the universe, the champion of liberty, the conqueror and subduer of necessities.[4]

In *The New Atlantis*, Bacon advocated a scientific foundation, Salomon's House, to engage in discovering 'the knowledge of causes, and the secret motions of things; and the enlarging of the bounds of the Human Empire'. In 1660, thirty-four years after Bacon died, the Royal Society was founded; its motto – *Nullius in verba*, take no one's word – an exhortation to healthy scepticism. For Bacon, knowledge about nature was not to be acquired for its own sake, but for the purpose of enabling man to use nature to better his material conditions of life – the invention of invention, as Whitehead put it. *Nam et ipsa scientia potestas est*, Bacon wrote. Knowledge is power.

Karl Marx called Bacon the prophet of the Industrial Revolution, his ideas looking forward to 'an alteration in the form of production and to effective control of nature by man' – as Marx wrote in *Das Kapital*. Karl Popper called Bacon the spiritual father of modern science and the creator of the Industrial and Scientific Revolutions with his idea of 'a *material* self-liberation through knowledge'.[5] It made Western civilization different from all others, past and present. 'European civilization is an industrial civilization … It uses engines, sources of energy which are non-muscular. In this, European and American civilizations differ fundamentally from all other great civilizations, which are or were mainly agrarian and whose industry depended on manual labor.'[6]

Power has a political as well as physical or scientific dimension; power over people as well as power over nature. Plato's republic was ruled by the wise. Scientists were to govern the New Atlantis. In a pre-democratic society where monarchs ruled by divine right, Bacon's political idea was revolutionary, especially from the pen of someone who had been Lord Chancellor of England, the highest political office under the crown.

Popper described Bacon and scientists generally as epistemological optimists. They believed humans can know with certainty the truth about the natural world. 'Once the truth stands revealed before us, it is impossible for us not to recognise it as truth,' is the way Popper characterized epistemological optimism. As Descartes put it, God does not deceive us, making explicit science's debt to religion. According to Whitehead, the scientific endeavour was driven by what he called 'the unexpugnable belief that every detailed occurrence can be correlated in a perfectly definite manner, exemplifying general principles.'[7] This belief, Whitehead thought, came from the prior epoch and 'the medieval insistence on the rationality of God'.

The mental framework science inherited from religion also explains why some people rejected the manifest truth. Such people were, in Popper's words, 'epistemological sinners' who erred because they stubbornly clung to their prejudices, in contrast to those who saw truth when it was set before them. 'It is the unbiased mind, the pure mind, the mind cleansed of prejudice, that cannot fail to recognize the truth.'[8]

These pre-scientific traits – that the truth of a rational, predictable nature could be apprehended by the human mind and that non-believers were not just ignorant but, in some way, morally corrupted – would become leitmotifs in debates on the science of global warming. In November 2009, the Australian prime minister Kevin Rudd denounced climate change deniers. 'They are a minority. They are powerful. And invariably they are driven by vested interests,' Rudd warned. Small in number, they were 'literally holding the world to ransom'.

Popper's own theory of scientific knowledge was provoked by his experience in Vienna after the First World War. The truth of Marxism, of Freudian psychoanalysis and Alfred Adler's psychology explained everything in their respective fields, including why people didn't believe them, because of their class interest or because of unanalyzed repressions crying out for treatment.

A confrontation with a young, uniformed member of the Nazi party armed with a pistol might, Popper thought, have planted the seed for his classic book, *Open Society*: 'What, you want to argue? I don't argue: I shoot.'[9] An open society is one that not merely tolerates dissenting opinions, but respects them: 'Democracy (that is, a form of government devoted to the protection of the open society) cannot flourish if science becomes the exclusive possession of a closed set of specialists.'[10]

By contrast, what Popper termed the 'critical optimists' based their views on the Socratic insight that to err is human. Less numerous, Popper numbered Locke among them. In this respect, the framers of the American constitution followed Locke. Checks and balances and the separation of powers are an implicit repudiation of the assumption that 'the people', or, at any rate, a majority of them, cannot err, a view made explicit in the *Federalist Papers*.

Bacon died in 1626, a decade and a half before the outbreak of the English Civil War, a conflict that was to lead to the development in England of the classical liberal political philosophy and find its most perfect expression in the constitution of the American republic. Thomas Hobbes' *Leviathan* was published in the last year of that war when John Locke, whose development of Hobbes' ideas inspired both Montesquieu and America's Founding Fathers, was not yet twenty years old.

The debate Bacon and Locke could not have, whether society should be governed by scientists or whether it should be based on popular sovereignty mediated by a liberal constitutional order, is one that, some three hundred years after Locke's death, sparked into life in debates on global warming. Writing in September 2009, *New York Times* columnist Thomas Friedman asked whether China's one-party system was better at tackling an issue like global warming because, Friedman argued, it could impose 'the politically difficult but critically important policies needed to move a society forward in the twenty-first century'.

In his 2004 book on globalization, Martin Wolf argues that after a millennium or more of flat-lining the decisive technological shift from human and animal power to inanimate energy, most importantly hydrocarbons, was the key factor in the sudden and extraordinary take-off in economic performance and living standards within the last two and half centuries. 'What is called the industrial revolution is better named the energy revolution.'[11] Its first stirrings were apparent in Bacon's lifetime. Production of pit coal in England rose from an average of two hundred and ten tonnes a year in the decade 1551-60 to nearly two million tonnes in the decade 1681–90, a growth rate of over seven per cent a year.

Why coal? Why fossil fuels? Why did this energy revolution depend on the combustion of hydrocarbons and, as an unavoidable by-product, the emission of carbon dioxide into the atmosphere?

Hydrocarbon molecules are rich in energy. When fully oxidized as carbon dioxide (CO_2) there is no more energy to be extracted from a carbon atom. Carbon dioxide can be re-energized by exposure to sunlight and water in the presence of an enzyme catalyst, a process known as photosynthesis. Plant leaves are able to do this in abundance, producing energy-rich molecules like sugar and cellulose. Cellulose, as wood, is still the principal fuel for many in the developing world. The remains of plants of the carboniferous era (around three hundred million years ago) are the source of fossil carbon which is the result of the geological processes of time, heat, and pressure, increasing their energy density with the loss of their oxygen atoms.

At the start of the Industrial Revolution, it was more convenient to mine coal which, as a solid, could be easily transported and stored. Oil was obtained from plants (olive oil, linseed oil) or from animals like whales. Plants and animals, including humans, convert carbohydrate to hydrocarbons (fatty acids) to store energy efficiently. In using fossil fuels, mankind unlocked a store of energy used by plants and animal and, from the time of the Industrial Revolution, started to apply it on an industrial scale.

By the end of the eighteenth century, Britain was becoming the world's first industrialized economy. The Promethean Revolution was underway.

If Bacon was the prophet of man's material liberation through the advance of science and technology, Malthus was its Jeremiah – prophesying that mankind's future was to be trapped in an agrarian past which the Promethean Revolution was already making history.

If ever there was an inflection point in the economic history of mankind, this was it. It was a spectacularly inapposite moment to be writing a treatise on economic development and population based on the assumption of the static technological endowment of pre-industrial societies when industrialization was taking mankind out of the Malthusian trap.

In his 1798 *Essay on the Principles of Population*, Malthus warned his readers that his view of human life had a melancholy hue. The great question of his age was, Malthus said, whether mankind was accelerating down a path of illimitable and hitherto unforeseen improvement (which it was) or instead whether humanity's destiny was to be condemned to perpetual oscillation between happiness and misery.

Malthus's argument proceeded from two propositions: the first, that food is necessary for the existence of man; the second, that, as this celibate Anglican cleric put it, 'the passion between the sexes is necessary and will remain nearly in its present state'.[12] From these two propositions – as trite as they are true – Malthus launched the core of his case: 'The power of population is indefinitely greater than the power in the earth to produce subsistence for man.' Summarizing the dilemma mathematically – any proposition that can be put into a mathematical form automatically improves its propagation – human population tended to expand at a geometric rate, whereas the means of subsistence expanded arithmetically. The two lines could not diverge for any length of time. The gap could only be closed by the population line moving back to the subsistence line through war, pestilence or famine.

Human progress would, Malthus surmised, be neither rapid nor unlimited. Instead, humanity would be subjected to repeated checks, forcing a reversion of human population growth back to the trend of the subsistence line. To Malthus, this appeared to be decisive proof against the possibility of a society 'all the members of which should live in ease, happiness, and comparative leisure; and feel no anxiety about providing the means of subsistence for themselves and their families'.

As an ordained clergyman, Malthus had to square checks on population growth with the existence of a benevolent God. Misery and vice – the twin mechanisms by which human population growth would be corrected – were natural and therefore sanctioned by God. As Malthus explained, 'the ordeal of virtue is to resist temptation'. The religious dimension of Malthus's theory of population growth was summarized by the nineteenth-century French economist Frédéric Bastiat. 'If you multiply inconsiderately, you cannot avoid the chastisement which awaits you in some form or other, and always in a hideous form – famine, war, pestilence etc.'

Economic history demonstrates conclusively that Malthus was wrong to believe that, without population controls, no human society could completely escape a downward pull towards subsistence. If Malthus's *Essay* demonstrated anything, it is that the accuracy of forecasts of doom is inversely related to their political impact.

The *Essay*'s impact was immense, running to six editions in Malthus's lifetime. It speaks volumes for the potency of the scare Malthus created that in the middle of the Napoleonic wars, fighting France, a country whose population of nearly thirty million was nearly double Britain's (over one third of which was Irish and of doubtful loyalty), a growing population was seized on in Britain as a cause of alarm. Such was its effect in creating 'alarmist' views about the future (the word used by Britain's Office of National Statistics in describing the episode) that in 1800, Parliament passed the Census Act. The first official census took place in 1801, subsequent ones every ten years afterwards except during the Second World War.

The 1801 census found that England and Wales had a population of 8.9 million. At the 1851 census, it had doubled to 17.9 million, rising to 32.5 million in 1901.

Had this nearly four-fold population increase led to greater disease and pestilence and put pressure on living standards?

While infant mortally rates remained almost unchanged during the nineteenth century, recent research suggests that the trend of increased life expectancy from around forty years in Malthus's day to between seventy and eighty years at the end of the twentieth century began around 1850.

Living standards rose as well.

Farming cash wages rose by twenty-eight per cent. Cash wages in industry rose by over fifty per cent. These rises understate the rise in living standards because prices fell and the purchasing power of money more than doubled. Living standards benefited even more from falling prices than rising wages.

None of these trends is consistent with Malthus's prediction of population growth triggering reversions to subsistence. Instead, high population growth was associated with a sustained move from subsistence to becoming a society which Malthus claimed could not possibly exist.

How could the economy support a larger population that nearly quadrupled in one hundred years? Industrial production increased twelve-fold and the output from Britain's coalmines rose tenfold between 1815 and 1901.

In the preface to the *Essay*, Malthus wrote that, even if in theory he could be shown to have been wrong, he would 'gladly retract his present opinions and rejoice in the conviction of his error'.

Although he was to live thirty-four years into the new century, the economic evidence did not lead him to retract and rejoice. Neither did it lead to his views being quietly forgotten. The power of the Malthusian substructure of sin, punishment and redemption overwhelmed the contrary evidence to become a recurring feature of the consequences of man's relationship with the Earth and with nature. Modern man's escape from the Malthusian trap is either illusory or temporary.

Take for example Maurice Strong, the Canadian environmentalist who was secretary of first United Nations (UN) conference on the environment in Stockholm in 1972 and the Rio Earth Summit twenty years later. The first chapter of Strong's 1999 autobiography *Where on Earth are We Going?* is set in 2031. It foretells humanity's fate unless, that is, we are 'very, very lucky' or 'very, very wise'.[13] Nation-states have imploded; the international order completely broken down; there are food shortages, energy shortages, more people perishing from severe weather than in the two world wars of the previous century; a Great Earthquake strikes in 2026; Americans are dying like flies from excessive heat (there was not enough electricity for air conditioners).

A mystic figure by the name of Tadi emerges to synthesize all the main world religions into one. 'In this Time of Troubles God must call all to a new and transcendent unity,' Tadi concluded. There was, however, a presentiment of a New Dawn. The human population was falling to what it had been at the beginning of the twentieth century, 'a consequence, yes, of death and destruction – but in the end a glimmer of hope for the future of our species and its potential for regeneration'.

Sin, punishment, redemption.

Malthus's population theory's most lasting impact was not in economics but in biology. In his autobiography, Charles Darwin wrote about how it helped catalyze his theory of evolution:

> In October 1838, that is, fifteen months after I had begun my systematic inquiry, I happened to read for amusement Malthus on *Population*, and being well prepared to appreciate the struggle for existence which everywhere goes on from long-continued observation of the habits of animals and plants, it at once struck me that under these circumstances favorable variations would tend to be preserved, and unfavorable ones to be destroyed. The results of this would be the formation of a new species. Here, then, I had at last got a theory by which to work.[14]

In contrast to the struggle-for-existence models of Malthus and evolutionary biologists, modern economics has incorporated the greatest finding of his friend and rival, David Ricardo.* 'Though an awareness of the benefits of specialization must go back to the dim mists of antiquity in all civilizations,' according to the *New Palgrave Dictionary of Economics*, 'it was not until Ricardo that this deepest and most beautiful result in all economics was obtained.' Specialization and the increasing division of labor distinguish advanced societies from primitive ones. It was trade that enabled Britain to specialize in manufacturing and coal mining,

---

* Not all economists elevated Ricardo above Malthus. Keynes wrote in the 1930s, 'if only Malthus, instead of Ricardo, had been the parent stem from which nineteenth-century economics proceeded, what a much wiser and richer place the world would be today' – not that Keynes subscribed to fears about running out of resources.

importing food from the Americas and other parts of the world which had a comparative advantage in agriculture.

In debates on the environment and global warming from the late 1960s to our own day, biologists and other natural scientists tend to see economic processes through Malthusian spectacles. Most economists follow Ricardo. Because the Malthusian narrative is about man's relationship with nature, the voices of natural scientists are generally given more weight in these debates.

Nature misleads when transposed to human society. It offers food chains, at the top of which are carnivores where the winner takes all and the loser forfeits their life. Nature also provides examples of symbiotic relationships (the closest to us physically being the flora lining our gut). But these latter relationships hardly compare to the conscious intent inherent in economic bargaining and to the specialization of activities within a single species which exchange both enables and rewards.

There is nothing comparable in nature to Ricardo's elucidation of comparative advantage. Trade depends on arguably man's greatest invention – money. Trade is voluntary; the parties to an exchange only undertake it if each of them believes it will make them better off. Thus trade generates positive sum outcomes.

Natural scientists' thinking about economic issues is also conditioned by the first law of thermodynamics. This states that energy cannot be created or destroyed, only transformed. How can mankind's numbers grow and consumption increase, like an economic perpetual motion machine, without incurring some equivalent loss somewhere else? Economic activity must therefore have a limit because it consumes what it depends upon, so the argument goes. This leads scientists and environmentalists (often they're the same people) to worry about resource depletion and the planet's carrying capacity.

The analogy with physics does not hold because the driver pushing outwards the boundary of economic potential is the expansion of human knowledge. In this respect, the market economy has always been the 'knowledge economy'. Knowledge is not like one of Paul Ehrlich's rotting bananas. As Bacon put it, knowledge is power.

3
ANTECEDENTS

Most of the population-theory teachers are Protestant pastors.

Karl Marx

So far then as our wealth and progress depend upon the superior command of coal we must not only stop – we must go back.

William Stanley Jevons

By the beginning of the second half of the nineteenth century, Malthus's prediction – that it was impossible for any human society to escape subsistence without some form of population control – was no longer tenable. His theory had to be reformulated or discarded.

The critical responses of leading economists of the day prefigure those that accompanied the emergence of environmentalism at the end of the 1960s and the debates on global warming two decades later.

The first response is an ancestor of the global warming party. William Stanley Jevons was born in 1835, the year after Malthus died. The economist Joseph Schumpeter, a tough assessor of reputation – he thought Adam Smith much overrated – lauded Jevons for his 'brilliant conceptions and profound insights'. Jevons was 'without any doubt one of the most genuinely original economists who ever lived', Schumpeter wrote.[1] Praise in economics does not come higher.

'His definitive breakthrough came with the publication of *The Coal Question* in 1865, which predicted a decline in Britain's prosperity due to the future exhaustion of cheaply extractable coal,' Jevons' biographer Harro Maas wrote.[2]

The Coal Question begins by rehearsing Malthus's key argument: although human numbers tended to increase in a uniform ratio, the supply of food cannot be expected to keep up. 'We cannot double the produce of the soil, time after time, *ad infinitum*.'[3] Conceding that innovation would 'from time to time' allow a considerable increase, this would only buy time. 'Exterior nature presents a certain absolute and inexorable limit,' Jevons maintained. Although Malthus's fundamental insight still held, Jevons argued that the growth of manufacturing and free trade 'take us out of the scope of Malthus's doctrine'. But this would not free mankind from resource constraints. The inability to grow enough food was no longer the check on human progress. Now it was coal – 'the Mainspring of Modern Material Civilization', as he called the carbonaceous rock. 'With coal almost any feat is possible or easy; without it we are thrown back into the laborious poverty of early times,' Jevons claimed. How Britain should respond to this challenge was not merely an economic issue; it was a question of 'almost religious importance'.

The transfer of the check on civilization's progress from farm to coal mine had actually worsened humanity's material predicament. 'A farm,' Jevons argued, 'however far pushed, will under proper cultivation continue to yield for ever a constant crop. But in a mine there is no reproduction, and the produce once pushed to the utmost will soon begin to fail and sink to zero.' Jevons thus anticipated both the idea of 'sustainability' that was to emerge in the 1970s and the rationale for renewable energy.

Whereas exporting farm products – 'the surplus yearly interest of the soil,' as Jevons put it – could be unalloyed gain, Jevons argued that to export coal was to be 'spendthrifts of our mineral wealth.' 'Are we wise in allowing the commerce of this country to rise beyond the point at which we can long maintain it?' Jevons asked.

His answer was unequivocal. Britain should understand that any increase in its prosperity and its power in the world was temporary.

> If we lavishly and boldly push forward in the creation and distribution of our riches, it is hard to over-estimate the pitch of beneficial influence to which we may attain in the present. *But the maintenance of such a position is physically impossible. We have to make the momentous choice between brief greatness and longer continued mediocrity.*

The second response provided the most vigorous counter-attack to Malthus. If Malthus was right, Marx and Engels had to be wrong. So they deployed some of their most cutting invective against him. In 1865, the same year as *The Coal Question*, Marx called Malthus's essay a 'libel against the human race'. Twenty years earlier, Engels described Malthus's law of population as 'the most open declaration of war of the bourgeoisie upon the proletariat'. His *Essay* was 'nothing more than a schoolboyish, superficial plagiary', Marx said, ridiculing Malthus's vow of celibacy.

'Where has it been proved that the productivity of the land increases in arithmetical progression?' Engels asked in his 1844 essay *The Myth of Overpopulation*. True, the area of land was limited. Even if it was assumed that additional labour did not always yield a proportionate increase in output, there was, Engels argued a third element, which 'the economists, however, never consider as important' – science. 'What is impossible for science?' Engels asked.

Jevons was emphatic. Science could not free mankind from resource constraints. 'A notion is very prevalent,' Jevons wrote, 'that, in the continuous progress of science some substitute for coal will be found, some source of motive power, as much surpassing steam as steam surpasses animal labor.'[4] He attacked a popular scientific writer of the time for spreading such notions as 'inexcusable.' The potential of electricity was based on 'fallacious notions', comparable to belief in perpetual motion machines.

What about petroleum? While superior in some respects to coal, it was nothing but the essence of coal. Besides, there wasn't very much of it. 'Its natural supply is far more limited and uncertain than that of coal,' its high price already reflected its scarcity. According to Jevons, 'an artificial supply can only be had by the distillation of some kind of coal at considerable cost.' The future, Jevons asserted, lay in the development of the steam engine and the possibility of multiplying by at least threefold its fuel efficiency. 'If there is anything certain in the progress of the arts and sciences it is that this gain will be achieved, and that all competition with the power of coal will then be out of the question,' Jevons wrote.

Perhaps it needed someone with the imagination of H.G. Wells to envisage a world transformed by the internal combustion engine (which was being developed in the 1860s) and the gas turbine (1930s). However, by the 1860s, the dynamo, discovered by Michael Faraday in 1831, was being commercialized. During the 1870s, dynamos were generating electricity cheaply enough to power factories and begin to replace steam on railways and tramways.

Marx and Engels displayed a much deeper grasp of the dynamic power of capitalism than Jevons. In the same year Jevons was making his assertions about the future's dependence on coal and steam, Engels received a letter commenting on the similarity between Darwin's account of plant and animal life and Malthusian theory.

'Nothing discredits modern bourgeois development so much as the fact that it has not yet succeeded in getting beyond the economic forms of the animal world,' Engels replied:

> We start from the premise that the same forces which have created modern bourgeois society – the steam engine, modern machinery, mass colonization, railways, steamships, world trade – these same means of production and exchange will also suffice … to raise the productive powers of each individual so much that he can produce enough for the consumption of two, three, four, five or six individuals.[5]

The third antecedent is Frédéric Bastiat. In Schumpeter's unkind estimation, Bastiat was like the swimmer who enjoys himself in the shallows but drowns when he swims out of his depth. 'I do not hold that Bastiat was a bad theorist,' Schumpeter commented, 'I hold that he was no theorist.'

Theorist or not, when it came to enquiring why Malthus was mistaken, Bastiat asked the right question: Why did Europe no longer suffer from periodic famine? The answer, Bastiat thought, had been provided by another French economist, Jean Baptiste Say. As civilization advances, the means of existence – the living standards people at any given time think are the minimum needed to maintain themselves and their families – diverge from the means of subsistence, the bare minimum needed to keep body and soul together.

> The means of existence, by reason of social progress, have risen far above the means of subsistence. When years of scarcity come, we are thus enabled to give up many enjoyments before encroaching on the first necessities of

life. Not so in such countries as China or Ireland, where men have nothing in the world but a little rice or a few potatoes. When the rice or potato crops fail, they have absolutely no means of purchasing other food.[6]

Malthus's population principle should therefore be amended so that population growth is no longer linked with the means of subsistence, Bastiat argued, but with the means of existence; 'the point where the [population] laws of *multiplication* and *limitation* meet, is removed, and elevated.'[7] People will have as many children as they can afford and maintain a certain standard of living, one that tends to rise over time.

Bastiat's insight is relevant to the debate on global warming a century and a half later. The greater the gap between the means of existence and the means of subsistence, the greater a society's resilience to climatic disaster, however caused. Australia is the world's driest inhabited continent. In the early 1980s, it suffered an intense drought, causing an estimated A$3 billion in losses. If a similar drought hit sub-Saharan Africa, the issue wouldn't have been the scale of economic losses but the extent of the humanitarian disaster. The difference is a function of economic development. In rich countries, people don't die from drought and crop failure.

Bastiat's ameliorist position reflects a different cast of mind from the pessimistic outlook of Malthus and Jevons. Malthus, he thought, had 'fixed his regards too exclusively on the somber side. In my own economical studies and inquiries, I have been so frequently led to the conclusion that *whatever is the work of Providence is good*, that when logic has seemed to force me to a different conclusion, I have been inclined to distrust my logic'.

Bastiat's view of the harmony of class interests was in complete contradiction to Marxism's class warfare analysis of history. Marx and Engels viewed Bastiat, a liberal, bourgeois economist, with even greater disdain than Schumpeter did. But in responding to Malthusian views on environmental limits to population growth and economic activity, they were on the same side of the argument. Writing in 1895, towards the end of his life and of a century that had witnessed the greatest increase in production up to that point in history, Engels remarked:

> I do not understand how anyone can speak today of a completion of the Malthusian theory that *the population presses against the means of subsistence* at a time when corn in London cost twenty shillings a quarter, or half the average price of 1848–70, and when it is generally recognized that *the means of subsistence are pressing against the population* which is not large enough to consume them![8]

Capitalism would collapse because it produced too much too cheaply, Marxists used to argue.

That couldn't be said of the collapse of communism in 1989. For sure, the environmental degradation of the communist regimes of Eastern Europe revealed that communism had been faithful to its founders' Promethean ideology of man's subjugation of nature, just as it required the subjugation of mankind. However, communism did not fail because it had poisoned the Earth, polluted the skies or drained inland seas. Neither did it fail because it had run out of natural resources. As an economic system, it failed because it could not produce.

The functional extinction of Marxism as a living ideology was to have a profound impact on the success of the idea of global warming and its ascendancy in the early 1990s. The decline of Marxism removed one of the two economic antecedents from the nineteenth century that would have opposed environmentalism and alarmism about global warming. From the 1960s onwards, the growth of the environmental movement would expand to occupy the space on the political spectrum vacated by classical Marxism. It left the ameliorists, Bastiat's successors, to fight the battle alone against the depletionists, the descendants of Malthus and Jevons.

Jevons' forecasts, or prophecies as Keynes called them, also tell a story relevant to our day. We cannot definitively verify economic forecasts to justify calls to tackle global warming. Unlike Jevons' contemporaries, we can see whether the fame *The Coal Question* earned him was justified by events.

Jevons simply took a three and a half per cent annual growth rate and extended it for a century. For the first decade, the forecast wasn't too bad with a 3.1 per cent average rate of growth a year. But by 1881, the divergence was unmistakable, with actual output of coal nearly twelve per cent less than Jevons had reckoned. The divergence kept growing and coal output peaked and started to decline in the second decade of the twentieth century.

Overall, Jevons calculated that Britain would need to produce one hundred and two billion tonnes of coal in the period 1861–1970 with annual production in 1961 projected to be 2.2 billion tonnes. It was this colossal number, which Jevons argued was beyond Britain's physical resources, that led him to conclude that Britain's 'present happy progressive condition' was of limited duration. In reality, total coal production in the hundred years to 1965 was a shade under two billion tonnes, less than the *annual* amount Jevons had forecast for 1965. The forecast by one of the most brilliant economists of his or any age was out by a factor greater than fifty. By 2007, British coal consumption had fallen to around sixty-three million tonnes, of which some twenty million was produced domestically – less than one per cent of what Jevons had projected for the last year of his series.

Even more spectacular than Jevons' over-estimation of the importance of coal was his dismissal of petroleum. Here is the Jevons coal curve again, this time with the rising curve of petroleum imports.

Keynes said that Jevons wrote *The Coal Question* to shock and establish his public reputation. The book was a bestseller. It led to Jevons meeting Gladstone who told Jevons that his book was masterly. John Stuart Mill argued that because Britain's prosperity was limited, the National Debt should be paid down – a proposition that Keynes argued should have been dismissed with only 'a little reflection'. If demand for coal was to increase indefinitely at a geometric rate, future national income would be so much greater than present national income that the dead-weight represented by the National Debt would become of little account.[9]

How did Jevons get it so wrong?

One of his economist contemporaries said that not only was Jevons simply a genius, he was also a brilliant logician. Jevons believed that forecasting the future was a matter of logic. He once wrote in his journal of waking one sunny morning with the sure sensation that logic had disclosed the future to him, only for it to slip his grasp – epistemological optimism taken to a delusory extreme. Then there's the role of character, as explained by Keynes: 'There is not much in Jevons' scare which can survive cool criticism. His conclusions were influenced, I suspect, by a psychological trait, unusually strong in him, which many other people share, a certain hoarding instinct, a readiness to be alarmed by the idea of the exhaustion of resources.'

Out-turn vs. Jevons forecast – coal

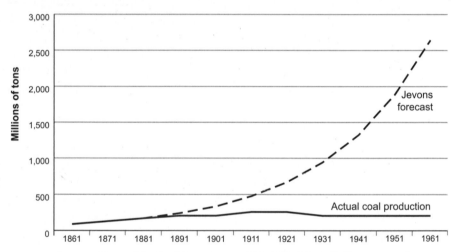

Sources: Jevons, The Coal Question, p. 213; B.R. Mitchell, International Historical Statistics: Europe 1750–2000, Tables D2, D12, H1, J1
Note: 1921 average of 1920 and 1922

Keynes went on to relate that he'd been told by Jevons' son, an economist of some note, about how his father

> held similar ideas as to the approaching scarcity of paper as a result of the vastness of demand in relation to the supplies of suitable material … He acted on his fears and laid in such large stores not only of writing paper, but also thin brown packing paper, that even today, more than fifty years after his death, his children have not used up the stock he left behind of the latter; though his purchases seem to have been more in the nature of speculation

than for his personal use, since his own notes were mostly written on the backs of old envelopes and odd scraps of paper, of which the proper place was the waste-paper basket.[10]

The rationality of Jevons' response to his fear of resource depletion provides a contrast with the manner of his end. Ignoring the advice of his doctor that he avoid swimming, on holiday one August, Jevons drowned at sea.

Out-turn vs. Jevons forecast – oil

Sources: Jevons, p. 213; BR Mitchell, Tables D2, D12, H1, J1

In December 2007, NASA climate scientist James Hansen wrote to the British Prime Minister Gordon Brown, sending a copy of his letter to the Queen. Britain, along with the United States and Germany, had contributed more carbon dioxide emissions per capita than any other country, Hansen claimed. Accompanying the letter was a short analysis of 'basic fossil fuel facts'. There was a graphic totalling carbon dioxide emissions in the period 1751–2006. Coal was the single largest culprit. 'Fully half of the excess CO_2 in the air today (from fossil fuels), relative to pre-industrial times, is from coal,' the analysis said.

Is this a bad thing? As the economist Ronald Coase reminded us, to answer the question, we need to know the value of what is obtained as well as what is sacrificed in obtaining it.

Suppose that at the beginning of the nineteenth century, as the Industrial Revolution was gaining momentum, a climate scientist had found a link between carbon dioxide emissions and global temperatures. And suppose the politicians of the time, invoking the precautionary principle that society should not do things that might involve unknown and unquantifiable risk, had followed the advice of today's leading proponents of the global warming consensus. As a result, fossil fuel extraction would have been capped when Britain, the world's leading coal producer, was mining less than twenty million tonnes of coal a year.

Posing the counter-factual provides a reality check. If today we implemented deep emission cuts, we cannot be sure how different the future would be in terms of economic development or how the climate of the future might be. But we can be fully confident that if the combustion of coal and other hydrocarbons had been severely restricted from the start of the nineteenth century, the economic take-off of the Industrial Revolution would not have happened: we would all be a lot poorer, our lives would be shorter and most of us would be earning our living working in the fields.

We can also see that the benefits of the Industrial Revolution were not out-weighed by the costs of any resultant change in the climate, insofar as such changes can be attributed to industrialization. The Maldives, or any other in-habited islands, did not sink beneath the oceans – if Darwin was right about the formation of coral atolls, they would not have anyway.* Neither is Bengal inun-dated. There hasn't been a mass extinction of species due to climate change. It is quite possible that such changes in the climate as a result of industrialization might, in fact, have been benign. Would a colder world have made us better off – with no coal, no electricity, no gas-fired central heating?

Around the time that Marx, Engels, Bastiat and Jevons were debating Mal-thus, a scientific breakthrough was occurring with the first experimental demon-stration of the warming effect of carbon dioxide. Before it could happen, scientists needed to have identified and isolated carbon dioxide. And to do that, they had to discard one of chemistry's most cherished theories.

It was in the 1750s, during the early stirrings of the Industrial Revolution, that Scottish chemist Joseph Black demonstrated what he called 'fixed air' had funda-mentally different properties from ordinary atmospheric air. In the last three dec-ades of the century, Joseph Priestley prepared and differentiated some twenty new 'airs.' At this point, scientific understanding encountered a block in the form of

---

*   In 1836, during the last year of his voyage on HMS Beagle, Darwin hypothesized that coral atolls had been formed by subsidence of the ocean bed of "extreme slowness." Witnessing the unrelenting power of the Indian Ocean, Darwin reflected that an island built of the hardest rock would ultimately be demolished by such irresistible forces. "Yet these low, insignificant coral islets stand and are victorious: for here another power, as antagonist to the former, takes part in the contest. The organic forces separate the atoms of carbonate of lime one by one from the foaming breakers, and unite them into a symmetrical structure. Let the hurricane tear up its thousand huge fragments; yet what will this tell against the accumulated labour of myriads of architects at work night and day, month after month." Charles Darwin, *Voyage of the Beagle* (first published 1839; Penguin 1989), p. 346 & 338.

the phlogiston theory of combustion. This explained chemical changes caused by combustion in terms of a substance, phlogiston, being lost into the atmosphere.

The invention of phlogiston wouldn't be the last time that scientists invented something to explain something else. In the nineteenth century, scientists invented ether because their materialistic assumptions required something through which light could undulate and electromagnetic occurrences happen. 'If you do not happen to hold the metaphysical theory which makes you postulate such an ether,' Whitehead wrote, 'you can discard it. For it has no independent vitality.'

The overthrow of the phlogiston theory is a *locus classicus* in Thomas Kuhn's *The Structure of Scientific Revolutions* which describes how a scientific paradigm defines scientists' field of study, then experiences a crisis to be supplanted by a new paradigm in a process similar to a political revolution. Kuhn challenged the idea that scientific knowledge proceeded through cumulative breakthroughs. The depreciation of historical fact and context was deeply ingrained in the ideology of the scientific profession, Kuhn wrote. Science still needed heroes, so it revised or forgot their works that did not fit the present. The outcome was 'a persistent tendency to make the history of science look linear and cumulative'.[11] History, Kuhn argued, did not bear out this linear view of scientific progress: 'Cumulative acquisition of unanticipated novelties proves to be an almost non-existent exception to the rule of scientific development.'[12]

The phlogiston theory broke down because it demanded greater and greater contortions to explain how the escape of a substance was consistent with the increased weight of what was left behind: phlogiston was incorporeal; it was the lightest known substance; it had negative weight. This enabled the Frenchman Antoine-Laurent Lavoisier, who had established his oxygen theory of combustion seven years earlier, to administer the *coup de grâce*. In an extraordinary passage anticipating Popper's theory of science, Lavoisier wrote in 1785:

> Chemists have made phlogiston a vague principle, which is not strictly defined and which consequently fits all the explanations demanded of it. Sometimes it has weight, sometimes it has not; sometimes it is free fire, sometimes it is fire combined with an earth; sometimes it passes through the pores of vessels, sometimes they are impenetrable to it. It explains at once causticity and non-causticity, transparency and opacity, colour and the absence of colours. It is a veritable Proteus that changes its form every instant![13]

In 1800 Frederick William Herschel, a German astronomer settled in England, found that sunlight passing through a prism produced heat just beyond the red end of the visible spectrum. Herschel had stumbled upon infrared radiation. In the 1820s, Jean-Baptiste Fourier took Herschel's discovery as the basis of his speculation that the Earth's temperature could be 'augmented by the interposition of the atmosphere', because, as Fourier hypothesized, 'heat in the state of light find less resistance in penetrating the air, than in re-passing into the air when converted into non-luminous heat'.[14] Had Fourier 'discovered' the greenhouse effect? The self-styled Newton of heat wasn't sure. It was difficult to know how far the atmosphere influenced the average temperature of the globe. Fourier, a formidable mathematician, despaired of solving this problem: 'We are no longer guided by a regular mathematical theory.'

Whether or not Fourier could be said to have discovered the greenhouse effect, it had not been demonstrated experimentally. In 1859, a few months before Charles Darwin published *The Origin of the Species*, the Irish scientist John Tyndall discovered that different gases absorb radiant heat of different qualities in different degrees. Provoked by an interest in glaciers and Alpinism, Tyndall set up a series of experiments to, as he said, put these questions to nature. After seven weeks of intense experimentation in his basement laboratory at the Royal Institution, Tyndall declared 'the subject is completely in my hands'.

Three weeks later at a lecture attended by Prince Albert, Tyndall demonstrated his findings and concluded:

> The bearing of this experiment upon the action of planetary atmospheres is obvious … the atmosphere admits of the entrance of the solar heat, but checks its exit; and the result is a tendency to accumulate heat at the surface of the planet.[15]

Changes in the composition of the atmosphere, Tyndall wrote two years later, might have produced 'all the mutations of climate which the researches of geologists reveal'.

Towards the end of the century, the Swedish scientist Svante Arrhenius was also attracted to the idea that changes in the composition of the atmosphere could explain what caused successive glacial cycles. After a colleague pointed out to him that industrial processes were releasing carbon dioxide into the atmosphere and would gradually alter its composition, Arrhenius did some calculations to quantify the effect. In 1896 Arrhenius produced a paper estimating that a doubling of carbon dioxide in the atmosphere would increase temperatures by 5-6°C.

As an explanation of glacial cycles, Arrhenius' paper did not attract much scientific support. In the 1920s, the Serb mathematician Milutin Milanković proposed a competing theory relating glacial cycles to changes in the Earth's orbit, which scientists found more plausible. By the end of the twentieth century, the Milanković cycles had fallen out of favor. Scientists now favored theories that explained climate change in terms of changes in the atmosphere.

A new paradigm held scientists in its grip.

4
FIRST STIRRINGS

Was it for this
That one, the fairest of all rivers, loved
To blend his murmurs with my nurse's song,
And, from his alder shades and rocky falls,
And from his fords and shallows, sent a voice
That flowed along my dreams? For this, didst thou,
O Derwent! winding among grassy holms
Where I was looking on, a babe in arms,
Make ceaseless music that composed my thoughts
To more than infant softness, giving me
Amid the fretful dwellings of mankind
A foretaste, a dim earnest, of the calm
That Nature breathes among the hills and groves.

William Wordsworth

At the beginning of the twentieth century, the rudiments of a mechanism of man-made global warming had been documented. And as if to confirm it, temperatures rose through the first decades of the new century. According to a 1922 newspaper report, the Arctic seemed to be warming up. Fishermen, seal hunters and explorers sailing the seas around Spitsbergen and the eastern Arctic pointed to 'a radical change in climatic conditions, and hitherto unheard-of high temperatures ... Ice conditions were exceptional. In fact, so little ice has never before been noted'.[1]

The 1930s was a hot decade in North America. 1934 would vie with 1998 as the warmest year of the twentieth century (NASA's Goddard Institute, one of the custodians of the global temperature records, would have difficulty deciding which was the hotter). Six of the fifteen hottest Julys since 1895 were in that decade – each year from 1930 to 1936 except 1932. The 1936 heat wave, which killed five thousand people, set temperature records, some of which have not been exceeded. Dust bowl storms swept the American prairie, caused by severe drought and soil erosion, described by a NASA climate scientist as 'the major climatic event in the nation's history'.

The Central England temperature series, the longest reliable temperature series in the world, shows temperatures rising from the middle of the 1890s, a partial decline in the first decade of the twentieth century, followed by a rising plateau over the next three decades. Data from Nordic countries also conforms to a broad pattern of rising temperatures to peak around 1940.

Might this have been caused by industrialization?

One man thought so. Guy Stewart Callendar was born in 1898, two years after Arrhenius's 1896 paper. The son of a distinguished Cambridge physicist, Callendar was a talented scientist specializing in the thermodynamics of steam and kept copious notes on temperature and climate records. In 1938, Callendar published a paper on *The Artificial Production of Carbon Dioxide and Its Influence on Temperature.*

Temperatures, as measured by two hundred weather stations worldwide, had risen and the five years 1934–38 were easily the warmest such period at several stations whose records started up to one hundred and eighty years earlier. Humans had added one hundred and fifty billion tonnes of carbon dioxide to the atmosphere since the end of the nineteenth century, around three quarters of which was still in the atmosphere. Callendar estimated that the higher level of carbon dioxide was a major factor explaining the temperature rise, accounting for two thirds of the warming trend of 0.005 degrees centigrade a year experienced in the first half of the twentieth century.

Global warming was first known in scientific circles as the Callendar Effect. If the 1930s' Dust Bowl had happened fifty or sixty years later, it doesn't require much imagination to guess what it would have been blamed on.* Why did it take half a century for the Callendar Effect to be transformed from a scientific curiosity into a planet-threatening crisis?

In the 1930s, there were more pressing issues – the Depression, recovery from the First World War and the threat of a Second, the rise of fascism and communism and the retreat of the nineteenth-century liberal order.

Something else had to change.

Attitudes towards man's interference in nature underwent a profound transformation over the course of the twentieth century. Nowadays, it is (pre-)judged as bad, manifested in opposition to genetically modified food crops; foregoing the opportunity to wipe-out malarial mosquitoes in Africa with the use of DDT and the banning of polychlorinated biphenyl (PCB), a non-toxic chemical used in a wide variety of applications believed by some to be a cancer-causing toxin. At the beginning of the twentieth century, science was allied with industrial progress in benefiting mankind.

This transformation is reflected in changed attitudes towards global warming. Arrhenius thought that burning fossil fuels would accelerate a virtuous cycle in preventing a rapid return to the conditions of the ice age, removing the need for a forced migration from temperate countries to Africa. It might even inaugurate a new carboniferous era of enormous plant growth. Callendar shared this belief. Carbon dioxide emissions would extend agriculture northwards, stimulate plant growth and, quite possibly, indefinitely delay return of the 'deadly glaciers.' Callendar was to have direct experience of the benefits of intervening in nature. During the Second World War, Callendar designed the FIDO fog dispersal system for the Royal Air Force. 'There is no need,' Churchill said, 'to fight the enemy and the weather at the same time.'

---

*   The 1930s drought is now attributed to a cooler Pacific and warmer Atlantic diverting and weakening the jet stream blowing from the Atlantic away from the Midwest (http://www.nasa.gov/centers/goddard/news/topstory/2004/0319dustbowl.html).

America was first to experience the impact of concern about nature on politics. There, it reflected the cultural values of its leaders, especially Theodore Roosevelt.

After the First World War, Britain imported environmental ideas from Germany, where they were grafted on to an indigenous strain of urban nostalgia for the countryside. Far from America being an inspiration, British environmentalists saw America's commercial culture as embodying the features they most detested in their own. For many of them, Germany showed the path back to nature.

> It was still the Wild West in those days, the Far West, the West of Owen Wister's stories and Frederic Remington's drawings, the West of the Indian and the buffalo-hunter, the soldier and the cow-puncher. That land of the West has gone now, 'gone, gone with lost Atlantis', gone to the isle of ghosts and of strange dead memories. It was a land of vast silent spaces, of lonely rivers, and of plains where the wild game stared at the passing horseman.[2]

Theodore Roosevelt's need for nature never left him. As president, Roosevelt would often take his companions on lengthy point-to-point walks, walking in a dead straight line.

> On several occasions we thus swam Rock Creek in the early spring when the ice was floating thick upon it. If we swam the Potomac, we usually took off our clothes. I remember one such occasion when the French ambassador, Jusserand … was along, and just as we were about to get in to swim, somebody said, 'Mr Ambassador, Mr Ambassador, you haven't taken off your gloves,' to which he promptly responded, 'I think I will leave them on; we might meet ladies.'[3]

It is hard to imagine the Marquis of Salisbury, the inhabitant of Downing Street when Roosevelt took office, leading such an expedition. Salisbury took no exercise other than riding a tricycle around his estate at Hatfield House and grew immensely fat.

The vigor espoused by Roosevelt contrasts with the solitary contemplation of Henry David Thoreau, who spent two years living on the edge of Walden Pond in the 1840s. Thoreau's was a radical rejection of civilization. He pitied the young men who inherited farms and houses and cattle. Better to have been born in open pasture and suckled by wolves, so they might see their fields made them 'serfs of the soil.' To Roosevelt, it was 'right and necessary' that the way of life of the Wild West should pass – 'the safety of our country lies in its being made the country of the small home-maker.' Both men shared what Ralph Waldo Emerson called an original relation to the Universe. 'I am no worshipper of Hygeia,' the Greek goddess of health, Thoreau wrote, 'but rather of Hebe … who had the power of restoring gods and men to the vigor of youth,' a sentiment that accords with Roosevelt's philosophy of the strenuous life.

As a child, Roosevelt seemed destined to become a naturalist. He assembled a collection of rare birds and mammals that he later donated to the Smithsonian. He had an astonishing ear for birdsong. On a walk in England's Itchen Valley with Lord Grey, the British foreign secretary, having heard a bird sing, all he needed to know was its name, and it was unnecessary to tell him more. 'He knew the kind of bird it was, its habit and appearance. He just wanted to complete his knowledge by hearing the song,' Lord Grey recalled.

'More was accomplished for the protection of wild life in the United States than during all the previous years, excepting only the creation of the Yellowstone National Park,' Roosevelt wrote of his time in the White House.[4] He acted, he said, 'to preserve from destruction beautiful and wonderful wild creatures whose existence was threatened by greed and wantonness', overseeing the creation of five national parks, four big game refuges, fifty-one bird reservations and the enactment of laws to protect the wildlife of Alaska.

Roosevelt also subscribed to the depletionist scare first popularized by Jevons the previous century. 'The idea still obtained,' Roosevelt later wrote, 'that our natural resources were inexhaustible.' He invited state governors and the presidents of societies concerned with natural resources – the environmental NGOs of the day – to a three day conference held in May 1908 at the White House. 'It is doubtful whether, except in time of war, any new idea of like importance has ever been presented to a Nation and accepted by it with such effectiveness and rapidity,' Roosevelt wrote.

The conference established the National Conservation Commission, led by Gifford Pinchot, one of the leading conservationists of the era. Published six months later, the commission's report estimated that America's natural gas fields would run out within twenty-five years and its oilfields by the middle of the century. Coal was less of a worry. Reserves appeared adequate to the middle of the twenty-first century. When it was sent to Congress in January 1909, the report was described as 'one of the most fundamentally important documents ever laid before the American people'.[5]

With less than three months of his presidential term left, Roosevelt decided to convene a North American Conservation Conference. It met in February (some things happened faster at the beginning of the twentieth century than at its end). The conference decided that the issue needed to be elevated from a continental to a global level – 'all nations should be invited to join together in conference on the subject of world resources, and their inventory, conservation and wise utilization'. Before February was out, the US government had sent invitations to forty-five governments. On March 4th, Roosevelt's term was over and the project lapsed. It would take another sixty-three years before the first global conference on the environment in Stockholm in 1972 and not, as Roosevelt had planned, in 1909 at The Hague.

Theodore Roosevelt's leadership had brought nature into American politics, uniting the nascent environmental movement already prone to division. Conservationists, led by Gifford Pinchot, were depletionists following in Jevons' footsteps and believed in taking a utilitarian approach to the use of natural resources. Preservationists such as John Muir, the first president of the Sierra Club, founded in 1892, rejected the Conservationists' utilitarianism. For Muir, as for Emerson and Thoreau, nature was a transcendental reality. Nature was needed for men's souls, not for meeting their material needs. Roosevelt vaulted over Muir's breach with Pinchot. Pinchot had advocated allowing sheep – hoofed locusts, Muir called them – to graze in forest reserves. Roosevelt spent three days with Muir in Yosemite and, acting at Muir's prompting, expanded the Yosemite National Park.

The First World War heightened American concerns about energy supplies, a 'case of national jitters' after the narrow escape from acute energy scarcity in 1917–18. The 1917 Lever Act put the coal industry under government control, setting prices, allocating supply and requisitioning firms not acting in the public interest, all of which helped trigger America's first energy crisis – lightless nights, heatless Mondays, shutdowns of factories producing non-essential items, and fuel riots.

Those jitters led to the creation of the United States Coal Commission in 1922. Its three-thousand-page report, assembled by a five-hundred-strong team, sounded the alarm of impending scarcity – the output of natural gas had begun to wane and the production of oil could not long maintain its present rate. As for coal, rather than analyze the impact of government regulation in restricting supply (the geology of the United States hadn't changed in the fourteen years since the findings of Roosevelt's National Conservation Commission), it argued that output of coal would struggle to meet demand.

Appointed by President Harding, the commission's report was received by his successor, Calvin Coolidge. The new president responded to the commission's report in his first annual message to Congress in December 1923. Acknowledging that the price of coal was 'unbearably high', Coolidge gave the barest of nods to its depletionist case and made plain he was not going to intervene in the coal industry. 'The Federal Government probably has no peacetime authority to regulate wages, prices, or profits in coal at the mines or among dealers, but by ascertaining and publishing facts it can exercise great influence,' Coolidge declared. The real threat to coal supplies was not from imminent resource depletion. It was the coal industry's poor labor relations. On this front, Coolidge promised 'uncompromising action', urging passage of legislation to settle labor dispute as 'exceedingly urgent'.[6]

Coolidge's stand stemmed the depletionist tide for much of the Roaring Twenties. Although Pinchot contemplated running against Coolidge for the Republican nomination in the 1924 election, he decided not to. The Depression in the 1930s reversed the tide. The assumptions underpinning much of Franklin Roosevelt's Depression-era economic policies ran in the opposite direction. High prices and the dangers of scarcity gave way to low prices and glut, which were blamed for worsening the Depression.

In Europe, calls for a return to nature between the two world wars were a reaction to modernity rather than an articulation of modern man's need for it – with one major exception. Often they embodied a call to return to the ways of the past; sometimes they were linked to questions of national identity and, on occasion, more than tinged with blood-and-soil nationalism.

The few voices from the political Left tended to be outside the mainstream. The Marxist belief in scientific socialism and the coming dictatorship of the proletariat made communists such as Lenin ideologically hostile to attempts to battle the laws of history. The British Labour party, in the words of Ernest Bevin, emerging out of the bowels of the trade union movement, was hardly in the business of turning the working class into peasants. Elsewhere in Europe, peasant-oriented movements and parties were not about recovering man's relationship with nature, but pressing claims for land redistribution from landlord to tenant.

In its attitude to nature, National Socialism, together with its British offshoot, was profoundly different from Italian fascism and similar movements elsewhere in Europe, which placed culture and nature as conflicting entities, elevating culture above nature. This led to tension between Italy and Germany when Mussolini embarked on a program of cutting down trees in the South Tyrol. According to Anna Bramwell in her courageous book, *Ecology in the 20th Century*,

> European fascism, where it had a program, emphasized forward-looking technological planning and urban development … Germany was the exception, with a tradition, in practice as well as in theory, of looking to nature for philosophical guidance.[7]

That tradition pre-dated Hitler's rise and was appropriated rather than created by Nazi ideologists. Indeed, the word ecology, *Oekologie*, was invented in 1866 by Ernst Haeckel as the science of relations between organisms and their environment.

In the 1920s and 1930s, German ideas permeated Britain. They added to an array of ideas championed by prominent literary figures of the age: D.H. Lawrence and the distinctive German brand of serious nature and sun worship and healthy exercise; the Catholic socialism of Hilaire Belloc and G.K. Chesterton (advocates of giving every worker three acres and a cow); T.S. Eliot's revival of the Christian society and medieval parishes. These disparate voices shared a common loathing of (sub)urban bourgeois culture and hostility to liberal industrial capitalism.

In their criticisms of capitalism and its exploitation of nature, there was little to distinguish Left from Right. Just six months before the outbreak of the Second World War, T.S. Eliot wrote about 'the evil in particular institutions at particular times'. These included 'problems such as the hypertrophy of the motive of Profit into a social ideal, the distinction between the use of natural resources and their exploitation, the use of labor and its exploitation, the advantages unfairly accruing to the trader in contrast to the primary producer, the misdirection of the financial machine, the iniquity of usury, and other features of a commercialized society'.[8]

These ideas were seeded in the fertile political soil of nostalgic patriotism that located the essence of England in the English countryside, a country where the desire of people to have their own garden is visible to anyone flying over London into Heathrow airport. 'The sounds of England, the tinkle of hammer on anvil in the country smithy, the corncrake on a dewy morning, the sound of the scythe against the whetstone, and the sight of a plow team coming over the brow of a hill, the sight that has been in England since England was a land,' words spoken by Stanley Baldwin in a nostalgia-laden St George's Day address in 1924.

British politicians began to respond to the consequences of the rapid expansion of England's cities. Electric trains had extended the reach of the suburbs and, in the interwar years, the motorcar was further extending it. In 1935, Baldwin's government passed the Restriction of Ribbon Development Act, designed to limit unplanned suburban spread. Protecting the countryside from expanding cities was an outgrowth of nineteenth-century agitation for social reform of curing urban problems through town planning. Ebenezer Howard proposed the garden city as a peaceful path to reforming industrial society. 'The key to the problem,' Howard wrote in 1898, 'is how to restore the people to the land, that beautiful land of ours, with its canopy of sky, the air that blows upon it, the sun that warms it, the rain and dew that moisten it, the very embodiment of Divine love for man.'

In 1926, Sir Patrick Abercrombie, one of the leading town planners of the age, founded the Council for the Protection of Rural England. Was this an English version of Muir's communion with nature in Yosemite and the foundation of the Sierra Club in California? Peter Hall, the leading authority on the English planning system, detects a different motive. Rural England was a place of segregation and social stratification.

> The more prosperous members of the old county society, joined by selected newcomers from the cities, have sought to defend a way of life which they regarded traditionally as their right. The weapon they have used, and it has been a powerful one, is conservationist planning. The result has been to segregate the less affluent newcomers as firmly as ever the medieval cottagers were.[9]

By contrast, George Orwell's England embraced the urban England where Labour drew its support, 'the clatter of clogs in the Lancashire mill towns, the to-and-fro of the lorries on the Great North Road, the rattle of pin-tables in the Soho pubs'.[10] There was a lively debate within the Left's intelligentsia between urban socialists and rural distributists, the former pushing for the nationalization of the means of production, the latter for the expropriation of landowners and the redistribution of land to the rest of society.

In 1928, Hilaire Belloc chaired a debate between George Bernard Shaw, speaking for socialism, and G.K. Chesterton championing distributism. Belloc, himself a distributist, summed up. Industrial civilization, he prophesied, 'will break down and therefore end from its monstrous wickedness, folly, ineptitude' or it will lead 'the mass of men to become contented slaves'.[11] Before the First World War, Chesterton had written of the need for a Peasant Proprietorship in *What's Wrong with the World*. 'If we are to save property, we must distribute property, almost as sternly and sweepingly as did the French Revolution,' a sentiment closer in spirit to Robespierre than Edmund Burke.

Orwell was dismissive of the distributists, writing that a return to small-scale peasant ownership was impossible. In *The Road to Wigan Pier*, Orwell wrote of what he called the Chesterton-type of writer, who wanted 'to see a free peasant or other small-owner living in his privately-owned and probably insanitary cottage; not a wage-slave living in an excellently appointed Corporation flat and tied down by restrictions as to sanitation'.

After the First World War, the younger generation espousing return-to-the-soil ideas broke off in an entirely new direction. In 1920, a leading member of Baden-Powell's Boy Scout movement, the twenty-six-year-old John Hargrave founded the Kindred of the Kibbo Kift, the name purportedly an ancient Kentish dialect meaning proof of great strength. Hargrave's movement had a program of open-air education, camping and naturecraft and the Brotherhood of Man. A Quaker and a pacifist, Hargrave – White Fox to his followers – believed society could only be regenerated by training a cohort of the strongest few. Inspired by North American Red Indians (the movement adopted totem poles), the Kibbo Kift combined Nordic sagas and Saxon dress.

How could anyone have taken Hargrave and his movement seriously?

An idea seizing the imagination of the leading minds of one age can appear bizarre or absurd to a later one. Within three years of its founding, Hargrave had signed up some of the luminaries of the 1920s to his Kibbo Kift advisory council: Norman Angell, author of the celebrated pre-war book, *The Grand Illusion*, a future Labour Member of Parliament and winner of the 1933 Nobel Peace prize; the evolutionary biologist Julian Huxley; the Belgian playwright Maurice Maeterlinck and winner of the 1911 Nobel Prize in literature; the Indian poet Rabindrath Tagore and H.G. Wells. D.H. Lawrence was put off by Hargrave's impractical ambition, but, for all that: 'He's right and I respect him for it … If it weren't for his ambition and lack of warmth, I'd go and Kibbo Kift with him.'

In 1923, Hargrave first met Major C.H. Douglas, the theoretician of the Social Credit movement. Social credit, a now forgotten monetary explanation of the economic difficulties of the inter-war years, was based on Major Douglas's cranky A+B theorem. It earned Douglas a mention in Keynes' 1936 *General Theory of Employment, Interest and Money* – 'a private, perhaps but not a major in the brave army of heretics'. Eight years later, the Kibbo Kift became the Green Shirt Movement for Social Credit. Now the Green Shirts had a comprehensive solution to society's problems, an economic prong and an environmentalist prong together providing a critique of capitalism and urban industrial society. Accompanied by a corps of drummers, Green Shirts marched through English towns and cities, singing the Anthem of the Green Shirt movement:

> *Dead Men arise*
> *From the catacombs of Death!*

went the refrain, the anthem ending with the stirring cry,

> *Wake, now, the Dead!*
> *By numbers held in thrall;*
> *On this fruitful Earth*
> *There is Earth-wealth for all!*

While Hargrave and the Green Shirts looked to the past and to Red Indians and Saxons, for inspiration, others looked across the English Channel. 'England is dying, capitulating to the forces which she herself has set in motion,' reads the preface to a book published in 1928.

> Germany is the sole country where there is a positive challenge to the mechanism and commercialism which we associate with America, but which we in England take lying down, without real protest or power to discover an alternative. If contact with Germany cannot stir the faint-heart hopes to a new quickness which will attune us with the forgotten genius of our own Britain, some of us might realise that Germany is fertile soil in which we may plant the seeds of our experience with the surety of their having increase.[12]

Britain and Germany: A Frank Discussion Instigated by Members of the Younger Generation, was co-edited by Rolf Gardiner, who had briefly been a member of the Kibbo Kift and had originally introduced Hargrave to Major Douglas and social credit.

In a chapter, 'Have the Northern Peoples a Common Destiny?', Gardiner repudiated sterile internationalism and rejected Hargrave's pacifism. 'Should some polity arise which calls forth a genuine star-like quality in the blood of men we should be prepared to fight and die for that quality,' Gardiner wrote.[13] Europe's artificial states should be dissolved to reveal the real frontiers of 'race and culture, and they are made manifest in the form and effluence of landscapes, of regions inhabited by past generations.'

Gardiner disdained the British Union of Fascists for being too lower middle class and urban for his taste. Instead, national regeneration would come about through an alliance of aristocracy and yeomanry. Together with the Earl of Portsmouth, Gardiner was the leading advocate of organic farming, the pair paying a visit to Walther Darré, the Nazi agriculture minister. Their journal, *The New Pioneer*, was edited by an ex-member of the National Socialist League, an organization that criticized Moseley's fascists for being insufficiently anti-Semitic, and devoted much of its space to Gardiner's back-to-the-land program and organic farming.

The complexity of the various cross currents of the non-traditional Right can be seen with Henry Williamson, author of *Tarka the Otter*, published in 1927, and one of the greatest nature writers of the twentieth century. Having fought in the First World War, Williamson ended it strongly sympathetic to Germany and a Bolshevik. On returning from the 1935 Nuremburg Rally, Williamson expressed his revulsion of the rootless civilization he saw in London, 'hoardings, brittle houses, flashiness posing as beauty, mongrel living and cosmopolitan modernism.'[14] Two years later, he joined Moseley's British Union of Fascists. Agricultural reform was central to Moseley's plan to recreate British society. Under the Fascists, land use throughout Britain was to be centrally planned and organized, with programs of re-afforestation, sewage disposal and building motorways to create jobs and re-cycle materials.

Brown Shirts and Green Shirts clashed on the streets of Britain's cities. According to Bramwell, 'the Green Shirts had an element of Quaker niceness, of world unity pacificism' completely at variance with the violence and anti-Semitism of Oswald Moseley's fascists.[15] In 1937, the Public Order Act swept the Brown Shirts and Green Shirts off the streets. The Second World War rendered environmentalism irrelevant, putting some of its more vocal pro-German supporters behind bars. In 1944, the Green Shirts were being written off by Labour MP Tom Driberg as a 'small, fantastic cult of nature worshippers'.[16] Hargrave dissolved the organization in 1951.

Gardiner and his circle found a more successful route into the post-war era. In June 1945, he co-founded the Soil Association and for many years sat on its council. Until the early 1960s, its journal, *Mother Earth*, was edited by the former British Union of Fascists' agricultural secretary, who had spent part of the war in prison. Sixty years after it was founded, eighty per cent of the organic food sold in Britain's supermarkets was being certified by the Soil Association.

Thus British environmentalism emerged from an alien gene pool from the Progressive policies of Theodore Roosevelt a decade or two earlier, with a strong anti-democratic virus and narrative of national decay. To British environmentalists, American culture exemplified all that was wrong with the modern world. 'The English have now reached a point in their history where they must seek a new focus,' Gardiner wrote in 1928; 'the saga of our nationhood and Empire is finished.'[17] Britain had to reject materialism and accept where its true destiny lay: 'Germany is once again a young and puissant nation, pulsating with life and hope, and, like Faust, striving for possession of her own soul.'[18]

In German culture, nature – especially trees and forests – is part of national identity as in no other European country. The Nazi seizure of power in 1933 means that this aspect of German culture became entwined with National Socialist ideology. And here one comes to an awkward truth for the post-1945 environmental movement. Were it not for its crimes, the Nazi record on the environment would have been praised for being far in advance of its time. Nazi Germany introduced anti-vivisection laws. It was the first country in Europe to create nature reserves. In 1934, the Nazis required that new tree plantations should include deciduous trees as well as conifers. In 1940, the Nazis passed hedgerow and copse protection ordinances to protect wildlife as its armies were laying waste to Europe. Following the annexation of part of Poland, one sixth of the new territory's arable land was reserved for new forests and woodland. At the height of the war, Hitler vetoed Ministry of Agriculture schemes to drain and reclaim moorland.

How far could these environmental policies be described as specifically Nazi rather than reflecting more generalized German values and priorities? Put another way, how central was protecting the environment to Nazi ideology?

The story of organic farming in the Third Reich illustrates the Nazis' real priorities. Walther Darré, in charge of the agriculture ministry until 1942, was a convinced Nazi; the titles of two of his books give a flavour: *The Peasantry as Life Source of the Nordic Race* (1928) and *A New Nobility of Blood and Soil* (1934). He was also a supporter of bio-dynamic (organic) farming that had been advocated by Rudolf Steiner in the 1920s, even though Steiner had been an early opponent of the Nazis before his death in 1925.

Rudolf Hess, Hitler's deputy, was a follower of Rudolf Steiner's teachings on the threat to human health posed by artificial fertilizers and Himmler, the head of the SS, encouraged organic farming. Others in the Nazi leadership, such as Goering, were strongly opposed. Following Hess's flight to Britain in 1941, Darré was marginalized, the Gestapo led a crack down on organic farmers and the technocrats in the regime focused on maximizing what farms could produce. The Nazi project was not about saving the planet, but about world conquest, the extermination of the Jews and the enslavement of peoples deemed racially inferior.

To become an international movement and make a global impact in the post-war world, environmentalism needed what its European versions in the pre-war period would have abhorred; mass international travel and communications, multinationals and global consumer brands, a transnational entertainment culture – all the things that go to create a global village. Hitler's defeat in 1945 and the depth of Nazi barbarism meant that post-war environmentalism wouldn't speak German as its first language. In Britain, environmentalism's high conservative lineage and the stain of its association with National Socialism also meant that the next burst of environmentalism would be led from a different social class and from a different part of the political spectrum.

Environmentalism needed to be Americanized. In the post-war world, environmentalism spoke with an American accent – descended from Thoreau and Emerson, Pinchot and Muir and the environmental politics of Theodore Roosevelt.

5
TURNING POINT

Human beings are now carrying out a large scale geophysical experiment of a kind that could not have happened in the past nor be reproduced in the future ... This experiment, if adequately documented, may yield a far-reaching insight into the processes determining weather and climate.

Hans Suess and Roger Revelle, 1957

In 1962, when Silent Spring *was first published, 'environment' was not even an entry in the vocabulary of public policy.*

Al Gore

After the Second World War, Callendar exchanged ideas with the younger generation of climate scientists such as Gilbert Plass and Charles Keeling. In 1956, Keeling obtained funding for an observatory at Mauna Loa in Hawaii to record atmospheric concentrations of carbon dioxide. Over time, results from Mauna Loa would show rising levels of carbon dioxide, the Keeling Curve as it became known.

Five years later, Callendar finished the last of his essays. In his 1938 paper, he wrote that 'the course of world temperatures during the next twenty years should afford valuable evidence as to the accuracy of the calculated effect of atmospheric carbon dioxide'.[1] His 1961 paper concluded that the trend toward higher temperatures was significant. Yet Callendar's confidence was undermined by the downturn in global temperatures in the 1950s and 1960s. England's three skating Christmases of 1961, 1962 and 1963 – the coldest since 1740 – shook Callendar's belief in the Callendar Effect.

The post-war era began by subscribing to Francis Bacon's belief that the progress of science and technology aid mankind's mastery of nature and improve his material wellbeing. That belief was to be challenged and then recede over the course the second half of the twentieth century. During it, the prime source of ideological hostility to the Baconian view of man's relationship with nature shifted from the Right to the Left of the political spectrum; the conservationism of the century's first half would be supplanted by environmentalism and the public role of scientists would change from illuminating and enabling mankind's progress along the path of greater material wellbeing to issuing warnings about economic growth and telling governments how they should mitigate the effects of such progress. For a development of such enormity, it all happened with surprising speed.

The end of the Second World War and the start of the Cold War saw the return of fears that the US might run out of raw materials, exacerbated by the Korean War. In January 1951, President Truman appointed William Paley, president of the CBS network, to head a Materials Policy Commission. 'We cannot allow shortages of materials to jeopardize our national security nor to become a bottleneck to our economic expansion,' Truman wrote to Paley.

When its eight-hundred-and-nineteen-page report, *Resources for Freedom*, appeared the following year, it was unmistakably a product of its time.

> The United States, once criticized as the creator of a crassly materialistic order of things, is today throwing its might into the task of keeping alive the spirit of Man and helping beat back from the frontiers of the free world everywhere the threats of force and of a new Dark Age which rise from the Communist nations. In defeating this barbarian violence moral values will count most, but they must be supported by an ample materials base. Indeed, the interdependence of moral and material values has never been so completely demonstrated as today, when all the world has seen the narrowness of its escape from the now dead Nazi tyranny and has yet to know the breadth by which it will escape the live Communist one.[2]

Soaring demand, shrinking resources, upward pressure on costs, the risk of wartime shortages all threatened to stop or reverse Americans' rising standard of living. It sounded as if *Resources for Freedom* would be the latest product of the Jevons cookie cutter. Instead, *Resources for Freedom* turned out to be the most comprehensive official statement of the ameliorist position made by any government or intergovernmental body before or since. It was a popular fallacy, the report concluded, 'to regard our resource base as a fixed inventory which, when used up, will leave society with no means of survival'. American history showed why. 'In developing America,' the commission asserted, 'our forebears consumed resources extravagantly, but we are certainly better off in materials than they were. It would be unreasonable for us, their posterity, to suggest that they should have consumed less so that we might consume more.'

> Using resources today is an essential part of making our economy grow ... Hoarding resources in the expectation of more important uses later involves a sacrifice that may never be recouped; technological changes and new resource discoveries may alter a situation completely.[3]

To act in this way, the report suggested, would be rather like the early settlers of New England conserving bayberries to provide light for a generation that lives by kilowatts.

At times, the report reads like a point-by-point refutation of Jevons. On recycling, it sided with Keynes, who, as we saw in Chapter Three, had criticized Jevons for hoarding scraps of paper. It was another popular fallacy, the commission said, to equate physical waste with economic waste, an attitude which can lead to 'devoting a dollar's worth of work to 'saving' a few cents' worth of waste paper and old string.' The waste paper bin does have an economic function after all.

The commission had three fundamental convictions. Growth is good ('it seems preferable to any opposite, which to us implies stagnation and decay,' was the common sense justification); belief in private enterprise, the profit motive and the price system; and internationalism. Trade was good for America and America was stronger when its allies were stronger. If the US did not work to raise living standards in the rest of the free world, it would 'hamper and impede further rise of our own, and equally lessen the changes of democracy to prosper and peace to reign the world over.'[4]

Resources for Freedom made two fundamental recommendations – a do and a don't to maximize economic efficiency. 'There is no magic formula which will yield the right answers ... Yet there is one basic economic principle which, if applied to the limit of available facts and injected consciously into each judgment, can provide a basic thread of consistency. THE LEAST COST PRINCIPLE.'[5] The commission recognized the risk of interest groups exploiting political concern about the availability of resource supplies to cut legal privileges and special deals for themselves.

The second recommendation was a plea to avoid protectionism and autarky. 'The United States must reject self-sufficiency as a policy and instead adopt the policy of the lowest cost acquisition of materials wherever secure supplies may be found.' The idea of self-sufficiency, the commission observed, 'amounts to a self-imposed blockade and nothing more.'[6] It urged that tariff and other barriers to trade be removed, as these were adding to the free world's material problems. 'By interfering with market pressures of supply and demand, they prevent normal development of the tendency to move toward the lowest cost sources of materials – a movement which, as the commission has pointed out, is essential in promoting the most rapid economic growth of both the United States and the less developed countries.'[7]

The US should unilaterally eliminate import duties on any commodity if the US was likely to become dependent on it. In the case of petroleum – 'the great enigma of future energy supplies' – the commission envisaged with equanimity that by 1975 the US would have increasingly turned to foreign supplies, although it suggested stockpiling in case of war. It went on to urge repeal of the Buy American Act 1933 and similar protectionist legislation, 'a relic of depression years and depression psychology'.

In the history of concern about industrialized societies running out of resources, which started in 1865 and continues to the present, *Resources for Freedom* stands apart. Based on its examination of history, the commission rejected the depletionists' assumption that there was a fixed lump of resource which, when used up, was gone forever. Compared to the drumbeat of depletionist reports that punctuated the 1970s prophesying resource-induced and environmental catastrophes before the end of the twentieth century, the view of the world set out in *Resources for Freedom* turned out to be right. The world reached 1975 (the time horizon adopted in the report) and the beginning of the twenty-first century closer to the state envisaged in *Resources for Freedom* than the predictions made by the depletionists. For that reason alone, *Resources for Freedom* demands attention and a respectful hearing.

It would have been easy for the Paley Commission to have lapsed in to paranoia. The international situation in 1952 was much riskier than fifty years later. The Soviet Union had acquired the Bomb, the Iron Curtain divided Europe, and the US and its allies were fighting a war in Korea in which the American commander had requested the use of atomic weapons against China.

The policy optimism of *Resources for Freedom* reflected the outlook of a governing elite that a few years earlier had led America to victory in the Second World War and had just put in place the architecture that would win the Cold War. It was confident about America's future and the capacity of its economy to keep growing. It had the self-confidence to be internationalist. Far from falling for the illusion of 'energy independence', *Resources for Freedom* recognised that American prosperity and security were enhanced by an open trading system.

The same year as the Paley Commission produced its report, Congress passed the Domestic Minerals Program Extension Act. It requires the federal government to reduce American dependency on foreign sources of critical or strategic materials and remains in force to this day. In 2003 congressional testimony, the director of the US Geological Survey admitted that shortly after passage of the Act, 'minerals were in surplus rather than shortage ... by 1956, even uranium was in over-supply'.[8] Thus within months of the Paley Commission's report, the market had solved the problem it had been established to examine, insofar as there had been one in the first place. It would be another twenty years before politicians and policymakers thought that they, rather than markets, were needed to solve the problem of resource scarcity.

Long-term, the commission foresaw little risk of resource depletion. As land-based resources became more costly, and as technology improved, man would increasingly turn to the sea. 'Covering seventy-one per cent of the earth's surface, the sea is a great reservoir of minerals. In its three hundred million cubic miles of water, the sea probably contains, in solution, all the elements found in the earth's crust.'[9]

Such a view would have confirmed the worst fears of Rachel Carson, an aquatic biologist working for the US Fish and Wildlife Service, whose book, *The Sea Around Us*, had been published the previous year. 'We will become even more dependent upon the ocean as we destroy the land' – a view which reflected what one of her admirers has called her aesthetic of ocean-centrism.[10]

Although trained as a scientist, Carson had always assumed she was going to be a writer. She majored in English, switching to zoology. 'I thought I had to be one or the other; it never occurred to me, or apparently anyone else, that I could combine the two.'[11] Her favorite writers were Henry Thoreau and Henry Williamson. She kept Thoreau's journal beside her bed and her first book, *Under the Sea Wind*, was inspired by Williamson's *Salar the Salmon*.[12]

'I never predicted the book would have a smashing success,' Carson said of *Silent Spring*, which was published in 1962.[13] It spent much of that autumn at the top of the *New York Times* bestseller list and sold over one hundred thousand copies in the week before Christmas. In August, *Silent Spring* sparked a question at President Kennedy's press conference on the dangers of DDT, leading to a report by the President's Science Advisory Committee. It concluded that Carson had been right to warn about the dangers of pesticides. In his address to the UN General Assembly in September, President Kennedy tried to move international relations beyond Cold War divisions to the common challenge of improving the conditions of mankind. Plunder and pollution were the foes of every nation and Kennedy proposed a worldwide program of conservation.

At one level, Carson's crusade against pesticides failed. According to an official Environmental Protection Agency appreciation of *Silent Spring* sixteen years on, 'Americans now apply more than twice the amount of pesticides they did before *Silent Spring* was published.' What *Silent Spring* changed was vastly more consequential. In a review for the Book-of-the-Month Club, Supreme Court Justice William O. Douglas compared *Silent Spring* to *Uncle Tom's Cabin* for its revolutionary implications.

'You are a poet not only because you use words so well,' wrote one of her fans in a letter that evidently meant much to Carson as she copied it to a friend, 'but because by describing non-human life, you make us readers understand our place on earth so much better. As I drive home along the Hudson tonight I'll feel more human for having read your lovely, loving words today.'[14]

Silent Spring's title was a masterstroke, suggested to Carson by her editor. Discarded titles included 'Man against the Earth' and 'The War against Nature', which better describe the larger dimension of the book's purpose at the cost of the book's capacity to provoke the kinds of reaction (those lovely, loving words) if its purpose had been made plain in its title. Spring has an almost religious quality about it. 'In a pleasant spring morning all men's sins are forgiven,' Thoreau wrote in *Walden*. 'Through our own recovered innocence we discern the innocence of our neighbors ... There is not only an atmosphere of good will about him, but even a savor of holiness groping for expression.'[15] For modern man to destroy spring is tantamount to extinguishing the possibility of redemption.

The book opens with what Carson described as a fable for tomorrow, the silent spring. A rural town in the heart of America hit by a 'strange blight' casting an 'evil spell'; sudden deaths which doctors couldn't explain – children dead within a few hours; apple trees that yielded no fruit; birds that trembled violently and couldn't fly: 'No witchcraft, no enemy action had silenced the rebirth of new life in this stricken world. The people had done it to themselves.'[16] Although the town did not exist, Carson wrote that every one of the disasters had actually happened somewhere, allegedly silencing the voices of spring in countless towns across America.

The book proceeds to catalog a string of disasters. A dog and baby exposed to an insecticide. Within an hour the dog is dead and the baby little more than a vegetable. Two children in Florida dead after touching a bag that contained a Parathion – a favorite means of committing suicide in Finland, we're told. An unfortunate chemist who swallowed 0.0424 ounces of Parathion to assess its effects – paralyzed before he could reach the antidotes. The housewife scared of spiders who sprayed her house three times with an aerosol containing DDT and petroleum distillate. She felt unwell, saw a doctor and was found to be suffering from acute leukaemia – dead the next month.

The threat of cancer lurked everywhere. People were living in a sea of carcinogens. Worse still was the prevalence of cancer among children. Twenty-five years previously, cancer in children had been rare. 'Today more American school children die of cancer than from any other disease.'[17] Attempts to cure cancer were doomed, Carson wrote, 'because it leaves untouched the great reservoirs of carcinogenic agents which continue to claim new victims faster than the as yet elusive 'cure' could allay the disease'.[18]

After her death in 1964 (the same year as Callendar), Carson acquired an iconic status as a lonely female scientist speaking up for nature against an entrenched male scientific establishment financed by vested interests. While working on *Silent Spring*, Carson was being treated for breast cancer, sanctifying *Silent Spring* as the product of a green martyrdom. Questioning its scientific basis is considered bad form or worse, especially from the wilder shores of eco-feminism.*

Although framed as an indictment of pesticides, Carson's general theory was that flora and fauna were threatened by any substance which they had not previously encountered.

> With the dawn of the industrial era the world became a place of continuous, even accelerating change. Instead of the natural environment there was rapidly substituted an artificial one composed of new chemical and physical agents, many of them possessing powerful capabilities for inducing biologic change.[19]

Each year, five hundred new chemicals were being introduced to which 'the bodies of men and animals are required somehow to adapt'. Man's speed was nature's enemy. 'The rapidity of change and the speed with which new situations are created follow the impetuous pace of man rather than the deliberate pace of nature.'[20]

---

* In *The Recurring Silent Spring*, Patricia Hynes, a former official of the EPA, wrote of her anger 'at living in a world in which nature and women are presumed to exist for the use and convenience of men, so that the destruction of nature and violence against women are interconnected, increasingly technologized, and infect all corners of the earth.' H. Patricia Hynes, *The Recurring Silent Spring*, Elmsford, (1989), p. 2.

This scientific hypothesis at the core of *Silent Spring* is not one that was subsequently developed with any success. Even her admirers do not ascribe pre-eminence to the science of *Silent Spring*. For them, its significance lies elsewhere. Linda Lear's 1997 biography describes *Silent Spring* as a 'fundamental social critique of a gospel of technological progress.'[21] A leading textbook of environmental issues in American history argues that *Silent Spring* was more important for what it said about modern society and its relationship to nature than about chemistry and biology.

> For all the debate about the specific merits of pesticide regulation that Carson's work ultimately produced, at bottom her profoundly radical and most enduring argument was that corporate and governmental institutions of power – whether out of arrogance or ignorance, or for base profit motives – had disrupted the balance of the natural world and now threatened human health.[22]

As social theory, the arguments in *Silent Spring* are decidedly patchy. Who was to blame? 'The authoritarian temporarily entrusted with power.'[23] But elsewhere, Carson castigated 'the mores of suburbia' which led to the use of herbicides to remove unwanted crabgrass. Power mowers, Carson complained, had been fitted with devices to disseminate pesticides, 'so to the potentially dangerous fumes from petrol added the finely divided particles of whatever insecticide the probably unsuspected suburbanite has chosen'.[24] Then there were farmers exceeding the prescribed dosages and using chemicals too close to harvest time or more insecticides than were necessary and in other ways displaying 'the common human failure to read the fine print.'[25] If Carson's reputation had rested on the power and originality of her analysis of social causation, she would have been a marginal figure.

During her life, Carson was accused by some of her critics of being unmarried, which was true but irrelevant – although not everyone thought so ('I thought she was a spinster. What's she so worried about genetics for?' exclaimed a member of the Federal Pest Control Review Board) – and a communist. Classical Marxism was about reorganizing industrial societies along socialist lines. Carson was against industrial societies altogether. Carson was not a communist; she was an ecologist. 'What is important is the relation of man to all life, when through our technology we are waging war against the natural world,' she said in 1963.[26] Writing to a friend, she put it more baldly: 'In truth, man is against the earth.'[27]

Carson summarized her credo in a 1964 CBS interview:

> I truly believe that we in this generation must come to terms with nature, and I think we're challenged as mankind has never been challenged before to prove our maturity and our mastery, not of nature, but of ourselves.[28]

In a statement written not long after *The Sea Around Us* was published, Carson gave voice to an aesthetic, her belief system. 'I believe that whenever we destroy beauty, or whenever we substitute something man-made and artificial for a natural feature of the earth, we have retarded some part of man's spiritual growth.'[29]

The ecologism espoused by Carson conforms to what Nietzsche had anticipated at the end of the nineteenth century. Since the time of Copernicus, science had been pushing man from the centre of the universe and would push him to its periphery. 'It destroys my importance,' Kant observed. According to Nietzsche, science was destroying the self-respect humans once had. The faith in their dignity, their uniqueness, their irreplaceable position in the chain of being, their position as children of God, had gone. The human being was becoming a mere animal.

Biologists in the second half of the twentieth century point out that *homo sapiens* is one species among the countless that have existed since life began, and, in their eyes, not an especially significant one. Palaeontologist Stephen Jay Gould liked to counter the received picture of evolution working from simple forms of life toward complexity, with man at the summit. 'The most outstanding feature of life's history is that through 3.5 billion years this has remained, really, a bacterial planet. Most creatures are what they've always been: They're bacteria and they rule the world,' Gould told an interviewer. 'We need to be nice to them.'[30]

Carson told the viewers of CBS that man still talked in terms of conquest. 'We still haven't become mature enough to think of ourselves as only a tiny part of a vast and incredible universe.'[31] Carson's ocean-centrism overwhelmed mankind's significance. 'He cannot control or change the ocean as, in his brief tenancy of earth, he has subdued and plundered the continent,' she wrote in *The Sea Around Us*.[32]

Ecologism differs from the transcendentalism of Emerson and Thoreau. With the latter, the individual's openness to nature elevates him and shows him his uniqueness. By contrast, ecologism preaches man's intrinsic insignificance. Not that its followers find their own lives devoid of importance. Towards the end of her life, Carson was drawn toward belief in the immortality of her soul. When consideration moves from a particular individual to the relationship of human society to the natural world, ecologism brings a radically altered perspective, one of a disruptive, out-of-control species, putting its survival and that of other species at risk by upsetting natural balances.

Ecologism was a child of the 1960s given the breath of life by Rachel Carson. It would be the philosophy used to evaluate man's impact on nature. The lasting effect was the presumption that human interventions in the natural world were potentially harmful. New ones should be avoided unless it could be demonstrated that they were harmless, giving rise to the precautionary principle. The logic of the precautionary principle runs counter to the development of human society, especially its acceleration with the Industrial Revolution. As environmentalism reached a peak in 1972, the Canadian economist Harry Johnson countered that 'man's whole history has been one of transforming his environment rather than accepting its limitations'.[33]

In focusing exclusively on mankind's impact on nature, ecologism – the ideo-logical core of the environmental movement – opposed the claims of the poor to seek better lives for themselves. There were too many of them and meeting their demands would put further pressure on fragile ecological balances and accelerate resource depletion.

To succeed in its mission to save the planet, environmentalism depended on global participation. Its ability to persuade developing countries that doing so would be in their interests would be environmentalism's greatest challenge. Its success would determine the fate of its most ambitious project – saving the world from the perils of global warming.

6
SPACESHIP EARTH

We came all this way to explore the moon, and the most important thing is that we discovered the Earth.

Bill Anders on Apollo 8, December 24th 1968

On March 23rd 1964, the first session of the United Nations Conference on Trade and Development (UNCTAD) was held in Geneva. Four thousand delegates from one hundred and nineteen countries attended, making it the largest international event held up to that date. The United Nations had designated the 1960s the Decade of Development and the creation of UNCTAD reflected the growing weight of Third World countries in the UN General Assembly. In the 1950s, Latin America had twenty out of the fifty-one seats in the General Assembly. Decolonization meant that in the 1960s, the Third World constituted the majority. The Geneva conference saw the formalization of the G77 group as the largest voting bloc in the UN.

Only two speakers were given a standing ovation, the improbably pinstriped Che Guevara and the secretary-general of the conference, Raúl Prebisch. Like Guevara, Prebisch was an Argentine exile. Although the CIA had kept him under surveillance in the 1950s, Prebisch was no revolutionary. Born in 1901, he had a meteoric rise. Appointed under-secretary of finance at the age of twenty-nine, he quickly rose to become general manager of Argentina's central bank.

On the outbreak of the Second World War, realizing it would lead to American economic leadership, Prebisch went to Washington to negotiate a trade deal. A three-minute courtesy call with Franklin Roosevelt turned into an hour-long meeting. Roosevelt urged Prebisch to nationalise Argentina's British-owned railways, at the same time warning him that railways were a poor investment.

Following a coup in 1943, Prebisch was sacked. There was no place for him in the new regime, so he decided to develop the economic policies Argentina should adopt after the war. Prebisch believed that Argentina's reliance on agriculture condemned it to boom-bust cycles and decline relative to the industrialized world. At the centre of his economic thinking was a structural rift between the core and the periphery of the international trading system. Trends in commodity prices relative to prices of manufactured goods between 1873 and 1938 provided evidence, Prebisch argued, that the commodity-producing periphery was at a structural disadvantage to the industrialized core.

A UN conference at Havana in 1949 set the stage for Prebisch's new ideas. Tensions between the US and Latin America had been rising, fuelled by resentment of the US. Prebisch's speech electrified the conference. 'I gave consistency to ideas that were in the air. I did not create *ex novo*: I reflected an intellectual reality,' he explained towards the end of his life.[1] Prebisch's Havana Manifesto demonstrated why Third World countries were condemned to relative decline and provided their governments with policies that overturned the traditional ones prescribed by classical economics, just as Keynes had for economies during the Depression of the 1930s. 'One of the conspicuous deficiencies of general economic theory from the point of view of the periphery,' Prebisch told the conference, 'is its false sense of universality.'[2]

Exhilarated by the instant acclaim and attention, Prebisch suddenly disappeared. According to his biographer, 'His sensational public success had unleashed an unbridled sexuality heretofore contained by a life of disciplined work and family.'[3] Three weeks later, he re-emerged to be acknowledged as the intellectual leader of a new path for the economic development of the Third World, one which emphasized the need for governments to actively stimulate industrialization.

The path of economic development advocated by Prebisch ran in diametrically the opposite direction to the demands of the environmental movement in the West unleashed by *Silent Spring*. Thus two new forces were emerging to re-shape the post-war world. Their inner logic was antithetical to each other. Unless they could identify a common foe or develop a new synthesis, only one of them could prevail.

One aspect of Prebisch's economic analysis meshed with the growing environmentalist movement in the West in the 1960s. Both were based on an assumption that markets failed. For Western environmentalists, markets led to environmental degradation and resource depletion. For Prebisch and his followers in the Third World, the benefits from trade between the core and the periphery – North and South as it would become in the UN development debates of the 1970s and 1980s – were distributed unequally.

But in other ways, the two clashed. Environmentalists blamed the ills of society and threats to the environment on industrialization. Prebisch argued that Third World governments should intervene to accelerate industrialization. Western environmentalists believed resources were running out, implying higher commodity prices. Prebisch argued that prices of commodities would fall relative to prices of manufactured goods. Who was right? A 1992 study of twenty-six individual commodity prices over the period 1900–1983 suggests neither of them were. Of the twenty-six, sixteen were trendless, five had statistically significant negative trends and the remaining had positive trends.[4]

Whatever its objective merits as an economic blueprint for developing countries, the Havana Manifesto was a political fact, one institutionalized with the creation of UNCTAD and the formation of the G77. In forming the most numerous voting bloc in the UN General Assembly, the Third World had succeeded in carving out a bargaining position in international negotiations wherever their agreement was required. Whenever it could, the G77 would insist that international issues be addressed through the General Assembly.

Environmentalists in the West were slow to recognize the dilemma this posed for environmentalism. During the 1960s, environmentalism went global – its mission became planetary. In July 1965, five days before a fatal heart attack, Adlai Stevenson, America's ambassador to the UN, spoke of humanity as passengers on a little spaceship dependent on its vulnerable reserves of air and soil.

> We cannot maintain it half fortunate, half miserable, half confident, half despairing, half slave – to the ancient enemies of man – half free in a liberation of resources undreamed of until this day. No craft, no crew can travel safely with such vast contradictions. On their resolution depends the survival of us all.

The following year, economist Kenneth Boulding picked up Stevenson's theme. Entitled 'The Economics of the Coming Spaceship Earth', Boulding argued that past civilizations had imagined themselves to be living on an illimitable plane. 'There was almost always somewhere beyond the known limits of human habitation, and over a very large part of the time man has been on earth, there has been something like a frontier.'[5] This was the 'cowboy economy', symbolizing 'the reckless, exploitative, romantic, and violent behavior, which is characteristic of open societies'. In the cowboy economy, consumption was regarded as a good thing, as was production. Now mankind was transitioning to the spaceman economy. Man's material wellbeing would be constrained by the economics of the closed sphere. Everyone in the world shared the same spaceship. On spaceships, supplies are rationed; 'The less consumption we can maintain a given state with, the better off we are.'

The metaphor of the spaceship economy carried a strong presumption in favor of global governance. What anyone did anywhere was the concern of everyone, everywhere.

Stevenson's speech had been drafted by Barbara Ward, a British intellectual who, alongside Rachel Carson, has a strong claim to be the most consequential environmentalist of the twentieth century. In 1966, Ward gave a lecture, *Space Ship Earth*, in which she argued that mankind's survival depended on developing a government of the world.

The longevity of China's government, Mao being the latest dynasty, demonstrated that world government was possible. If two thousand years of rule can work for twenty-five per cent of the world's population, 'we can hardly argue that the task of government becomes *a priori* impossible simply because the remaining three-quarters are added', Ward argued.[6] Would a world government have to be authoritarian, she asked? No, look at the continental federation of the United States. 'If a free continent is possible, why not an association of free continents?'[7]

The expansion of environmentalism to embrace the future of the planet had huge implications for global politics and environmentalism. Whereas socialism within one country was a viable political doctrine, one-country environmentalism would be nothing more than symbolism. Saving the planet demanded global policies.

On Christmas Eve 1968, nearly a quarter of a million miles from Earth, the three crew members of Apollo 8 orbited the moon. Each took turns to read the first ten verses of the book of Genesis, ending their reading with 'God bless all of you, all of you on the good Earth'. The photographs were even more arresting; the first of the whole Earth, a blue disc suspended in a black void, and of the Earth rising above the curve of the moon's horizon, an unforgettable contrast between the only known planet that supported life and the grey, pitted emptiness of the lunar landscape. Environmentalism had acquired as its icon one of the most powerful images ever produced.

Back on Spaceship Earth, a growing number of the crew were gaining their independence from declining imperial powers. The newly independent states had little interest in replacing an antiquated form of imperialism with a newer version and were adopting economic policies that ran counter to the anti-industrialization tenets of Western environmentalism. How could the irreconcilable be reconciled?

One man above all developed and popularized the notion that the solution to the environmental and spiritual crisis in the First World could be found in the Third World. Hailed by Jonathon Porritt, a leading British environmentalist and adviser to Prince Charles, as 'the first of the 'holistic thinkers' of the modern Green Movement', Fritz Schumacher was born in Bremen three years before the First World War. Revulsion at the Nazis led Schumacher to flee Germany and settle in Britain. He quickly developed a reputation as an able economist, engaging Keynes in discussions about post-war currency reform and working on the 1944 Beveridge Report on full employment.

His input into the Beveridge Report impressed Hugh Gaitskell. When he became minister of fuel and power in the 1945 Labour government, Gaitskell suggested that the newly nationalized National Coal Board appoint Schumacher as its chief economist. Like Jevons before him, Schumacher became alarmed at the prospect of resource depletion, ironically just as post-war demand for coal started to fall. In a 1954 lecture, Schumacher articulated his concern by placing the depletion of finite energy resources in the context of man's relationship with nature.

> We are living off capital in the most fundamental meaning of the word. Mankind has existed for many thousands of years and has always lived off income. Only in the last hundred years has man forcibly broken into nature's larder and is now emptying it out at a breathtaking speed which increases from year to year ... The whole problem of nature's larder, that

is the exhaustion of non-renewable resources, can probably be reduced to this one point – Energy.[8]

Schumacher's work at the Coal Board went hand in hand with a growing preoccupation with organic farming. He bought a house in Surrey with a four-acre garden, promptly joined the Soil Association and became an avid reader of its journal *Mother Earth*, edited by the former fascist Jorian Jenks. He invited an expert from the Soil Association to give a lecture to the Coal Board's gardening club. 'To listen to a lecture on food production in the Headquarters of fuel production,' was, Schumacher told his colleagues, 'the most significant concentration on the essential that I could imagine.'[9] He also developed an interest in Eastern mysticism and astrology, working out his children's horoscopes. If used 'correctly and wisely', astrology could be a useful instrument in understanding one's fellow men, Schumacher claimed.

The event that pulled these diffuse strands together was an invitation to be an economic adviser to the government of Burma. The trip was the final stage in Schumacher's transformation. He left for Burma as an economist and returned as a soothsayer and guru. The Economic and Social Council of Burma, which had invited him, was not thrilled with its economic adviser seeking answers to economic questions from orange-robed monks.

Stopping off in New York, 'this American madhouse', the contrast with Burma led him to declare that he had found the cure for Western civilization – Buddhist economics. A Buddhist approach would distinguish between misery, sufficiency and surfeit. 'Economic "progress" is good only to the point of sufficiency, beyond that, it is evil, destructive, uneconomic,' Schumacher said. Economics was based on materialism and the religions of the East offered an alternative, the new guru thought. Mahatma Gandhi had laid down the foundation for a system of economics that, Schumacher believed, would be compatible with Hinduism and Buddhism too.

'Buddhist economics' enabled Schumacher to recast the narrower depletionist arguments developed by Jevons ninety years before into a new language mixing ecologism and Eastern mysticism:

> A Buddhist economy would make the distinction between 'renewable' and 'non-renewable resources'. A civilization built on renewable resources, such as products of forestry and agriculture, is by this fact alone superior to one built on non-renewable resources ... The former co-operates with nature, while the latter robs nature. The former bears the sign of life, while the latter bears the sign of death.[10]

All this was poured into Schumacher's classic, *Small is Beautiful*. It gave him a huge cult following. The one thing it wasn't was what Schumacher claimed it to be: 'A study of economics as if people mattered.' Answering critics who asked what economics had to do with Buddhism, Schumacher replied. 'Economics without spirituality can give you temporary and physical gratification, but it cannot provide an internal fulfilment.'[11] Economics without Buddhism is like sex without love. The remark illustrates Schumacher's flawed premise. Economics is not a science of happiness. Economic welfare is only a part of human welfare; it has no claims to explain what falls outside its domain or resolve the problems of the spirit, ethical problems or indeed instruct people on how to find happiness.

Instead *Small is Beautiful* is a collection of *pensées* and injunctions, predominantly of a religious and philosophical character – many of them trite, trivial or bizarre. Why is small beautiful? 'Man is small, and, therefore, small is beautiful,' said the Sage of Surrey.[12] Technology must therefore be redirected to the actual size of man. Buddhist economics means simplicity and non-violence. What might this mean in practice? 'It would be highly uneconomic, for instance, to go in for complicated tailoring, like the modern west, when a much more beautiful effect can be achieved by the skilful draping of uncut material,' Schumacher opined.[13]

What has this to do with economics? As Humpty Dumpty explained to Alice, 'When I use a word, it means what I choose it to mean' – an attitude that is the mark of a guru, not a philosopher or economist.

Schumacher liked trees. If people followed the Buddhist injunction to plant a tree every few years, the result would be a high rate of genuine economic development. 'Much of the economic decay of south-east Asia (as of many other parts of the world) is undoubtedly due to a heedless and shameful neglect of trees,' Schumacher wrote.[14] Anticipating a deeply held belief of the Green Movement, he argued that Buddhist economics was hostile to international trade, 'production from local resources for local needs is the most rational way of economic life, while dependence on imports from afar and the consequent need to produce for export to unknown and distant peoples is highly uneconomic and justifiable only in exceptional cases and on a small scale'.[15]

Similarly, Schumacher favored renewable resources. 'Non-renewable goods must be used only if they are indispensable.'[16] As with his opposition to international trade, Schumacher did not provide anything that might be described as a chain of economic reasoning supported by evidence. Even the Buddhist tag was a misnomer. The Sermon on the Mount's blessing of the meek was re-interpreted to mean 'we need a gentle approach, a non-violent spirit, and small is beautiful'.[17] A militant atheist as a young man, Schumacher was received into the Catholic Church in 1971. Responding to a questioner on his final trip to the US, Schumacher said: 'It was what I call Buddhist economics. I might have called it Christian economics, but then no one would have read it.' Schumacher was really the successor to Hilaire

Belloc and G.K. Chesterton for the 1960s flower-power generation, reaching an audience of millions.

For all Schumacher's popularity among the impressionable and the credulous in the First World – in 1977, he visited the White House, where he gave a delighted Jimmy Carter a copy of *Small is Beautiful* – the reaction to ideas in the Third World was distinctly cool, perhaps reciprocating his own. 'India is a sewer,' he remarked in 1973.[18] His doctrine of 'intermediate technology' would have locked the Third World into permanent inferiority to the West and was patronizing to boot:

> All development, like all learning, is a process of stretching. If you attempt to stretch too much, you get a rupture instead of a stretch, or you will lose contact and nothing happens at all.[19]

He was an ardent supporter of the Soil Association, bequeathing them the royalties from *Small is Beautiful*. 'Their methods bear the mark of non-violence and humility towards the infinitely subtle system of natural harmony,' absolving it of the Nazi sympathies of its founders, and was strongly opposed to the use of pesticides and chemical fertilizers to increase crop yields.[20]

Application of Schumacher's ideas to the Third World would have been disastrous. It was Norman Borlaug's Green Revolution, not Schumacher's intermediate technology or the Soil Association's organic farming, which fed the Third World and avoided the mass starvation and the deaths of millions of Indians confidently predicted by Paul Ehrlich in his 1968 bestseller *The Population Bomb*. As Borlaug, winner of the 1970 Nobel Peace Prize, pointed out in a 1972 newspaper article, wheat yields in India and Pakistan doubled in the six years from 1965, with similar progress in rice yields. When Borlaug died in 2009, the Indian agriculture minister paid tribute. India and many other nations owed a debt of gratitude to 'this outstanding personality' for helping to forge world peace and saving the lives of two hundred and forty-five million people worldwide.[21]

Gratitude in the Third World, but what about in the West? For American radicals, the implications of the Green Revolution were quite different. Borlaug's achievement wasn't about feeding the hungry and saving lives; the Green Revolution had damaged the prospects of a Red Revolution and had extended capitalist agriculture to the tropics.[22]

From his perch at Berkeley in California, the leading ideologist of the New Left Herbert Marcuse re-formulated classical Marxism – a belief system entirely indifferent to nature – to capitalize on the surge in environmentalism. 'Violation of the Earth is a vital aspect of the counter-revolution,' Marcuse told a 1972 conference on ecology and revolution.[23] Monopoly capitalism is waging war against nature, 'the more capitalist productivity increases, the more destructive it becomes. This is one sign of the internal contradictions of capitalism.' Marcuse yoked together the two issues bringing young people out onto the streets: 'The genocidal war against people is also "ecoside".'

The passengers and crew of Spaceship Earth were nothing if not disputatious.

7
LIMITS TO GROWTH

*The crux of the matter is not only whether the human species will survive,
but even more whether it can survive without falling into a state of worthless
existence.*

Executive Committee of the Club of Rome, 1972

April 22nd 1970 – Americans celebrated the first Earth Day. A quarter of a million
people gathered in Washington, DC. Altogether twenty million people took part
in peaceful demonstrations across America. President Nixon's 'Silent Majority',
who sided with him over the Vietnam War against the anti-war protestors, was
marching with them on the environment.

In preparing his first State of the Union Message in January 1970, Nixon
noted that public concern on environmental issues had risen from twenty-five
per cent in 1965 to seventy-five per cent at the end of 1969.[1] Nixon decided to
pre-empt the Democrats on the environment and prevent them using it against
him in the 1972 presidential election. His decision made, Nixon went the full
distance: 'The time has come for a new quest – a quest not for a greater quantity
of what we have, but for a new quality of life in America.' The question America
would answer in the 1970s was whether it would make peace with nature. Using
the language of environmentalism, Nixon spoke of making reparations to nature.
'Through our years of past carelessness we incurred a debt to nature, and now
that debt is being called.' He would propose the 'most comprehensive and costly'
environmental program in America's history.

In private, Nixon told one of his aides, 'In a flat choice between smoke and
jobs, we're for jobs … But just keep me out of trouble on environmental issues.'[2]

Cynical? The *New York Times* gave him the headline he wanted:

NIXON, STRESSING QUALITY OF LIFE, ASKS IN THE STATE OF
THE UNION MESSAGE FOR BATTLE TO SAVE ENVIRONMENT;
OFFERS 'NEW ROAD'

Congress passed laws to regulate air and water pollution and in July, Nixon
created the Environmental Protection Agency by carving out functions from existing federal departments. One of the Cabinet secretaries losing staff in the reor-
ganization asked Nixon whether it was all right for a Cabinet member to say, 'No
comment.' Nixon shot back, 'Yes. And it's about time.'[3] He appointed William
Ruckelshaus, a lawyer, to be the EPA's first administrator. Ruckelshaus then led
a seven-month hearing into DDT. Despite the thinness of scientific evidence of
DDT's threat to human health, in 1972 the EPA banned its use – ten years after
publication of *Silent Spring* – in a highly political reaction to the over-use of the
pesticide by American farmers.

America's second most important elected Republican approached environmental politics differently. In his campaign for governor in 1966, Ronald Reagan had caused consternation with his remark about California's redwoods: 'A tree is a tree – how many more do you need to look at?'[4] Environmental sensitivities in California had been raised by the blowout of an oil platform six miles off the coast of Santa Barbara in January 1969, leading to a drilling ban in federally controlled waters off the central Californian coast.

Once elected, Reagan's environmental policies turned out to be a lineal descendant of Teddy Roosevelt. He created a fifty-eight-thousand-acre Redwood National Park. He reached agreement with Nevada to restrict development of Lake Tahoe. Reagan also supported legislation to ban the registration and sale of automobiles that did not meet Californian auto emission standards. In what his long time biographer Lou Cannon called his finest environmental achievement, Reagan stopped the Dos Rios Dam construction project. 'Ike, I hate to see a beautiful valley destroyed,' he told Ike Livermore, the state's secretary for resources. In 1973, Reagan saddled up and, followed by one hundred packhorses, rode to a meadow underneath Minaret Summit in the eastern Sierra. Waving a white hat, Reagan dismounted and made a speech opposing the Trans-Sierra highway project. It would have been the only road cutting the two-hundred-and-eleven-mile John Muir Trail between Yosemite and Mount Witney. The highway was never built.

Unlike the localized surges in environmental politics earlier in the twentieth century, environmentalism in the late 1960s and early 1970s was synchronized across the Western world. It was powered by a common concern that man's activities – the pollution of air and water – were poisoning the biosphere (Carson had called *Silent Spring* her poison book). Governments across the Western world responded in similar ways. A 1976 UN survey found that seventy countries had created agencies or departments at national level with environmental protection as a core function.

The wave would crest in 1972. Then, even more swiftly than it rose, it broke. Fifteen years would pass before the second wave began to roll, propelled by global warming.

Both drew their energy from fear that industrialized societies were destroying the balance of nature on which humans depended. The first wave was more extreme in its claims, the second more consequential in its effects. Both started during a long period of prosperity. The first began towards the end of what economic historian Angus Maddison called the Golden Age and the second during the growth phase of what Maddison dubbed the Neoliberal Order.

They had similar cheerleaders – NGOs and a cast of experts comprising scientists, mostly physicists and biologists, acting as Jeremiahs warning that time was running out to avert catastrophe. These experts, possessing the gift of seeing the future, were blind to the past – ignorant of the recurring pattern of alarmist forecasts from Malthus and Jevons onwards and incurious as to why similar prophecies in the past had all been wrong. Compared to those earlier efforts, in the sixties and early seventies, forecasts of collapse and doom were high decibel, Technicolor productions for worldwide release.

Might they share the misconceptions that pre-determined the predictive failure of earlier ones? A persistent feature of the environmentalist position is to ignore economic history and fail to ask how or why industrial societies had escaped the Malthusian trap in the past.*

More often, the boot was on the other foot, as if questioning forecasts of imminent collapse was irresponsible. Responding to the charge of being Micawbers for believing that society would continue to find solutions to problems of pollution, would increase food supplies and find sufficient raw materials for future growth, the economist Wilfred Beckerman wrote in 1974:

> Our predictions are firmly based on a study of the way these problems have been overcome in the past. And it is only the past that gives us any insight into the laws of motion of human society and hence enables us to predict the future.[5]

At the same time, environmentalists called for alternatives to modern industrial society – the only economic structure known to have kept widening the gap between subsistence and people's living standards. 'We can free our imagination from bondage to the existing system and realize that twentieth-century civilization is only one, and not necessarily the best, of the many possibilities among which mankind is free to choose.' Not the words of a dope-smoking drop-out, but the view an official report to the British government in 1972.[6]

The first environmental wave is close enough to enable ageing baby boomers to look in the mirror and see a 1960s' student. It is sufficiently distant for its predictions of eco-doom – a term coined at the time – to be verified. These forecasts assumed economic growth would be ultimately destructive and cause man's collision with nature. It should therefore be deliberately halted though government action.

In addition to its alleged harm on the environment, it was argued that growth didn't make people happy; that the way GDP numbers were put together meant

---

*   Insofar as human rather than climatic history was used as a source for argument, it was of pre-industrial societies that had suffered the effects of resource depletion and natural climate variation. Thus Al Gore in *Earth in the Balance* (1993) writes of the impact of volcanoes on the climate and the disappearance of the Minoan civilization (p.58), on the lead up to the French Revolution (pp. 59–60); on the ending of warm periods and the fall of the Roman Empire (p. 64) and in triggering the Black Death and wiping out the Viking settlers on Greenland towards the end of the Middle Ages (p. 67); other climatic changes, such as changes in rainfall, leading to the disappearance of the Mycenaean civilization in 1200 BC (p. 65) and the abandonment of Fatepur Sikri in India in the sixteenth century (p. 64); the onset of warmer periods, possibly causing the collapse of the Mayan civilisation around AD 950 and the Irish potato famine in the middle of the nineteenth century (p. 69).

that the costs of growth were not fully accounted for; that growth led to private affluence and public squalor. The president of the European Commission, Dr Sicco Mansholt, called on European governments to change their economic policies. 'We should no longer orient our economic system towards the pursuit of maximum economic growth,' he wrote in a 1972 newspaper article.

The most famous anti-growth tract was produced by the Club of Rome. Founded in 1968, the Club described itself as an 'invisible college' of some seventy experts. They were united in the belief that 'the major problems facing mankind are of such complexity and are so interrelated that traditional institutions and policies are no longer able to cope with them'.[7] The Club commissioned a study based on a computer model developed by Jay Forrester, a computer scientist and management professor, that purportedly demonstrated how the world economy would develop over the two hundred years to 2170.

The results were summarized in *The Limits to Growth*, which appeared in 1972. Its headline finding generated global publicity, helping the book sell four million copies. On an unchanged trend, the limits to growth would be reached in one hundred years. 'The most probable result will be a rather sudden and uncontrollable decline in both population and industrial capacity,' the authors claimed.[8] Similarly to Jevons' call in 1865 that Britain should shrink its economy before it was forced to, the Club of Rome authors argued that precipitate decline could be avoided if immediate steps were taken to stop further growth and thereby establish a condition of 'ecological and economic stability' which they asserted would be sustainable far into the future.

Their doom-laden advice rested on a simple misconception which in turn led the authors to mis-specify the problem they were analyzing as 'growth in a finite system'. Resources are used up, not by growth, but by production and consumption. Given its premise of finite resources, the desired state of 'ecological and economic stability' would also run out of resources and collapse.* The only difference was timing; it would just happen a bit later. The two-hundred-year model runs illustrated in the book conveniently ended before they would have showed this happening. Catastrophe postponed, not avoided, by limiting growth.

Like Jevons, the Club of Rome authors expressed a particular animus towards those who reasonably thought technological progress could overcome the limits to growth. It was, they wrote 'the most dangerous reaction to our findings.'[9] Technology, they argued, had adverse side-effects. For sure, the Green Revolution

---

*   The authors evidently realized their error, because the argument was shifted to a different issue from resource depletion to the system's responsiveness to feedback loops. 'Delays in the feedback loops of the world system would be no problem if the system were [sic] growing very slowly or not at all.' Dennis L. Meadows et al, *The Limits to Growth: A Report for the Club of Rome's Project on the Predicament of Humanity* (1972), p. 144. No evidence was offered for this to conflict with the common sense observation that dynamic societies are more adaptable than static ones.

had increased food production and the number of agricultural jobs. It had also widened inequality because larger farms adopted the better methods first. Therefore an extreme version of the precautionary principle was required to deal with the threat posed by technology. Any mass-produced new technology should be forbidden unless all physical and social side effects could be demonstrated beforehand. If a new technology lifted one limit to growth, then it would have to be shown that the next check was more desirable than the one being lifted.

It was Malthusianism for a technocratic age; its vision of a post-industrial society, where change would proceed at a pace the Pharaohs of ancient Egypt would have been comfortable with. Reduced to its essentials, the logic of *The Limits to Growth* implied that human beings were the fundamental problem because humans consume resources.

The conclusions of *The Limits to Growth* were by no means extreme for the time, but fairly representative of intellectual opinion. The Organisation for Economic Cooperation and Development (OECD) set up a study group chaired by Harvey Brooks, a Harvard engineering professor. In a 1971 report, the group argued that developed societies were fast approaching a condition of near saturation. Even in higher education, people were suffering from information overload which risked stifling the production of new knowledge.

The January 1972 edition of the *Ecologist* was devoted to 'A Blueprint for Survival'. It was endorsed by thirty-seven eminent experts, including five Fellows of the Royal Society and sixteen holders of science chairs at British universities, two Nobel laureates and Sir Julian Huxley, who was a subscriber to virtually every environmental cause, from the Kibbo Kift in the 1920s, eugenics and population control in the 1930s, the first director-general of UNESCO in 1946, co-founder of the World Wildlife Fund (WWF) in 1961, and writer of the preface to the British edition of *A Silent Spring*.

Environmental problems were not accidental malfunctions of modern society, the 'Blueprint' stated. They were 'warning signs of a profound incompatibility between deeply rooted beliefs in continuous growth and the dawning recognition of earth as a space ship.'[10] Industrial society, with its ethos of expansion, was unsustainable. Then came an unambiguous, cast iron prediction: 'Its termination within the lifetime of someone today is inevitable.'[11] A choice therefore had to be made between famine, epidemic and war (those three again) or initiating 'a succession of thoughtful, humane and measured changes' to spare our children the hardship and cruelty of the first option. Even the most fervent proponents of global warming might be nervous of tying themselves to such a tight timetable for civilization's rendezvous with catastrophe.

As with global warming, scientists weren't speaking from the fringes of the policy debate but given the role of defining what the problem was. In Britain, the Conservative government asked Sir Eric Ashby, a distinguished botanist and Fellow of the Royal Society, to chair a study ahead of the UN conference on the human environment in Stockholm. Its report, *Pollution: Nuisance or Nemesis?*, contains some of the most alarmist language presented to a British government in peacetime. Acknowledging that some members of the group considered that having less pollution and a better environment was a matter of political choice, like providing more hospitals, 'others among us felt such an attitude is dangerously complacent, and are convinced that a fundamental and painful restructuring of our industrial society is necessary if mankind is to survive'.[12] Growth had to be halted in 'a deliberate and controlled manner'. The sooner this happened, the better – 'the longer the change is delayed, the more the productivity of the biosphere will be damaged, and the lower will be the eventual sustainable level for our descendants'.[13] There was no analysis or evidence to support this dire conclusion.

Most alarming to the alarmist was the lack of alarm in the general population. 'The danger of entrusting the environment to the mandate of public opinion is that most people ascribe a higher priority to the present than the future,' the report said. 'Only in time of war do people willingly make such sacrifices.' While some members of the working party believed that 'the normal pace of politics' was adequate to solve environmental problems, 'others of us feel that there is no hope of action capable of grappling with the complex environmental crisis unless the issues are presented to the public in the same stark terms of national and even racial survival as maintained in war'.[14]

Wilfred Beckerman's experience of serving on Ashby's working group sharpened his criticisms of natural scientists' approach to economic and social matters. Two years later, he wrote that scientists who might be world authorities on phenomena pertaining in the physical world 'do not have a minimal understanding of the way that the world of human beings operates in general, or, in particular, the way that society reacts to problems such as pollution and demands for raw materials'. Understanding those problems required knowledge of human beings in a social context. Some scientists might have decided to become scientists, Beckerman suggested, 'precisely to shield themselves from these phenomena and to escape into a world where problems are not on a human scale, but microscopic or astronomic'.[15]

Beckerman argued against governments intervening to ensure the long-run supply of raw materials, as it would rest on the assumption that governments were better at forecasting future trends and developing new technology than private industry. 'There is no reason to make such an assumption,' Beckerman said. 'This does not mean private industry always gets it right; it doesn't, but it usually pays for its mistakes.'[16]

At that time, counter-arguments to environmentalism tended to come from the Left rather than the Right of political spectrum. Beckerman was an economic adviser to Tony Crosland, a senior Cabinet minister in the 1964–1970 Labour government and the Labour party's foremost intellectual. Environmental policies had a distributional impact because they were inevitably skewed towards meeting the wants of the better off in society at the expense of the poor, Beckerman argued. There was a natural hierarchy of human wants. As societies became richer, improved quality of life became more important. Collective policies favoring improved quality of life therefore favoured the better off. Crosland himself had a famous difference of opinion with the American economist J.K. Galbraith. In a brilliant put-down, he mocked Galbraith and then set out a point-by-point defence of economic growth.

In a 1971 pamphlet, Crosland warned that parts of the environmental lobby were hostile to growth in principle and indifferent to the needs of ordinary people, reflecting manifest class bias and a set of middle and upper class value judgments:

> Its champions are often kindly and dedicated people. But they are affluent and fundamentally, though of course not consciously, they want to kick the ladder down behind them … We must make our own value judgment based on socialist objectives: and that objective must … be that growth is vital, and its benefits far outweigh its costs.[17]

The triumph of environmentalism during the West's years of plenty was assured by the structural weakness of its opponents. Right-of-centre political parties represented class interests favouring environmental initiatives, but left-of-centre parties were divided between representing blue-collar workers (Crosland's position) and members of the middle-class intelligentsia, such as Galbraith. There was one obstacle to environmentalism that could not be so easily overcome: obtaining Third World support needed for any global agreement on the environment.

In 1972, a representative of a leading member of the G77 challenged the fundamental basis of ecologism. Brazil's Miguel Alvaro Osório de Almeida, who served as ambassador to the US and the UN and led Brazil's preparations for the Stockholm conference, delivered an insight of great subtlety:

> the problem to be solved in fact is not achieving an 'ecological balance', but, on the contrary, obtaining the most efficient forms of 'long-term ecological imbalance'. The problem is not to exterminate mankind now, in the name of ecological equilibrium, but to prolong our ability to use natural resources for as long as possible.[18]

A year earlier, de Almeida had shocked western delegates at a preparatory meeting for the Stockholm conference. 'To be many and to be poor is offensive to the sights and feelings of developed countries,' he told them. 'Most of their suggestions do not concern cooperation for increasing income, but cooperation to reduce numbers.'

Without a formula to reconcile rich countries' environmentalism and the developing world's hunger for economic growth, the Stockholm conference, due to convene on June 5th 1972, was heading for disaster. In fact, the germ of that formula was to be found in Adlai Stevenson's final speech, the passengers on the little spaceship, 'half fortunate, half miserable'.

The person who planted it there would do more than anyone else to fashion it into a political doctrine to form the basis of a treaty of convenience between environmentalism and the Third World. Her contribution was to be decisive in ensuring that the Stockholm conference would be a starting point rather than a showdown – something that could not be avoided, only postponed.

8
STOCKHOLM

Its real leader, the source of its inspiration and the directions it set for the future of our planet, was Lady Jackson, whom the world knows best as Barbara Ward.

Maurice Strong

Swedish scientists were finding an increase in the acidity of rain falling across Scandinavia. The Swedish government raised their concern at the UN. In 1968, the General Assembly passed a resolution to convene the world's first intergovernmental conference on the environment.

For all its enthusiasm, hardly any preparations had been made for the conference, to be hosted by Sweden in 1972. Through a mutual friend, Sweden's ambassador to the UN approached Maurice Strong to head the conference secretariat. Undersecretary of state Christian Herter for the Nixon administration lobbied Strong to take the assignment. A formal offer from U. Thant, the UN secretary-general, followed. Strong did not need much persuading. It was the opportunity he had been preparing for all his adult life.

Born in Canada in 1929, as a boy growing up in Depression-era Manitoba, Strong remembers asking his mother: 'If nature could be so right, how could human society be so wrong?'[1] Inspired by Churchill and Roosevelt's vision of a global organization to ensure world peace, 'I knew at once that I wanted to be part of that endeavor.' Strong lived side by side with Inuit people in the Canadian Arctic working for the Hudson's Bay Company. At seventeen, he got his first job at the UN as a junior clerk in the passes office.

Reading *Silent Spring* deepened his concern about the environment, Carson's 'cry of alarm' confirming what he already believed. It also reinforced his belief in global governance. The environment is supranational and transcends the nation state – 'one of the great underlying truths of environmental politics', according to Strong. The establishment of effective global governance and management was the single most important challenge of the next generation, Strong said in a lecture marking the fifteenth anniversary of the Stockholm conference. 'I find it hard to conceive that civilization could continue through the coming century if we fail to do this.'[2]

In a career straddling politics and business, Strong worked in the oil and mining industries before becoming head of the Canada's international aid agency in 1966. Where some saw a contradiction between his environmentalism and digging minerals out of the ground, Strong saw 'positive synergy' between mankind's economic and environmental needs.

> It's through our economic life that we affect our environment, and it's only through changes in economic behaviour, particularly on the part of corporations, which are the primary actors in the economy, that we can protect and improve the environment.[3]

Strong's sole foray into elective politics was not, he admitted, his finest hour. In 1977, Pierre Trudeau persuaded him to become a parliamentary candidate. Within weeks, Strong asked Trudeau if he could withdraw. It was the making of him. As an unelected politician, Strong could make a far greater impact on global affairs.

Strong's lack of educational qualifications, he believed, barred him from a traditional ascent in international politics, so he pursued a business career to get noticed and get on. Power is augmented by influence derived from extensive and diverse networks, he found. Through them, Strong became powerful.

The list of his positions takes up four and a half pages of his autobiography. They include membership of the advisory boards of Toyota and Harvard's Centre for International Development, UN undersecretary and adviser to the UN secretary-general, senior adviser to the president of the World Bank (he talent-spotted the bank's president, Jim Wolfensohn, in the 1950s), Chairman of the Earth Council, trustee of the Rockefeller Foundation, a board member of the Davos World Economic Forum and of the regents of New York's Episcopalian cathedral St John the Divine, as well as being a member of the Vatican's Society for Development, Justice and Peace, Fellow of the Royal Society and member of the Club of Rome (*The Limits to Growth* was on the right lines but ahead of its time, he thought). He knew how to obtain political access and hedge his bets. In the 1988 American presidential election, Strong donated $100,000 to Mike Dukakis's campaign, became a trustee of the Democratic National Committee and fundraised for the Republican National Committee at the same time.

He believed the best form of conflict resolution was conflict avoidance. 'I had learned never to confront but to co-opt, never to bully but to equivocate, and never to yield … The oblique approach can often be the most direct one.'[4] Strong was the right man for the job, as the Stockholm environment conference was threatened by two sets of global divisions, between East and West and between North and South. Strong could not overcome the former. If East Germany was invited, the US and West Germany wouldn't attend. If East Germany wasn't invited, the Soviet Union and its main Warsaw Pact allies wouldn't show up, which is what happened. Nonetheless, Strong insisted on having a Soviet on his staff (he ended up with two) and at the conference itself he met the Soviet ambassador to Sweden almost daily to brief him on the conference.

The division between North and South was a bigger threat. In March 1971, Strong received a signal from Yugoslavia of a potential boycott by developing countries. It galvanized him. He decided he had to win over India and Brazil as the most important members of the G77. In June he went to New Delhi hoping to arrange a meeting with Indian Prime Minister Indira Gandhi. It was a long shot as rising tension with Pakistan would lead to war six months later. He got his meeting and persuaded Mrs Gandhi to deliver the keynote address to the conference.

Brazil felt its sovereignty threatened by environmentalists claiming the Amazon as a global preserve. Although Strong started a dialog with the Brazilian government, it continued to lobby the rest of the G77 to back its hard line text at the UN emphasizing national sovereignty, more additional aid and the curtailment of Strong's authority. A Swedish text supportive of Strong was adopted by a large majority while Brazil's position received the support of a smaller majority (the US and the UK voting against).

A split was averted due to Strong's recognition that bridging the divide required more than diplomatic skill. It needed something he did not possess: the intellectual creativity to devise a formula that would turn the conflict between the North's environmentalism and the South's ambitions for economic development into a new synthesis. Barbara Ward was the person he needed.

Although she did not have the instant fame Carson experienced with *Silent Spring*, it could be said that of the two, Ward's contribution was the more important. Few today take seriously the science in *Silent Spring* but Ward's concept of sustainable development has grown to be one of the dominant policy doctrines of our age.

The *New York Times* described Ward as a 'synthesizer and propagandist'. She was much more than that. Perhaps the closest comparison is Harry Hopkins' relationship with Churchill and FDR. Whereas Hopkins lived and worked in the White House, Ward's arena was global, projecting her ideas through the numerous political leaders who trusted her. At a White House meeting between the British Prime Minister Harold Wilson and Lyndon Johnson in February 1968, Johnson gave Wilson a note on the economic situation. Two and half pages of quarto, single spacing. 'It was drafted in terms which would appeal to him – the pure Roosevelt approach to the dangers of another 1931,' Wilson recalled. 'For the rest, I did not agree with the analysis. But, I said, I could trace a feminine hand in its drafting, English at that. He roared with laughter. No one knowing Washington would underrate the importance nor fail to recognize the handiwork of Barbara Ward.'[5]

Ward took her father's Quaker earnestness to improve the world and her Roman Catholicism from her mother. After graduating from Oxford, she lectured extensively and wrote her first book at the age of twenty-four. The next year she became assistant editor of the *Economist*, becoming its foreign editor four years later. She used to write the *Economist* leaders from her bed without effort or correction, what she called her 'fatal facility'.

Such was her remarkable talent as a speaker and broadcaster that in 1943 the British government sent her to the United States to win support for Britain's war effort. She stayed five months, lunching with Eleanor Roosevelt, dining with Vice President Wallace and amassing a circle of American admirers. In the 1945 election, she campaigned for Labour, escorting a young American naval officer, John F. Kennedy, whom Ward had come to know through his sister Kathleen. A speech she gave on full employment reduced the hard-bitten Ernest Bevin, Britain's most powerful trade union leader, to tears.

The next year she struck up a life-long friendship with J.K. Galbraith (a woman of rare and slender beauty, Galbraith recalled), who got her a post at Harvard's department of economics. Later she was appointed Albert Schweitzer Professor at Columbia (Nelson Rockefeller, Governor of New York, was also one of her circle) and her work received substantial backing from the Carnegie Foundation. Friendship with Adlai Stevenson opened doors to the leading figures in the Kennedy administration. She became a confidante of Robert McNamara, with whom she retained a strong influence during his time as president of the World Bank. Although she found Kennedy a cool personality who always kept his distance, the same was not true of his successor. Ward fascinated Johnson. Her books were the only ones he ever read, Johnson once said. She contributed to his speeches, including the 1964 Great Society speech. Two days before leaving the White House, Johnson wrote to her: 'Whatever mark we have made in these last five years clearly bears your stamp too.'[6]

In the 1950s, she lived in Africa and was friends with the first generation of African leaders – Kwame Nkrumah of Ghana, Tanzania's Julius Nyerere, Jomo Kenyatta of Kenya and Zambia's Kenneth Kaunda – whose economic policies helped ruin their countries. Visiting India in 1952, she met Prime Minister Jawaharlal Nehru, who had better things to talk about than five-year plans, and met his daughter, Indira Gandhi, who became a life-long admirer, a relationship crucial in persuading Gandhi to come to Stockholm.

She lobbied the Second Vatican Council on Third World development. In 1967, Pope Paul VI established the Pontifical Commission for Justice and Peace, with Ward as one of its members. The encyclical, *Populorum Progressio*, 'The Development of the Peoples', with its criticism of 'unbridled liberalism', its call for 'concerted planning' and the creation of a 'World Fund' are evidence of Ward's imprint.* It's no wonder that, for Strong, the prospect of working with Ward was 'a dream come true'.[7]

Like Strong, Ward believed world government was necessary and achievable. 'We live physically and socially in a post-national order. But we still worship the 'idols of the tribe," she wrote in a 1973 paper for the Vatican.[8] The urgent economic problems of the day required policies which could only be achieved at a 'planetary level'. Underlying the debate on Third World development was the ethical issue of world distribution of income and resources, a debate sharpened because of the unsustainability of 'the reckless economic expansion of the last three hundred years'.[9]

The belief that without massive aid transfers, developing countries could not improve their living standards was central to Ward's worldview.

---

*   The encyclical also gave the Catholic Church's sanction to the expropriation landed estates in the name of the "common good" – a position espoused by Catholic distributists such as G.K. Chesterton earlier in the twentieth century. Ward, like Schumacher, was a successor to the distributists.

The market alone cannot begin to accomplish the scale of readjustment that will be needed once the concept of unlimitedly growing wealth, mediated to all by a 'trickle down' process, ceases to be a rational possibility for tomorrow's world economy.[10]

In an *Economist* article on the eve of the Stockholm conference, Ward argued that the pursuit of 'destabilizing growth' was not the answer because it didn't reduce the gap between the rich and poor nations. 'It has no built-in tendency to redistribute the surpluses and tends on the contrary to skew still further the patterns of income it creates' – a view that has not withstood the test of time in a new era grappling with the implications of China's huge trade surpluses.[11]

The Marshall Plan should be the model for the whole world, a case she made in 1957 from the same podium on the tenth anniversary of General Marshall's speech, when she became the first woman to give the Harvard commencement address. Five years later in *The Rich Nations and the Poor Nations*, Ward argued that high income countries should allocate one per cent of their national income to aid programs. Selected by the Book of the Month club, the *New York Times* called it 'wise and inspiring'.

It was imperative, Ward argued, to manage the future. Collectively, human beings must know where they're going and what the world will look like in twenty years' time, she said in her 1966 paper *Space Ship Earth*. 'It is surely inconceivable that we should turn the whole human experiment over to forces of change which we can neither master nor even fully understand.'[12]

Marx and Engels have been heavily criticized for claiming to have discovered the scientific laws that govern history. Forecasts of environmental doom also depend on a deterministic view of history, in particular that the values and state of human knowledge of future societies are knowable by the present. In doing so, they presume the social organization of the anthill or the beehive and the curtailment of individual freedom and initiative – forces we can never fully understand or predict.

Strong commissioned Ward and the microbiologist René Dubos to write a book summing up the 'knowledge and opinions' of leading experts Strong had selected on 'the relationships between man and his natural habitat at a time when human activity is having profound effects upon the environment'.[13]

Only One Earth couldn't hide the divide between experts from First World and Third World. One from the former argued that societies should retreat from industrialization to agriculture. Representatives from the Third World argued the opposite. Industrial development should have priority over worries about future environmental damage and 'dreams of landscapes innocent of smokestacks', as one of them put it.

Global warming made a somewhat tentative debut in 1972. Small changes in the planet's balance of energy could change average temperatures by two degrees centigrade, 'downwards, this is another ice age, upwards to an ice-free age. In either case, the effects are global and catastrophic'.[14] The greenhouse effect could mean that temperatures rise 0.5°C by the end of the century. But if developing countries consumed energy at levels in the developed world, might that risk temperatures rising uncomfortably close to the catastrophic two degree centigrade threshold?

At that stage, global warming was not considered the main threat to mankind's survival. What was? 'Pockets of urban degradation in affluent countries provide us with a foretaste of what could become man's greatest environmental risk,' the authors of *Only One Earth* wrote. 'Spreading urban misery, city quarters of unrelieved ugliness and squalor in which imaginative life of young children may be as systematically starved as their bodies are undernourished.'[15] Catastrophe was just around the corner.

Although living things had survived ice ages, volcanic eruptions, earthquakes and typhoons of 'our unstable planet', they served as a warning. 'Like the giant reptiles of the Jurassic ages, some species have gone the "way to dusty death."'[16] What was the lesson of the extinction of the dinosaurs, given that it wasn't caused by the human activities that were risking the planet's viability? 'A need for extreme caution, a sense of the appalling vastness and complexity of the forces that can be unleashed and of the egg-shell delicacy of the arrangements that can be upset,' was the answer.[17]

The more immediate eggshell was Third World hostility to the promotion of environmental concerns at the expense of their development agenda. Strong convened a group from the development movement in New York. He would assess progress with Ward at one-to-one dinner meetings, always over a bottle of Dom Perignon, her favourite champagne. The message was simple and devastating. As expressed by Pakistan's Mahub ul Haq, industrialization had given developed countries disproportionate benefits and huge reservoirs of wealth, at the same time causing the very environmental problems that they were now asking developing countries to help resolve. Those who had created the mess in the first place should pay the costs of cleaning it up. Strong realized more needed to be done to save the conference.

In June 1971, Strong held a week-long seminar at Founex, fifteen minutes' drive from Geneva airport. 'One of the best intellectual exchanges I have ever participated in,' Strong recalled. 'It had a profound influence both on the Stockholm conference and on the evolution of the concept of the environment-development relationship.' Founex produced a three-part deal which gave developing countries what they sought. In turn, it gave Strong what he wanted – the presence of Third World countries at the conference.

The first part was the assumption that the Third World would not emulate the developed world's path to industrialization, expressed as a non-binding aspiration on the part of the Third World. Developing countries would 'wish' to avoid the patterns of development of industrialized countries.

The second related to national sovereignty. Suppose this turned out to be wrong? Where environmental objectives conflict with development objectives, each country had the right to decide the trade-off for itself.

The third part was about money: 'If the concern for human environment reinforces the commitment to development, it must also reinforce the commitment to international aid. It should provide a stimulus for augmenting the flow of resources from the advanced to the developing countries.'

The Founex deal was packaged in a formula that bound together a contradiction – the environmental problems of developed countries are caused by too much development but the Third World's environmental problems are caused by too little. A political compromise lacking internal consistency or empirical validation, its second half contains an important truth. As societies get wealthier, they can afford to spend more and place a higher value on a clean environment and unspoiled nature. When Bjorn Lomborg produced evidence suggesting that far from getting worse, environmental indicators had been improving as wealth had increased, he was attacked for demonstrating the validity of one part of the Founex formula – only it was the wrong half. As a political formula, Founex did the trick.

On June 5th 1972, the first session of the world's first environmental conference took place in Stockholm's Royal Opera House. Traffic delayed the opening ceremony. None of the communist states turned up apart from China, Romania and Yugoslavia. But the Third World nations did come. Kurt Waldheim, the new UN secretary-general, warned against man's unplanned, selfish and ever-growing activities. 'While the environment is an emerging, new and very serious problem, we must not forget that development is still the highest priority,' Waldheim said, acknowledging the position of the largest bloc in the UN.[18]

At times, the conference was chaotic. 'There is a strange sensation here of large groups of people wandering aimlessly about looking for someone else,' a reporter wrote.[19] In addition to a fleet of Saabs and Volvos, the Swedish government provided two hundred bicycles painted in the UN's blue and white. Not enough, delegates complained. By the conference's second day, half had disappeared, most ending up in hotel rooms.

The sense of disorganization was partly a result of one of Strong's masterstrokes. His most important allies in prosecuting the environmental agenda were the NGOs that had helped generate the political momentum for the conference in the first place. Despite resistance from the UN bureaucracy, Strong enlisted volunteers, led by the secretary-general of League of Red Cross Societies and Baron Axel von dem Bussche (a former member of Germany's anti-Hitler resistance) to encourage NGOs to come. More than four hundred NGOs did, attending a parallel Environment Forum at Hog Farm just outside the city.

Ward shuttled between Hog Farm and the main conference, arranging briefing sessions and tickets. According to one government insider, the NGOs had little discernible effect on 'the real action'. That wasn't why Strong wanted them there. It was to hold governments' feet to the fire after the agreements had been signed and everyone had gone home.

The NGOs and prominent experts also helped generate press interest. Biologist Paul Ehrlich showered praise on Strong. It was an 'absolute miracle' that Strong had said that poor countries could not close the gap with rich countries.[20] It was just ten years before the start of China's growth spurt.* For Ehrlich's fellow biologist Barry Commoner, the culprits were capitalism, colonialism and especially the US. The environmental crisis 'wrenched open' the brutality of racial competition for survival. Producing for the common good, not for private profit, would solve it. Colonialism had caused the population explosion, Commoner claimed. Rich countries should now pay reparations for it.[21]

Inside the conference, host Prime Minister Olof Palme also attacked the US, demanding the conference examine the environmental impact of the Vietnam War 'ecoside'. Even so, the Nixon administration managed the conference with considerable skill. On the conference eve, the EPA announced its DDT ban, leaving other countries to play catch up. Shortly before it ended, President Nixon announced a $100 million fund to finance new environmental activities, while France and China came under attack for jointly opposing a resolution calling for a halt to nuclear testing.

In her keynote address, Indira Gandhi blamed the profit motive for wrecking the environment and keeping people poor. The West's affluence had been achieved at the price of the domination of other countries, the wealth of the few coming about 'through sheer ruthlessness'. Modern man, she said, must re-establish an unbroken link with nature.[22] There were limits to how much should be done to protect the environment. India did not wish to impoverish the environment further, but could not forget its own people. 'When men feel deprived, how can we speak about preserving animals?'[23]

Barbara Ward, the conference's other star, addressed delegates in her sunglasses; it was unsustainable if two thirds of humanity stayed poor so that one third could stay rich. Robert McNamara, president of the World Bank, claimed that the evidence was 'overwhelming' that a century of very rapid growth had contributed to a 'monstrous assault on the quality of life in the developed [sic] countries'.

China, making its first appearance at a major international conference since taking its seat at the UN, used the conference to attack imperialism and colonialism. It tried to turn the proposed Preamble to the Declaration on the Human Environment into a Maoist diatribe. The environment had been endangered by 'plunder, aggression and war by colonialists, imperialists and neo-

---

*   Ehrlich has a record of making predictions that turn out spectacularly wrong. In 1970 he said that if he was a gambler, he would take even money that England would not exist in the year 2000. Quoted in Julian L. Simon, *The Ultimate Resource* (1981), p. 101.

colonialists'. Theories of over-population were attacked along classic Marxist lines ('the notorious Malthusian theory is absurd in theory and groundless in fact').[24]

As the conference drew to a close, it was touch-and-go whether it would end in failure. Exhausted delegates haggled about the final wording of a draft declaration and argued what had been achieved during the eleven days of the conference. Strong had to deal with the prospect of a walk-out by the Chinese, as Beijing hadn't instructed them how to vote on the final declaration. He astutely suggested that instead of leaving the room, they should just stand behind their seats. That way, they'd neither vote, nor abstain, but be reported as present. It enabled the conference chairman to declare the resolution passed by consensus.

'What came out of Stockholm is about what we expected – not much,' said Brazil's Carlos Calero Rodrigues. In his speech to the conference, Strong admitted that the draft declaration was less than the inspirational and comprehensive code of international environmental conduct that was needed. It was, he claimed, an indispensable beginning.

The Stockholm declaration enshrined the Founex contradiction, that the economic development of advanced nations caused environmental problems; in developing countries, the same process reduced them. It then went on to set out twenty-six principles. These included a condemnation of apartheid and all foreign and colonial domination (principle one); that natural resources should be subject to 'careful planning or management' (2); the benefits from finite resources should be shared by all mankind (5); the need for the transfer of 'substantial' financial and technological aid to developing nations (9); that schools and the mass media should disseminate information on the environment (19); and that international policy should be decided cooperatively by all countries 'on an equal footing' (24), i.e., making it more difficult for the West to impose its environmental priorities on the rest of the world.[25]

The conference also agreed nearly a hundred recommendations ranging from the need for genetic cataloging to measuring and limiting noise emissions. These included half a page of recommendations concerning the atmosphere. Governments should be 'mindful' of activities in which there is appreciable risk of effects on climate. Ten baseline stations should be established in remote areas to monitor changes in the atmosphere that might cause climatic changes. The World Meteorological Organisation should continue to carry out its Global Atmospheric Research Programme to understand whether the causes of climatic changes were natural or the result of man's activities.[26]

Compared to the half page given to global warming, the oceans and marine pollution took four and a half pages. Lured by the prospect of near limitless supplies of minerals on and beneath the seabed, in the first decade after Stockholm, the environmental and developing world agenda focused on the oceans. Attention only switched to the atmosphere and global warming from the mid-1980s.

The G77 backed Kenya's bid to host the new UN Environment Program based in Nairobi and Strong became its first head. Stockholm's most important legacy was the twinning of global environmentalism with the Third World's aid and development agenda as a way of managing their inherent contradictions.

The first environmental wave had risen with great suddenness and force. A decade separates *Silent Spring* and the Stockholm conference. Even more sudden was the speed of its apparent collapse.

When Barbara Ward died in 1981, obituarists ran out of superlatives, the *Guardian* calling her one of the most brilliant contributors to economic and political thought since the 1930s. Her contribution to the forging of global environmentalism and her role at the world's first conference on the environment barely rated a mention.

9
BREAKING WAVE

It is almost impossible for historians to understand how in advance of some of history's great revolutions, it was often not even realized that a change was about to take place.

Henry Kissinger

We have learned that more is not necessarily better, that even our great Nation has its recognized limits, and that we can neither answer all questions nor solve all problems.

Jimmy Carter, January 20th 1977

At 2pm on October 6th 1973, over two hundred Egyptian aircraft flew over the Suez Canal and two thousand artillery guns started a series of barrages. Twenty minutes later, the first Egyptian troops crossed the canal. Using high-pressure water pumps, they blasted passages through the Bar Lev line, a fortification running the length of the canal made of sand and concrete. Sixteen hours later, ninety thousand troops, eight hundred and fifty tanks and eleven thousand additional vehicles were on the east side of the canal. At the same time, the Syrians attacked Israel from the north.

A survey of polls in the US, Great Britain, and Germany suggest that 1972 marked the high point of concern about the environment. The circulation of the *Ecologist* more than halved between 1973 and 1976. Although membership of the largest environmental pressure groups continued to rise, only half a dozen of the environmental magazines started on or after Earth Day 1970 survived. Though in retreat, environmentalism broadened its reach. During this period, America's national security establishment became a functional ally of environmentalism as concerns about energy security coincided with environmentalists' campaign against fossil fuels.

The cause was the same event that led to the sudden disappearance of environmentalism from the world stage. Of all the wars of the twentieth century in which the major powers were not combatants, none had such an effect as the Yom Kippur war. To Henry Kissinger, it brought the curtain down on the post-war period.

The seemingly inexorable rise in prosperity was abruptly reversed. Simultaneously, inflation ran like a forest fire through the industrialized countries and recession left millions unemployed. … Transcending even the economic revolution was the emergence of oil as a weapon of political blackmail. The industrial democracies saw imposed on them not only an economic upheaval but fundamental changes in their social cohesion and political life.[1]

Environmentalists seized on this as vindication of their arguments. The party is over, Barbara Ward declared. Fritz Schumacher poured scorn on those who expected 'to get back onto the happy road of economic growth'.

> The present situation, I am certain, has nothing in common with any previous 'depression' or 'recession' ... It is not part of a cycle, is not a 'correction' or 'shake-out' or anything of this sort: It is the end of an era.[2]

Those who couldn't see this, Schumacher said, were indulging in the psychological exercise of 'the refusal of consciousness'.

One political leader decided to overcome the refusal of consciousness.

Barely mentioned in the 1976 presidential election, Jimmy Carter made the energy crisis the centrepiece of his presidency. In a June 1979 interview, it was put to Carter that a poll showed that sixty-five per cent of Americans did not believe there was a gasoline shortage. 'Is there a shortage?' the president was asked. 'Yes,' Carter replied.[3]

The facts suggested otherwise. In 1973 world oil production was 58.3 million barrels of oil a day. Output rose steadily to 65.7 million barrels a day in 1979. The energy crisis was created by governments, not geology. Shortages in the US were the result of government regulation rather than world supply. America's European allies were also enveloped in a refusal of consciousness. On the first day of the 1979 Tokyo G7 summit, Carter endured a bitter and unpleasant lunch when he was personally abused by German leader Helmut Schmidt. 'One of the worst days of my diplomatic life,' as Carter later called it.[4]

The world was not running out of oil. The major oil producers had formed a cartel and were exploiting their pricing power. But the circumstances of its sudden emergence enormously amplified its impact. The success of the initial Arab attacks was the single most traumatic event in Israel's history. Four days into the war, Israel decided to push deep into Syrian territory towards Damascus. To relieve the Syrians, the Egyptians launched further attacks, enabling the Israel Defence Forces to counter-attack. The Israelis crossed the canal and began to advance towards Cairo and cut off the Egyptian Third Army. Within a matter of days, Israel, Egypt and Syria had felt their existence at risk.

From mid October 1973, the US began a twenty-three-day airlift. It was meant to have been under the cover of darkness. When one of the first aircraft arrived in broad daylight, James Schlesinger, US defense secretary, said 'much of the population of Tel Aviv went out to the perimeter of the airport to cheer'. The airlift ricocheted around the Arab world. Two days later, Arab oil producers announced they were increasing the oil price from $3.01 to $5.01 a barrel, described by Kissinger as stunning and unprecedented. On October 20th, Saudi Arabia announced it was halting oil exports to the US. The next day Kuwait, Bahrain, Qatar and Dubai followed suit. The oil weapon had been unsheathed, triggering a wave of panic buying.

Throughout 1973, the Saudis had threatened to use oil as an economic weapon. In a TV interview in the late summer, King Faisal warned

> America's complete support of Zionism against the Arabs makes it extremely difficult for us to continue to supply the United States' petroleum needs and to even maintain our friendly relations.[5]

The threat had been discounted in Washington because of the failure of an attempted embargo in 1967. Such had been the Nixon administration's complacency that when, in 1969, the Shah of Iran offered to sell Washington one million barrels of oil a day for ten years for a strategic reserve, it was rejected out of hand.

By November 1973, President Nixon was telling Americans they faced the most severe energy shortage since the Second World War. They would have to use less heat, less electricity, less gasoline. 'The fuel crisis need not mean genuine suffering for any American, but it will require some sacrifice by all Americans.'[6] There would be a ten per cent reduction in flights and a fifteen per cent cut in supplies of domestic heating oil ('we must ask everyone to lower the thermostat in your home by at least six degrees') and requested highway speed limits be cut to fifty mph.

Invoking the Manhattan project and the Apollo space program, Nixon announced Project Independence: by 1980, America would meet all its energy needs without any imports. It marked a one-hundred-and-eighty-degree turn from the lowest cost, free trade principles of the 1952 Paley Commission and set a course which Jimmy Carter, George W. Bush and Barack Obama followed. It was also a complete flop. In 1973, the US imported 6.3 million barrels of oil a day. By 1980, this had risen to 6.9 million and, in 2007, the last year before the onset of recession, America imported 13.5 million barrels of oil a day, more than double the 1973 level.[7]

A December meeting of Persian Gulf producers in Tehran decided to increase the oil price from $5.12 to $11.65 a barrel. 'One of the pivotal events in the history of this century,' according to Kissinger.[8] The oil price increase from the beginning of the Yom Kippur War was unprecedented. Expressed in March 2009 dollars, it rose by $41.14 in four months and peaked $2.40 higher in June 1975, before trending down to a low of $46.33 in November 1978. Unprecedented but not unsurpassed. The oil price increase from January 2007 of $72.24 (in March 2009 dollars) to $123.73 in July 2008 was two thirds bigger than the 1973 oil price shock.[9] By then, the world had learned to live with high and volatile oil prices without turning it into a crisis.

In February 1974 Western foreign ministers met in Washington to take forward Kissinger's proposal for an oil buyers' group. France was strongly opposed. 'I won't be able to accept, no matter what conditions are established, a situation which requires us to forego Arab oil, for even a year,' President Pompidou told Kissinger.[10] Major oil importing countries such as West Germany and Japan decided to let their economies adjust to higher oil prices. German motorists were not going to wait in long lines to fill up their cars. 'We're going to recognise that if we want oil, we're going to pay the world price,' West Germany's economics minister said. The US went down the route of much greater regulation. The 1973 Emergency Petroleum Allocation Act authorized a raft of petroleum price, production, allocation and marketing controls, turning the oil price shock into an energy supply crisis.

The rift between Europeans and American policy became a defining feature of the Carter presidency. Carter was a true believer; he'd read *The Limits to Growth*. It had, he said, melded together scientific data 'to show a dismal prospect, even for the survival of mankind, if we fail to change some of those trends quickly'. Its language became a major theme of his inaugural address and his presidency.

Politically Carter's problem was that he saw limits where his fellow country-men did not. Convincing Americans about the energy crisis was like pulling teeth, Carter complained. By 1979, according to a briefing by his pollster Pat Caddell, Americans did not believe his warnings about future energy shortages. They were convinced that both the government and oil companies were either incompetent, dishonest or both.

'Tonight I want to have an unpleasant talk with you,' Carter told Americans three months into his presidency.[11] The energy crisis was the 'moral equivalent of war.' The alternative might be 'national catastrophe', Carter claimed. The lines at gas pumps had gone, but the energy problem was now worse; 'more waste has occurred, and more time has passed by without our planning for the future'. The immorality of waste was Carter's overarching theme. 'Our energy problems have the same cause as our environmental problems – wasteful use of resources.'

The next day, Carter presented his energy plan to Congress. He wasn't expect-ing any applause and didn't get any. It was 'torn to pieces in Congress in much the same way puppies might rip the stuffing out of a rag doll', critics noted.[12] A month after declaring war on oil and natural gas usage, Carter sent Congress a message on the environment: 'The transition to renewable energy sources, par-ticularly solar energy, must be made.'[13] He had solar panels installed on the roof of the White House (they got taken down for maintenance in 1986 and were never put back), backed gasohol (a mixture of ten per cent ethanol and ninety per cent petroleum), called for the creation of an energy security corporation financed by $5 billion of energy bonds, and a solar bank. Twenty per cent of America's energy should come from solar power by 2000, a target Carter said was crucial. The tar-get might as well never have existed. By 2007, renewable energy contributed less than seven per cent of America's needs while the contribution from solar energy was little more than a drop in the ocean – less than one tenth of one per cent.

The Nixon-era price controls inherited by Carter had a two-tier system so that the price of oil from developed reserves was capped at a lower price than oil from new fields. Selling gasoline at two different prices created local shortages and long lines at filling stations. During the 1976 primary campaign, Carter had pledged to remove price controls on natural gas prices, helping him win the crucial Texas primary. But he reversed himself once he got to the White House in what his domestic policy adviser Stu Eizenstat thought possibly the most fateful domestic decision of his presidency.

Deregulating energy prices wouldn't work, Carter claimed, because OPEC dominated the global energy market. Instead Carter got former defense secretary James Schlesinger, who had been a Rand Corporation systems analyst in the 1960s, to design a national energy plan. Its aim was to put the market to work to conserve oil and gas. 'Why not then, use the real market?' asked the plan's conservative critics. 'Even a systems analyst, however brilliant, would find it impossible to simulate the complexities of a fairly simple market system, let alone one as huge as the energy market,' they argued.[14]

Carter's energy policy was based on the (false) premise that energy supplies were soon going to be exhausted. 1973 world oil production was 58.5 million barrels of oil a day and rose to sixty-six million barrels a day in 1979.[15] Neither was it the case that America's reserves of energy were running low. In April 1977, shortly before Carter launched his energy plan, the Energy Research and Development Agency concluded that America's natural gas reserves could be expected to exceed its total energy needs well into the twenty-first century. The director of the US Geological Survey, Dr Vincent McKelvey, said that following a slew of recent discoveries America's natural gas reserves were about ten times the energy value of all previously discovered oil and gas reserves in the US combined. He left his post shortly after.

The more Carter talked about energy, the lower his poll rating sank. In 1977, he made three television addresses on the subject as well as one to Congress and countless appearances at public forums talking about the energy crisis. It left everyone confused, including Carter himself. In June 1979, Carter told an interviewer: 'I don't claim that the government has done a good job with the energy problem; it hasn't.'[16] When asked by Dan Rather in August 1980 to grade his performance, Carter gave himself a straight 'A' on energy. 'Very good, I think we've done better than we had anticipated, and that's one of the great achievements of our administration.'[17]

1979 was a particularly tough year for Carter's energy policy. In April he announced another energy package, including a windfall tax (preventing energy suppliers 'profiteering' from changes in energy prices was a constant theme). The public was cool and the attitude of Congress 'disgusting', Carter wrote in his diary. At the end of June came his mauling by Helmut Schmidt in Tokyo. To save face, the Europeans agreed to a formula that would meet Carter's objective of cutting oil imports by counting oil from the British North Sea as pooled with the other European members of the G7.

On his return home, Carter planned another televised address on energy for Independence Day. After reading the draft, he cancelled it and went to Camp

David to reflect. 'Some of the most thought-provoking and satisfying [days] of my presidency,' Carter wrote in his memoirs. He emerged to give what he thought was one of his best speeches, one that history knows as Carter's malaise speech.

By this stage of his presidency, the energy crisis had mutated into a moral and spiritual crisis. Inflation, unemployment and all the economic ills that ailed America were blamed on imported oil. 'This intolerable dependence on foreign oil threatens our economic independence and the very security of our nation,' Carter told America. 'It is a clear and present danger.' He asked Congress to give him authority to impose standby gasoline rationing and announced that he was setting import quotas on foreign oil. 'Beginning this moment, this nation will never use more foreign oil than we did in 1977 – never,' Carter emphasized.

For a time, it looked like the tide had turned. Oil imports had peaked in 1977 and continued to decline until 1985. Since the trough that year, apart from during the early 1990s recession, oil imports grew each year. Every year since 1994, oil imports exceeded 1977 and in 2009 were over one third higher.[18]

Carter's energy policies provide a test run for similar policies Western governments adopted in response to global warming. Energy policies based on the assumption of energy shortages end up creating shortages. As one historian of the Carter years has written, the energy crisis was a politically constructed artefact.[19] German car drivers did not queue for petrol because the German government decided to let the market work.

Invoking national security is a useful indicator of poor quality policy. Compelling countries to use higher cost, less efficient sources of energy than those available from the world market inflicts a continuing penalty on economic performance. The Paley Commission's lowest cost principle is better economics: obtain the lowest cost energy in whatever form, from whatever source, and let the economy adapt to changes in energy prices is likely to be more efficient than pre-emptively imposing high-cost energy on energy consumers. The Shah of Iran's suggestion to the Nixon administration of building a strategic energy reserve is a better answer to the threat of temporary supply disruptions than permanently hobbling the economy with mandates for high cost domestic energy.

Carter left another bequest to posterity. In his May 1977 environmental message, Carter announced a study on the implications of world population growth for natural resources and the environment. Work on *The Global 2000 Report to the President* took almost the rest of his term. In his farewell address, Carter said that population growth and resource depletion were one of the top three issues confronting America. 'The report is credible,' Madeleine Albright told colleagues on the National Security Council when it appeared in 1980.[20] Measured by the one and a half million copies sold, it was also a success.

'The world must expect a troubled entry into the twenty-first century,' the report predicted.[21] For sure, the report wasn't anticipating the possible effect of the millennium bug on computer programs. Neither did it anticipate what actually happened. Writing shortly after the century's end, Maddison found that

> the world economy performed better in the last half century than at any time in the past. World GDP increased six-fold from 1950 to 1998 with an average growth rate of 3.9 per cent a year compared to 1.6 per cent from 1820 to 1950, and 0.3 per cent from 1500 to 1820.[22]

As was the custom with such efforts, the study, led by Gerald Barney, a physicist, did not analyze why previous studies predicting a reversal of the benevolent economic trends since the Industrial Revolution had failed. Instead, the report's authors believed their methodology imparted an optimistic bias to their gloomy conclusions. In 2000 people would be poorer than in 1980 and life for most would be more precarious. Erosion of the Earth's 'carrying capacity' and the degradation of natural resources would give rise to global problems of 'alarming proportions'.

The most serious environmental impact would be on food production, caused by 'an accelerating deterioration and loss of resources essential for agriculture'. The report predicted that by 2000, more people, especially babies and children, would be dying from hunger and disease. Many of those (un)lucky enough to survive would be 'mentally and physically handicapped' by childhood malnutrition. In fact, life expectancy continued to lengthen. According to the UN, over the last two decades of the century global life expectancy at birth rose by four years to sixty-five and, for the least developed, countries it rose by three years, from forty-eight to fifty-one.[23]

Food production of developing countries, the report conceded, would increase, but not by enough to keep ahead of population growth. Wrong again. Compared to 1970, the supply of calories per head in developing countries rose by twenty-five per cent and the supply of nutrient per head grew by twenty per cent.[24] Inevitably the report predicted that oil would be running out. During the 1990s, oil production would approach its geological estimate of maximum production capacity. In 1980, world oil output, temporarily depressed by the Iranian Revolution, was 62.9 million barrels of oil a day. Ten years later, this had risen by four per cent to 65.4 million barrels a day and increased by fourteen per cent during the 1990s to reach 74.8 million barrels a day in 2000.[25]

In one respect, *Global 2000* was prophetic of a new era. 'Rising carbon dioxide concentrations are of concern because of their potential for causing a warming of the earth,' it said.[26] A doubling of atmospheric carbon dioxide by the middle of the twenty-first century would lead to a 2-3°C rise in temperatures in the Earth's middle latitudes and greater warming of polar temperatures which could lead to the melting of the Greenland and Antarctic ice caps, a gradual rise in sea levels and the abandonment of many coastal cities.

Had Carter been re-elected in 1980, might it have brought forward the age of global warming to the beginning of the 1980s? More likely, it would have meant a much more problematic launch than the one global warming was to have in

1988, when the stars were in perfect alignment. Environmentalism needs prosperity to thrive. Poor economic performance during the second half of the 1970s, especially in Britain and the US, meant that the economy came first.

Global warming also needs a benign global order. As *Global 2000* recognized: 'An era of unprecedented global cooperation and commitment is essential.' On Christmas Eve 1979, the Soviet Union invaded Afghanistan. In response, Carter stopped further grain exports to the Soviet Union and banned the US from competing in the Moscow Olympics. The Cold War needed to be won first.

There was another factor. The planet wasn't warming, or at least not many people thought it was. The opposite seemed to be happening. According to James Rodger Fleming, one of the leading historians of the science of global warming,

> by the mid 1970s global cooling was an observable trend. The US National Science Board pointed out that during the last twenty to thirty years, world temperatures had fallen, 'irregularly at first but more sharply over the last decade'.[27]

Global warming needed a warming world – economically, geostrategically and climatically. The world was not ready.

'How much do you miss dinosaurs? Would your life be richer if those giant pre-historic flying lizards occasionally settled on your front lawn?' Ronald Reagan asked a radio audience in 1977.[28] Reagan often discussed the energy crisis and environmental issues in radio talks during the Carter years. The question about dinosaurs was his way of suggesting that species had become extinct before man's appearance, but that feeling guilty about species loss had led to legislative over-reaction.

Reagan described himself as an environmentalist at heart who, like the majority of people, was somewhere in the middle between those who'd pave over everything in the name of progress and those who wouldn't let us build a house 'unless it looked like a bird nest'.[29] He didn't buy a CIA report cited by Carter that claimed the oil would be gone in thirty years. Past government reports had consistently under-estimated oil reserves. If they'd been right, the oil would have run out by now. Even if true, what was the point of Carter's conservation measures to reduce consumption by ten per cent if all it meant was that the oil ran out three years later? Gas-guzzling Californians were not to blame for the state's gasoline shortage, Reagan argued. The fault lay with government regulations that capped the price of oil and prevented Californian refineries from importing the low sulphur crude they were designed for. Having reduced oil production, the government was now proposing to reduce consumption. 'Why don't we try the free market again?' Reagan asked.

As president, that's what he did. One of Reagan's first acts was to lift all remaining domestic petroleum price and allocation controls. The price of oil peaked at $39 a barrel the month after Reagan took office. It fell to $18 a barrel five years later and ended the decade nearly fifty per cent lower than the peak after taking account of inflation. The 1980s turned out to be a decade of oil glut. The 1970s energy crisis had been solved.

After 1973, it had been an article of faith among environmentalists such as Schumacher that economic growth had gone for good, as a result of its collision with environmental limits. The pursuit of high standards of consumption in an advertising-crazed society would condemn societies to perpetual inflation, Barbara Ward argued in 1973. If pursued across the world, it would precipitate such an exhaustion of resources and such a toll of pollution that the technological system would simply crack under the strain.

And so, paradoxically, the renewed prosperity of the 1980s that environmentalism said could not happen became the prerequisite for environmentalism's own revival.

10
PUPATION

Does the flap of a butterfly's wings in Brazil set off a tornado in Texas?

Edward Lorenz, 1979

Before reaching their adult form, holometabolous insects such as butterflies undergo a complete metamorphosis. When they hatch, they are very different from the adult. Inside are the rudiments of the adult together with small blobs of tissue called imaginal disks. When the larva has grown enough, it stops moving to form a protective cocoon. The larva pupates and the imaginal disks grow into adult structures. What was a caterpillar becomes a butterfly.

In the decade and a half after the Stockholm conference, environmentalism underwent a form of pupation. It lost some features and gained others. Over time, the crusade against economic growth was replaced by talk of sustainable development and green growth. Environmentalists thought the sudden disappearance of growth was desirable. It took time for environmentalists to adapt to the new environment.

In 1974, the Club of Rome produced a sequel to *The Limits of Growth*. In some ways *Mankind at the Turning Point* was even more extreme. It too claimed to be a scientifically conducted assessment based on computer models. Mankind's 'crisis' was defined as two widening gaps, between man and nature and between North and South. 'Both gaps must be narrowed if world-shattering catastrophes are to be avoided,' although the book did not explain how a computer could measure the gap between man and nature.[1] Nonetheless, the authors argued that scientists had every right to make recommendations based on their subjective judgments, otherwise scientists would be barred from discussion of mankind's future, leaving the arena to those more ignorant about the world's future course.

Nearly four decades later, the hypothesis that scientists can see the future better than non-scientists can be tested. Comparing economic growth to cancer, the authors, Mihaljo Mesarovic, a professor of systems engineering, and Eduard Pestel, a professor of mechanics, predicted a series of regional catastrophes.* In the early 1980s, South Asia would be hit by a catastrophe which would peak around 2010. 'There is no historical precedent for this kind of slow destruction – the massive, agonising reduction of the population of an entire world region once inhabited by several billion people.'[2] Indeed.

---

* The impression of mankind as a form of malignant tumor was reinforced by a passage quoted at the beginning of the book's first chapter: "The world has cancer and the cancer is man." To add a Doctor Strangelove twist, in the second half of the 1970s Mesarovic and Pestel's "World Integrated Model" was used by the Pentagon for planning purposes. Julian L. Simon & Herman Kahn (ed.), *The Resourceful Earth: A Response to Global 2000* (1984), pp. 37–8.

Climate change was an additional threat. The continuous increase in carbon dioxide would lead to a rise in global temperatures, but the increase in particulate matter would cause falling temperatures. Either one would be disastrous, but cooling was more of a threat. 'Since 1945 the second trend has appeared to be prevailing. If it continues it will have grave consequences for [the] food production capacity of the globe.'

The *Guardian* thought *Mankind at the Turning Point* might be the most important document of the year. Otherwise it caused hardly a ripple. The world had moved on. At a conference in Philadelphia in 1976 to celebrate the bicentenary of America's independence the Club of Rome's founder, Aurelio Peccei, distanced the Club from its anti-growth crusade. *The Limits to Growth* had 'punctured the myth of exponential growth,' but further growth was needed to solve world poverty and threats to world peace. Even so, the book had served its purpose of 'getting the world's attention'. Finding solutions to the problems it had identified required puncturing a second myth – 'the myth of national competence'.[3] Peccei's view that individual nations were powerless to shape their economic future would take some hard knocks in the 1980s.

As little noticed was a conference Barbara Ward chaired between representatives of UNCTAD and Maurice Strong's UN Environment Programme in the Mexican town of Cocoyoc in October 1974. The Cocoyoc Declaration codified the basis of an alliance of convenience between the G77 and environmentalism in terms of their mutual enemy rather than what divided them. Thus it was a synthesis of Raúl Prebsich's Havana Manifesto and Barbara Ward's environmentalism.

> Large parts of the world today consist of a centre exploiting a vast periphery and also our common heritage, the biosphere. The ideal we need is a harmonized cooperative world in which each part is a centre, living at the expense of nobody else in partnership with nature and solidarity with future generations.[4]

A North American or a European child, on average, consumes outrageously more than his Indian or African counterpart, the Declaration said. The problem was not one of absolute physical shortage, but of the economic and social structures within and between countries, reflecting the legacy of colonialism. The market enabled powerful nations to exploit poor countries' natural resources at low prices and then sell manufactures back to them 'often at monopoly prices'. It demanded taxation of the 'global commons' as a first step towards a system of international taxation to generate automatic transfers from rich to poor countries.

The global commons that mattered most was the ocean. In 1973, Ward had written that the Conference on the Law of the Sea was *the* critical conference of the world's 'commons' of the twentieth century. There should be a special maritime authority to administer the ocean's mineral reserves. The Cocoyoc Declaration also said that a globally administered oceans regime had to be established 'with jurisdiction over a maximum area of the oceans'.

The Law of the Sea Treaty, negotiated through the 1970s, embodied the redistributionist principles of the New International Economic Order. It was one of the reasons that led President Reagan in 1982 to reject it. When Al Haig, Reagan's secretary of state asked him why, Reagan replied: 'Al, that's what the last election was all about … It was about not doing things just because that's the way they've been done before.'[5]

After Cocoyoc, the New International Economic Order gained momentum. Responding to a suggestion made by Robert McNamara in September 1977 the former German Chancellor Willy Brandt announced his readiness to launch an independent commission. The Brandt Commission on International Development marked a decisive move away from the militant anti-growth rhetoric of First Wave environmentalism. 'Many people in the North have questioned whether it is feasible, and even desirable, to maintain high rates of growth,' the commission acknowledged. Growth, the Brandt Report argued, helped fund environmental protection. It wasn't growth as such that was environmentally damaging, but particular technologies, lifestyles and industries which should be controlled by selective intervention.

The thrust of the report was that the gap between rich and poor was wide and getting wider, that it threatened the global economy and world peace, and could only be narrowed by a new architecture of global governance. 'Current trends point to a sombre future for the world economy and international relations,' the commission feared. Mankind was using up non-renewable resources. Pollution and exploitation of the atmosphere, sea and soil were all-embracing. 'Are we to leave our successors a scorched planet of advancing deserts, impoverished landscapes and ailing environments?' the commission asked.[6] 'The 1980s could witness even greater catastrophes than the 1930s,' it said. For an anti-Nazi who had adopted the pseudonym Willy Brandt to hide from the Gestapo, this was as bad a prognosis as it could possibly be, one that turned out to be totally ill-judged.

Among the twenty-five 'eminent persons' the commission invited to testify were Barbara Ward and Maurice Strong – and Raúl Prebisch. It was a close knit circle. One of the commissioners, former British Prime Minister Edward Heath, was a fervent admirer of Ward. Another commissioner, *Washington Post* owner Katharine Graham, was one of only three people McNamara would take a telephone call from before ten in the morning, the other two being his wife and Barbara Ward.

North-South: A Programme for Survival concluded that the world community must shape a new international economic order to transfer more resources from the North to the South. The welfare state had delivered social harmony within nations by protecting the weak and promoting the principles of justice. It had to be elevated to the global level. 'The world too can become stronger by becoming

a just and humane society. If it fails in this, it will move towards its own destruction.'[7] Rich countries should transfer 0.7 per cent of national income in aid to
poorer countries, rising to one per cent by 2000. There should be global system
of taxation, based on a sliding scale of national income, and taxes on exploiting
the seabed.

The Brandt Report was one of the items of the 1980 G7 in Venice. President
Carter also briefed fellow leaders on the *Global 2000* report. The summit
communiqué agreed that the G7 needed a better understanding of the long-
term effects of global population growth and economic development generally.
Formally welcoming the Brandt Report, the summit pointedly said that providing
aid must be equitably shared by the oil-exporting countries and the Soviet bloc,
which was tantamount to saying it wasn't going to happen. As Prebisch had
privately remarked about OPEC, 'the worst type of rich are the poor that have
been enriched'.[8]

At the 1981 G7 in Ottawa, host Pierre Trudeau prevailed on Ronald Reagan to attend the first North-South summit in Cancún later that year. Margaret
Thatcher wanted Reagan there for the opposite reason – to argue against the New
International Economic Order. At Cancún, Thatcher dismissed the Brandt Report
as fashionable talk and attacked its underlying idea of tackling poverty through
redistribution rather than wealth creation as wrongheaded. She and Reagan saw
off proposals to change the voting weights of the IMF and the World Bank. She
wasn't going to have British money in a bank run by those on overdrafts.

Cancún and Reagan's rejection of the Law of the Sea Treaty the following year
marked the end of North-South dialogue in response to the South's attempts to create
the New International Economic Order. When world leaders met in Rio de Janeiro in
1992, it was at the West's behest in the context of its fears about global warming. The
South's demands hadn't gone away.

The ideas championed by Carter and Brandt also began to attract the criticism
of academics. In the US, economist Julian Simon led a critique of *Global 2000* by
a group of academics which included one of Al Gore's favorite scientists, Roger
Revelle. Simon's studies of population growth and economic history had led him
to conclude that population growth, far from being a threat to prosperity, was
associated with rising prosperity. People were the ultimate resource, he argued.
Simon turned upside down *Global 2000*'s pessimistic end-of-century forecast: 'If
present trends continue, the world in 2000 will be less crowded (though more
populated), *less polluted, more stable ecologically*, and *less vulnerable to resource-
supply disruption* than the world we live in now.'[9]

On the other side of the Atlantic, David Henderson, a former World Bank
economist, wrote a searching analysis of the Brandt Report. Henderson argued
that the commission's view that inequalities between nations caused war had no
historical basis. War had been an endemic state of affairs throughout history,
whereas the condition of extreme inequality between nations had arisen only in

the recent past. The commission's understanding of why economies grew was naively political. Evidence of progress made by many poor countries had, in George Orwell's term, become an 'unfact'. Henderson's criticism centered on the commission's handling of uncertainty.

> It is a profound mistake to suppose that the issues of social and economic life are such that it makes sense to think in terms of 'solutions' to them, as though they were like the entries in a crossword puzzle, for which there can be found a recognized, uniquely correct and permanently valid set of responses.[10]

The belief that economic problems had determinate solutions, Henderson argued, embodied a definite magical element, 'so that the complexities and uncertainties of the world are wished away, and events are treated as though they could be made predictable and manipulable by formulae or spells'.

In December 1983, the UN General Assembly adopted resolution 38/161 – without a vote – to establish a special commission to propose 'long term environmental strategies for achieving sustainable development by the year 2000 and beyond'. To those who might have opposed it, the resolution appeared pretty innocuous. Member governments weren't bound by the commission's conclusions. Its costs were to be met by voluntary contributions from its sponsors, Canada, Japan, the Netherlands, Switzerland and the four Nordic nations.

The issues hardly appeared momentous. The previous year, the governing council of the UN Environment Programme (UNEP) had selected three topics of concern – hazardous waste, acid rain and the possible adverse environmental impact of large-scale renewable energy farms. The chairmanship was offered to the leader of the Norwegian Labour party and former prime minister, Gro Harlem Brundtland.

Like its forebears, the Brundtland Commission's report, *Our Common Future*, was predicated on impending doom. There was a trinity of crises: an environmental crisis, a development crisis and an energy crisis. 'They are one,' the report proclaimed.[11] According to Brundtland herself, a painful list of disasters had alerted 'all thinking people to the grave crisis facing our planet'.[12]

Hope was at hand. 'What we need is new concepts, new values and to mobilize will. We need a new global ethic.' The planet could be redeemed and the poor saved from sliding down a spiral of economic and ecological decline by embracing the doctrine of sustainable development.

What was the magic formula possessing such power? 'Sustainable development is development that meets the needs of the present without compromising the ability of future generations to meet their own needs.'[13]

It had the merit of superficial plausibility. But as the Paley Commission put it, the needs of future generations are unknowable – the syndrome of conserving bayberries for the electric age. In his review of the Brandt Report, Henderson argued that the distinction between essential and non-essential needs was alien to the conception of economic choice which underlies the case for using market modes of allocation in which 'there are no needs to be met regardless of cost, and to think in terms of a sharp distinction from essential to non-essential is meaningless'.[14]

At times, the search for the meaning of sustainable development gets caught in a loop of tautology. Living standards higher than 'the basic minimum' are only sustainable if consumption standards everywhere have regard for long-term sustainability.[15] Should the use of non-renewable resources be forbidden? No. 'The rate of depletion and the emphasis on recycling and the economy of use should be calibrated to ensure that the resource does not run out before acceptable substitutes are available' – a formulation that is meaningless because the world has never run out of a particular mineral. Species become extinct, not minerals.

In principle, the Brundtland formulation of sustainable development is consistent with having no policy at all other than to promote the efficient functioning of markets. This would, of course, clash with the report's presumption that markets were propelling the world towards some kind of planetary catastrophe.

So what does 'sustainable development' actually mean?

> In essence, sustainable development is a process of change in which the exploitation of resources, the direction of investments, the orientation of technological development, and institutional change are all in harmony and enhance both current and future potential to meet human needs and aspirations.[16]

In relation to the concept of needs, the Brundtland Report stated that the needs of the world's poor should have over-riding priority. Poverty was a major cause and effect of global environmental problems, the report asserted. Brundtland herself went further, telling the UNEP governing council that international economic inequalities were the root cause of poverty and environmental degradation.

The need for more resources for developing countries could not be evaded. 'The idea that developing countries would do better to live within their limited means is a cruel illusion.'[17] Thus sustainable development provided green packaging for egalitarianism and global income redistribution policies. 'Perceived needs are socially and culturally determined, and sustainable development requires the promotion of values that encourage consumption standards that are within the bounds of the ecologically possible and to which all can reasonably aspire.'[18]

Sustainable development is not an economic concept, but a political doctrine with far-reaching economic implications. In her speech to the UNEP at Nairobi in 1987, Brundtland herself described sustainable development as the report's 'overriding political concept.' It would help provide 'the key to open new doors of perception and entail inspiration for humankind in its quest for progress and survival … It requires fair access to knowledge and resources and a more equitable distribution within and among nations'.[19] Sustainable development, she claimed, turned on its head what she called the zero growth dogma of the early seventies. If done right, growth could be good and developing countries had no option but to seek to grow.

Yet the shedding of the limits-to-growth hair shirt was more in the nature of a re-branding statement than a substantive shift. There might not be absolute limits, but there were 'ultimate limits', the report said. The affluent would have to adopt lifestyles and live within 'the planet's ecological means'. 'Painful choices have to be made,' the report warned. 'In the final analysis, sustainable development must rest on political will.'[20]

A commission chaired by a Norwegian social democrat, financed by wealthy, high latitude, northern hemisphere countries was always going to advocate a form of globalized Scandinavian social democracy. Nearly nine hundred organizations and individuals gave their views to the commission or assisted it in some way. Seventy-seven came from Scandinavia and Finland, compared to thirty-five from the US and fifty-one from the Soviet Union, the last signifying an important step in the greening of international relations.

The Chernobyl nuclear accident in April 1986 and Mikhail Gorbachev's policy of *glasnost* had made discussion about Soviet attitudes to the environment permissible. 'No one even imagined the extent of our ecological disaster, how far we were behind the developed nations as a result of our barbaric attitude towards nature,' according to Gorbachev.[21] At one of the commission's public hearings in Moscow seven months after the Chernobyl accident, a member of the Soviet Academy of Sciences asked whether the development of civil nuclear power had not been premature, a position that before Gorbachev would have aroused the interest of the KGB (Gorbachev concluded that the Soviet Union could not dispense with nuclear power).

While the commission achieved balance between the industrial and developing nations, there was a marked imbalance within the latter, with Indonesia (ninety-six) and Brazil (eighty-eight) to the fore. Despite the 1984 Bhopal industrial disaster, there were only six from India and two from China.

One nation stood above all the rest: seemingly every organ of the Canadian government, its provinces and territories and a multitude of societies, students and individuals – two hundred and seven in all – were mobilized and made it their business to be involved. Maurice Strong was a commissioner and fellow Canadian Jim MacNeill, a 1971 Founex participant, was in charge of drafting its report. It is reasonable to say that the Brundtland Report and its aftermath represent Canada's most singular impression on world affairs.*

Political success required sustainable development to have something it lacked. The Brundtland Report and its antecedents made big claims about adverse trends harming the poor and the planet, which some day, would end in catastrophe, in some form. Yet these assertions were remarkably free of hard data. While

---

* There were no participants from Australia and New Zealand.

sustainable development implied limits, it couldn't say where they were or what exactly would happen if those thresholds were crossed. It was a doctrine in search of scientific authentication. As a political ideology, Marxism always claimed to be derived from scientific analysis. By contrast, sustainable development was an ideology, developed from a political formula, in search of science.

What transformed the impact of the Brundtland Commission was a joint conference of UNEP, the World Meteorological Organisation and the International Council of Scientific Unions in the Austrian town of Villach in October 1985. Called to assess the impact of increasing concentrations of carbon dioxide and other greenhouse gases on the climate, the conference concluded

> it is now believed that in the first half of the next century a rise of global mean temperature could occur which is greater than any in man's history.[22]

The estimated increase of between 0.3 and 0.7°C in global mean temperature during the previous one hundred years was, the conference said, consistent with the projected temperature increase attributable to the observed increase in CO_2 and other greenhouse gases, although the rise could not be ascribed in a scientifically rigorous manner to these factors alone.

An advisory group on greenhouse gases was formed, which liaised with Mac-Neill. Bert Bolin, a leading climate scientist and future chair of the Intergovernmental Panel on Climate Change, deemed the MacNeill channel to the Brundtland Commission essential in enabling the scientific community to get global warming onto the political agenda.

Global warming was one of four environmental threats identified by the Brundtland Commission (the others being urban pollution, acid rain and the risk of nuclear reactor accidents). If the meeting at Villach had not taken place, the thrust of the Brundtland Report and the need for the world to adopt sustainable development would have been the same. Brundtland herself did not mention global warming in her June 1987 prepared remarks to UNEP on the report.

Without global warming, sustainable development would not have shifted the world's political axis. With global warming, environmentalism had found its killer app. In turn, global warming became embedded in a pre-existing ideology, built on the belief of imminent planetary catastrophe – which many scientists subscribed to – with a UN infrastructure to support it and a cadre of influential political personages to propagate it.

A butterfly was ready to spread its wings.

11
ANNUS MIRABILIS

Global Warming Has Begun, Expert Tells Senate.

New York Times, June 24th 1988

It was hot. The date had been chosen after consulting the Weather Bureau. The air conditioning wasn't working and the windows had been left open overnight. The television lights made it worse.

June 23rd turned out to be a record-breaking day in a year when drought swept much of the United States. NASA scientist James Hansen wiped away the sweat as he told the Senate Energy and Natural Resources Committee: 'The greenhouse effect has been detected, and it is changing our climate now.'

If Philip Larkin could date 1963 as the year when sexual intercourse became a cultural phenomenon,

> Between the end of the Chatterley ban
> And the Beatles' first LP

then 1988 was the start of global warming as a political one. Parallels with the mid sixties were apposite. The rise of the Second Environmental Wave began towards the end of a decade of renewed prosperity after the lost decade of the 1970s.

> We got department stores and toilet paper
> Got Styrofoam boxes for the ozone layer

Neil Young sang in his 1989 'Rockin' in the Free World': 'Got fuel to burn, got roads to drive.' Britain experienced its Second Summer of Love, a return to the psychedelic 1960s with acid-fuelled rave parties and protests against road-building. Environmentalism needed ringing tills and gridlocked roads.

Global warming's arrival in the world was announced with a blaze of fanfares heralding potential catastrophe. Alarmism went hand-in-hand with predictions of temperature increases that turned out to be excessive. Although warnings that civilization was doomed because economic activity was destroying the biosphere had become something of a routine, two things were different this time. First, mainstream political leaders from across the political spectrum quickly joined and amplified the chorus. Second, an institutional apparatus was constructed to keep attention on the issue. Unlike the 1972 Stockholm conference and the creation of the UNEP far away from the centers of power, the Intergovernmental Panel on Climate Change was an inter-governmental body with close and pervasive relations with its sponsoring governments. The rhythm of the publication of IPCC assessment reports would help feed media interest and keep governments engaged.

And there was something else. It wasn't just one event. It was many. It was in the air.

Four days before Hansen's Senate appearance, the G7 summit in Toronto declared that 'global climate change, air, sea and fresh water pollution, acid rain, hazardous substances, deforestation, and endangered species require priority attention'. Climate change had first been mentioned in the 1985 Bonn summit, but the Toronto G7 went further. Canada had been the first western nation to support the Brundtland Report and host Brian Mulroney wanted to make sure environmental protection and the Brundtland Report were on the summit agenda.

The summit noted that the Brundtland Report's call that environmental considerations should be integrated into all areas of economic policymaking for the globe to continue to support humankind and formally endorsed 'the concept of sustainable development'. It had taken seventeen years from Founex to sustainable development being adopted by the G7. It was environmentalism's greatest international triumph up to that point.

Why did Ronald Reagan at his last G7 summit and Margaret Thatcher, then the longest-serving G7 leader, give their approval to a political doctrine that holds as its axiom that markets keep people in the developing world poor and devastate the environment? The answer is clearer with Thatcher. But there is little to suggest Reagan had changed his views. Most likely, sustainable development had been an issue the top levels of the administration had not considered that important. The first item on the summit communiqué had been the Soviet Union's withdrawal from Afghanistan and managing the de-escalation of the Cold War to a peaceful close.*

In his final year as president, Reagan continued to resist environmental regulations he thought might harm American economic interests. Earlier in 1988, Mulroney had visited Washington to push the Reagan administration into signing a treaty on acid rain. Reagan and secretary of state George Shultz ignored Mulroney's pleas, although Mulroney's meeting with Vice President Bush produced a promise that he would, a pledge that he kept after he became president. As Mulroney put it in his memoirs, 'unlike his decisive actions in East-West relations and other important international issues, Ronald Reagan, as I was to find out in April in Washington, was unable to fully seize the moment'.[1]

Eight days after the G7, Mulroney hosted a second Toronto conference. Organized by Gro Harlem Brundtland and Jim MacNeill, it was on global warming and its implications for global security. The language at the conference was alarmist, as were the temperature forecasts. The dangers from the growth of greenhouse gases were second only to nuclear war. Harmful consequences were already evident over many parts of the globe, although it wasn't specified where these were or

---

* When asked in 2010, former secretary of state George Shultz had no specific recollection of discussions on the Brundtland Report or sustainable development (email to author from Susan Schendel, October 14th 2010).

what they might be. Severe economic and social dislocation for present and future generations was predicted.

Average global temperatures were forecast to rise by between 1.5°C and 4.5°C before the middle of the twenty-first century. By 2009, one third of the way into 1988–2050 forecast period, average global temperatures, according to Britain's Hadley/Met Office temperature series, had risen by just over one quarter of a degree centigrade (0.257°C) compared to the half-degree rise implied by the lower band assumed by the Toronto conference. The top of the forecast band is way off the scale – out by a factor of nearly six.

The Toronto climate conference did make one correct forecast, anticipating the pattern of recommendations that should be adopted to combat global warming. Calling for a rapid reduction in North-South inequalities, the conference said that total emissions of carbon dioxide should be cut twenty per cent below their 1988 level by 2005, with all the cuts falling on the rich nations. Additional energy consumption by developing nations should be met by even steeper reductions on the part of the developed world. At some stage, rapid economic growth by developing nations implied that developed countries would have to cut their carbon emissions to less than zero.

In September came a powerful voice from an unexpected quarter. Echoing Revelle and Suess in their 1957 paper, Margaret Thatcher warned humans might have 'unwittingly begun a massive experiment with the system of the planet itself'.[2] Global warming now had a political champion of undoubted world stature. 'We are told,' although she didn't say by whom, 'that a warming of one degree centigrade per decade would greatly exceed the capacity of our natural habitat to cope,' she said in her speech to the Royal Society. Such a rapid rise might indeed have been a cause for alarm. The degree-per-decade rise alluded to by Mrs Thatcher implied that by 2010, the planet would have warmed a full two degrees centigrade compared to the 1.8°C to 4.0°C range predicted by the IPCC in its 2007 Fourth Assessment Report to the end of the twenty-first century.[3]

Two months later in Geneva, the Intergovernmental Panel on Climate Change met for the first time. Mustafa Tolba, executive director of the United Nations Environment Programme, asked Bert Bolin to be the IPCC's first chairman. An independent-minded meteorologist and sometime scientific adviser to Sweden's prime minister, Bolin was clear what the IPCC needed to do. It should provide a stringent assessment of the available scientific knowledge he thought had been lacking thus far. Bolin spared neither supporters nor opponents of anthropogenic global warming from criticism, whether it was Hansen's congressional testimony ('the data showing the global increase of temperature had not been scrutinized well enough'); the forecasts adopted by the Toronto climate conference ('not yet been generally accepted by the scientific community') or Mrs Thatcher ('seriously misinformed' on the scale of the warming effect).[4]

Bolin's opinions were not limited to scientific matters. He thought Mrs Thatcher's interest in global warming was motivated by a desire to close Britain's coal mines. More importantly, his opinions on the economics and politics of global warming, which would become a large part of the IPCC's work, were colored by his belief in sustainable development. For Bolin the scientist, there was no boundary demarcating the positivist world of science and the clash of normative values that characterizes political debate. Writing of the IPCC's work in developing policies to combat global warming, Bolin said, 'striving for sustainable development and an equitable world must be central features of any study of this kind'.[5]

In his role as IPCC chairman Bolin adjudicated technical disputes far outside his field of expertise. For example, he opposed the use of Purchasing Power Parity indices (used to compare GDP across frontiers) in the IPCC's economic scenarios because he (wrongly) thought they made assumptions incompatible with 'the basic goal of a future equitable world.'[6] From the outset, the IPCC and its first chairman were not going to limit themselves to diagnosis; they were going to set the parameters for the cure as well. Politicization of the IPCC's work was not an incidental risk that needed to be managed; it was inherent in the mandate the IPCC had devised for itself and Bert Bolin's expansive view of it.

The Geneva meeting also established IPCC's tripartite working group structure which has persisted to the present day; Working Group I to assess the available evidence, led initially by Sir John Houghton of Britain's Meteorological Office; Working Group II to assess the environmental and socio-economic impacts, led by Dr Yuri Izrael of the USSR's Hydro-Meteorological Service; and Working Group III to formulate response strategies, led by Dr Frederick Bernthal of the US State Department – and thought the most important of the three.

Twenty years to the day after his 1988 testimony, Hansen gave a reprise. 'Now, as then, I can assert that these conclusions have a certainty exceeding ninety-nine per cent,' Hansen testified.[7] There were already present, Hansen stated, the elements of a 'perfect storm' and a 'global cataclysm'. Even without further greenhouse gas emissions, the Arctic would soon be ice-free during summer. If emissions followed a business-as-usual scenario, Hansen claimed there would be two metre sea level rises by the end of the century and hundreds of millions of refugees. His call for the chief executives of oil and coal companies to be tried for 'high crimes against humanity and nature' caught the media's attention, as such inflammatory language was designed to do.

It also had the effect of diverting media scrutiny from other problematic aspects of Hansen's testimony. While complaining of a 'wide gap' between what the scientific community knew and the public believed, there was a widening gap Hansen was less keen to talk about – that between his 1988 temperature forecasts and what actually happened.

In his 1988 testimony, Hansen had presented long-term temperature projections based on three scenarios for future emissions of carbon dioxide and other greenhouse gases. These ranged from scenario A, which Hansen described as 'business as usual' through an intermediate scenario B to scenario C, which he described as illustrating the impact of 'draconian emission cuts' resulting in no net addition to concentration of greenhouse gases from 2000, i.e., by then, man-made emissions would be completely absorbed by natural sinks.

Because Hansen's team over-forecast the growth of other greenhouse gases apart from carbon dioxide, Hansen's intermediate scenario B provided the best fit with actual emissions, but not, interestingly, with the trend in global temperatures. By 2008, the trend in observed temperatures was well below scenario B and, in some cases, below even that for scenario C. Of Hansen's three scenarios, his temperature increase projections for scenario C provide the best fit with observed temperatures.

The message that the world didn't hear from Hansen's presentation was that it was getting the temperatures assumed by Hansen's no net emissions scenario without incurring any of its costs. Such a conclusion would have been consistent with the data, but would have undermined Hansen's characterization of the 'global warming time bomb' and demolished his claim of a better than ninety-nine per cent certainty.

Unlike the blanket TV coverage Hansen generated at his 1988 appearance, there were no cameras when Mrs Thatcher addressed the Royal Society on September 27th 1988. Told that the prime minister's speech was going to be on climate change, the BBC decided it wouldn't make the TV news.[8]

The speech had been a long time in the making. Flying back from visiting President Mitterrand in Paris in May 1984, Mrs Thatcher asked her officials if any of them had any new policy ideas for the forthcoming G7 summit in London. Sir Crispin Tickell, then a deputy undersecretary at the Foreign Office, suggested climate change and how it might figure in the G7 agenda. The next day, Tickell was summoned to Number 10 to brief the Prime Minister. The eventual result was to make environmental problems a specific item, and a statement in the London G7 communiqué duly referred to the international dimension of environmental problems and the role of environmental factors, including climate change. Environment ministers were instructed to report back to the G7 meeting at Bonn the following year, and duly did so.

Tickell's interest in climate change dated from the mid 1970s. Influenced by reading Hubert Lamb's book *Climate History and the Modern World*, Tickell took the opportunity of a one-year fellowship at Harvard to study the relationship between climate change and world affairs and wrote a book on the subject in 1977. Tickell recalls that people at the time thought his interest in the topic a bit eccentric, but he was the only non-American to have participated in the Carter administration's *Global 2000* project.

By 1987, Tickell had been appointed Britain's ambassador to the UN and informally was acting as Thatcher's envoy on global warming, his position at the UN making him privy to gossip from other nations.[9] On two occasions, Thatcher recalled him from New York to brief her. Tickell was always struck by her determined approach; in the world of politics, Thatcher was a woman in a man's world and someone with scientific training in a non-scientific world.[10] To meet the test, you had to know what you were talking about; if she challenged you, you needed to be sure of your ground; she could be remarkably vigorous, Tickell found. The prime minister wanted the government to grasp the importance of global warming. Ministers were called to Number 10 for briefings by climate scientists. 'You are to listen, not to speak,' the prime minister told them.

Returning to England for his summer holiday in 1988, Tickell called on Thatcher and suggested she make a major speech on global warming. She thought the Royal Society would be the perfect forum for it. She spent two weekends working on the draft with George Guise, one of her policy advisers.

In the speech, Thatcher addressed the Society as a scientist and a Fellow who happened to be prime minister. Environment policy was her main subject. Action to cut power station emissions and reduce acid rain was being undertaken 'at great and necessary expense', she said, building up to her main theme. 'The health of the economy and the health of the environment are totally dependent on each other,' implicitly rejecting the view of conventional economics of there being a trade-off between resources used for environmental protection which couldn't be used to raise output or increase consumption.[11] It was also clear that the G7's endorsement of sustainable development had not been an oversight or meant to be taken lightly, as far as she was concerned. 'The government espouses the concept of sustainable economic development,' she stated, although the new policy had not been discussed collectively by ministers beforehand or with Nigel Lawson, the Chancellor of the Exchequer.

Thatcher concluded her speech by referring to one of the most famous events in the Royal Society's history, when in 1919 Arthur Eddington displayed the photographic plates taken during the total eclipse of the sun earlier that year. The eclipse enabled Eddington to record whether light from distant stars was bent by the sun's gravity and verify a prediction of Einstein's theory of relativity.

Whitehead witnessed Eddington's demonstration. The scene, tense as a Greek drama, he wrote, was played out beneath the portrait of Isaac Newton, the Society's twelfth president, 'to remind us that the greatest of scientific generalizations was now, after more than two centuries, to receive its first modification'. In Vienna, reports of it thrilled the seventeen-year-old Karl Popper. What particularly impressed Popper was the risk implied by Einstein's theory, that light from distant stars would be deflected by the Sun's mass, because it could be subjected to a definitive test: 'If observation shows that the predicted effect is definitely absent, then the theory is simply refuted. The theory is *incompatible with certain possible results of observation* – in fact with results which everybody before Einstein would have expected.'[12]

These considerations led Popper to argue that the criterion for assessing the scientific status of a theory should be its capacity to generate predictions that could, in principle, be refuted by empirical evidence, what Popper called its fal-sifiability, or refutability, or testability. Every 'good' scientific theory is a prohibi-tion. The more a theory forbids, the better it is. Scientists should therefore devise tests designed to yield evidence that the theory prohibits, rather than search for what the theory confirms. If we look for them, Popper argued, it is easy to find confirmations for nearly every theory. 'Only a theory which asserts or implies that certain conceivable events will not, in fact, happen is testable,' Popper explained in a lecture in 1963. 'The test consists in trying to bring about, with all the means we can muster, precisely these events which the theory tells us cannot occur.'[13]

In 1988, proponents of global warming did not provide a similar black and white predictive test of the key proposition of global warming: the degree of warming with increasing levels of carbon dioxide in the atmosphere. It is therefore incapable of being falsified. The issue is not the capacity of carbon dioxide to absorb radiation in a test tube, which had first been demonstrated by John Tyndall in 1859, but the effect of increased levels of atmospheric carbon dioxide and other greenhouse gases on the temperature of the atmosphere. An answer can only be derived from empirical observation.

Revelle and Suess's characterization of mankind carrying out a large-scale geophysical experiment, further illustrates global warming's weakness as a scientific statement and its strength as a political idea. While prejudging the results of an experiment constitutes bad science, the proposition simultaneously generates powerful calls to halt the experiment before it is concluded. Yet questioning the science would inevitably be seen as weakening the political will to act. It created a symbiotic dependence between science and politics that marks 1988 as a turning point in the history of science and the start of a new chapter in the affairs of mankind.

12
TWO SCIENTISTS

There's no doubt that any man with complete conviction, particularly who's an expert, is bound to shake anybody who's got an open mind. That's the advantage of having a closed mind.

John F. Kennedy

Science is one of the very few human activities – perhaps the only one – in which errors are systematically criticized and fairly often, in time, corrected.

Karl Popper

We cannot know how the future will judge the era of global warming that dawned in 1988. We can listen to the voices from the past. The past can provide perspective. History is fluid, events are contingent, seemingly immutable categories that condition thought and action, evolve over time. Many scientists were environmentalists, subscribing to the tenets of environmentalism. Until the advent of global warming, environmentalism could not claim the support of science.

Before science became set in the concrete of the present, it is timely to introduce two scientists from the past and preview the terrain across which the science unfolded. The first died well before the global warming era and the main body of professional work of the second also took place before then. Neither, therefore, had any affiliation in the disputes about global warming.

Percy W. Bridgman was a Harvard physicist, winner of the 1946 Nobel Prize for his discoveries in the field of high pressure physics and one of nine signatories of Albert Einstein and Bertrand Russell's 1955 manifesto on nuclear disarmament. Bridgman's is a voice that speaks of the critical importance of verification in science and of the inherent problem in talking about the future, which for him assumed an entirely different character from studying the past. Writing of the 'inscrutable character of the future', Bridgman thought that statements about the future belonged to the category of pseudo-statements.

> I personally do not think that one should speak of making statements about the future. For me, a statement implies the possibility of verifying its truth, and the truth of a statement about the future cannot be verified.[1]

Why is verification important? 'Where there is no possibility of verification there can be no error, and "truth" becomes meaningless.'[2] Bridgman accorded greater confidence to the astronomer than to the man in the street to a verification of whether the sun had risen that day. If it had not, it would have involved suspension of the laws of mechanics which the astronomer had observed were manifest in every moving thing.

Global warming, of course, involves projecting increased levels of atmospheric carbon dioxide and other greenhouse gases into the inscrutability of the future. However it is not possible to determine global temperature in advance, solely by reference to the laws of physics and chemistry. The value of the climate sensitivity of carbon dioxide, the impact on average global temperatures from a doubling of the concentration of carbon dioxide, is unknown.

Though an upper limit can be inferred from global temperatures not increasing dramatically since the Industrial Revolution, there is no lower limit above zero.* Making an assumption more precise than this amounts to educated guesswork about what the forecaster believes *should* happen. If someone says, 'It will rain tomorrow,' what they mean is, 'I expect it will rain tomorrow.' Our predictions embody our expectations. As Bridgman put it, 'to "correctly-predict" is a verb which has only a past tense.'[3] Thus predictions of future temperature increases tell us less about whether the planet was heading for a climatic fever than the beliefs of the people making them.

Our second scientist was Britain's first leading climate scientist. Hubert Lamb was a friend of Guy Stewart Callendar. He had a long career in the Meteorological Office before being appointed founding director of the Climatic Research Unit in 1972 at what later became the University of East Anglia. Delving into weather records and other historical evidence such as diaries, pictures and population records, Lamb used his knowledge of weather patterns to argue that climate variability was an important but neglected factor in explaining the development and decline of various human societies.

In his last book, published in 1982, Lamb repeatedly used terms that no climate scientist would risk today. Warmer climates are genial, writing of the Medieval Warm Period, 'the genial climate of the high Middle Ages which coincided with the twelfth- and thirteenth-century climax of cultural development.'[4] The cooling trend of the fourteenth century is a 'climatic deterioration'. His research into that cooling (a factor he thought behind recurrence of the Black Death in Europe) led Lamb to be worried by the post-Second World War cooling trend. 1965 had been the year which halted the long recession of many Alpine glaciers and saw the return of Arctic sea ice around Iceland. The cooling trend in the northern hemisphere reversed an earlier upward trend, which had led Callendar to doubt the Callendar Effect, as global warming was known in those days. Lamb acknowledged that some mild winters in the 1970s had 'caused judgment to hesitate and produced the impression that the spate of writings in the 1960s about climatic change had overstressed the subject', adding: 'That was before the winters of record severity in parts of North America and Europe in the later 1970s.'[5]

---

*   In a 1991 paper, Canadian mathematician Christopher Essex demonstrated that the effect on global surface temperature of increased levels of carbon dioxide could be less than zero. Christopher Essex, 'What Do Climate Models Tell Us About Global Warming?' in *Pure and Applied Geophysics PAGEOPH*, Vol. 135, Issue 1 (1991), pp. 125–33.

Lamb therefore assessed the growing belief in anthropogenic global warming against his deep immersion in meteorological history and of post-war cooling. Claims of a warming of between two and eleven [!] degrees centigrade by 2100 were, he wrote carefully, 'an opinion, seemingly founded on firm scientific knowledge, which has to be taken seriously, even though we may notice some grounds for doubt and skepticism'.[6] One cause of Lamb's skepticism was his observation that the World Meteorological Organisation and other modellers assumed a range for natural fluctuation of plus or minus one degree, later reduced by the IPCC to as little as plus or minus half a degree. Lamb understood probably better than anyone that natural variations could greatly exceed the limited ranges assumed by later climate modellers.

It is only by assuming a small range of natural variation that modellers could project that warming caused by greenhouse gas emissions will 'swamp' all other elements of natural variation so the global warming 'signal' can emerge from the background 'noise'. Lamb responded

> There is a fallacy in this part of the case since it is impossible to define a figure for the range of natural variation of climate which is meaningful in this connection. The record of prevailing temperatures, whether over the past few centuries or over the much longer-term record of ice ages and interglacial periods, shows that the range of variation is itself subject to variation.[7]

To establish the amplitude of natural climate variability requires analysis of the climate record of past centuries and millennia. The instrumental record does not stretch sufficiently far – the oldest, the Central England Temperature record, goes back to 1659 on a monthly basis and to 1772 on a daily one. Creating a temperature record going back to the beginning of the current inter-glacial period over ten thousand years ago depends on the use of controversial proxy data. These include evidence from lake sediment, ice cores, pollen records and tree ring and coral ring studies.

What the bounds of natural variability might be – whether it is plus or minus one or half a degree – alters the significance of the estimated 0.6 degree (plus or minus 0.2) rise over the course of the twentieth century. If the rise falls within the bounds of natural variability, there is no need for man-made global warming to explain it. Having a version of the climatic past that showed a narrow amplitude of natural fluctuation was therefore essential in prosecuting the case for global warming.

There was a second reason. Part of the idea of global warming is that elevated temperatures are harmful. This invites comparison with previous climatic episodes. Are current temperatures unprecedented in human history? If not, how badly did human societies and ecosystems fare during previous warm periods? On the other hand, might such episodes, far from being harmful, have been benign and, to borrow Lamb's term, genial?

After 1988, climate scientists and the IPCC expended great effort on reconstruct-ing a global temperature record. In doing so, climate scientists descended from the impregnable heights of unverifiable assertion about the future to the arena of evidence of the past and the requirements of reproducibility and verifiability. 'Your statement that you have verified something is indifferent to me unless I believe that I could make the verification also,' was the way Bridgman described it.[8] It was on this field that the science of global warming suffered its first defeat, a strategic reversal from which it has not yet recovered.

A dramatic new version of the temperature record over the last thousand years was presented in 2001 in the IPCC's Third Assessment Report. The message appeared compelling: 1998 was the warmest year in the warmest decade in the last one thousand years. The Hockey Stick graph, which told a story of a declining temperature trend in the northern hemisphere for nine hundred years, then an abrupt uptick at the beginning of the twentieth century, could not meet a verification standard such as that required by Bridgman. Despite considerable obstruction, it was shown by independent researchers, notably Canadian mining analyst Steve McIntyre and his economist colleague Ross McKitrick, to be a statistical contrivance.

The counter-attack defending the Hockey Stick demonstrated even more clearly how rapidly scientific standards have changed. It was now argued that the universal principle of verification should be restricted to a select group of experts. Only papers in peer-reviewed literature that supported their views were to be considered legitimate in settling the dispute.

Popper issued a stark warning against such thinking with implications that go to the fundamental basis of Western civilization.

> An open society (that is, a society based on the idea of not merely tolerating dissenting opinions but respecting them) and a democracy (that is, a form of government devoted to the protection of an open society) cannot flourish if science becomes the exclusive possession of a closed group of specialists.[9]

A further danger with having a closed group is in reinforcing the natural tendency of the human mind to fit facts into preconceived theory. To explain why planets and stars were in positions they shouldn't according to the Ptolemaic system of the Earth being at the centre of the universe, medieval astronomers added geo-metrically complex and implausible epicycles. Adding 'epicycles' has come to be synonymous with adopting stratagems to avoid questioning the basic premise of a scientific proposition.

The hypothesis of dangerous global warming is sufficiently accommodating to be consistent with a wide spectrum of outcomes, even if these were not specifically predicted beforehand. The observed pause in the warming trend in the first decade of the twenty-first century was not predicted by climate scientists. After the event, they explained it away by invoking the Pacific Decadal Oscillation and not ques-tioning the premise of their models.

1988 came at the end of a first decade of warming, reversing the previous twenty-year trend of declining temperatures. This decline, which Lamb had thought indicated a long-term cooling trend, has since been interpreted as masking the rise that would have happened were it not for aerosols – microscopic particles formed in the atmosphere from coal-fired power station emissions – that reflect back the sun's incoming radiation.* As with the post-1998 cessation of warming, this rationalization is of the same type as that applied by medieval astronomers in invoking their epicycles. 'The physicist has, and the rest of us should have a temperamental aversion to *ad hoc* constructions in his theorizing,' Bridgman cautioned. For the use of ad hoc argument means there is no second method of arriving at a theory's terminus, and therefore, Bridgman argued, 'no method of verifying that the construction corresponds to anything "real"'.[10]

Instead of seeking to meet Popper's criterion of falsifiability, the activities of climate scientists conform to an earlier injunction pre-dating the Scientific Revolution: 'Seek, and ye shall find.' Climate scientists have followed this teaching from the Sermon on the Mount in their search for confirmatory evidence of global warming in shrinking ice caps, retreating glaciers and inferring past temperatures from tree rings.

The lack of a falsifiability test led scientists and governments into putting their faith in the existence of a scientific consensus as the guide to scientific truth. This too, according to Bridgman, is problematic. Consensus is one of the methods used to weed out individual error, but its role was strictly limited. He specifically rejected the notion of defining science as the consensus of all competent observers. Relying on verification by 'competent persons' during any epoch meant having a selection of people who had been subject to definite preconditioning. It might, Bridgman thought, signify little more than the discovery that under similar conditions, different human animals react the same way

> merely because there are so many common features in the construction of all human animals. Verification by consensus is always to be accepted with extreme reservation.[11]

In contrast to Bridgman's caution, consensus-building is at the heart of what the IPCC does. It rests on a highly questionable claim: the subjective opinions of

---

* Coal-fired power stations occupy a unique place in the demonology of global warming, being held responsible as the main culprits of global warming and the cause of the cooling in the third quarter of the last century. In 2008, Hansen compared trains carrying coal to power stations to those that carried Jews to the concentration camps. Thus a speculative view of the future has been accorded the same evidential status as historical fact, moreover the fact of the greatest crime of all time, and doubters about the former are subliminally bracketed with deniers of the latter (James Hansen, 'Global Warming Twenty Years Later: Tipping Points Near' www.columbia.edu/~jeh1/2008/TwentyYearsLater_20080623.pdf).

individual scientists can be transmuted into scientific knowledge through the process of being formed into an agreed judgment. In his 2009 book *Why We Disagree
About Climate Change*, climate scientist and IPCC author Mike Hulme cited the
IPCC's expressions of increasing confidence in attributing rising temperatures to
human activities as how consensus works to establish scientific knowledge; from
the caution of its First Assessment Report in 1990 to what Hulme called the bull-
ishness of its Fourth Assessment Report in 2007. According to Hulme,

> at each stage of the process, substantial judgment on the part of the
> scientist(s) is required; a judgment formed through dialog, dispute and
> compromise rather than through detached and disinterested truth-
> seeking.[12]

There was little alternative to relying on such procedures, Hulme argued. 'In
complex and uncertain areas of knowledge "objective methods" alone are rarely
adequate to establish what is known.'[13]

The purpose of consensus is to gain agreement with a view to taking collective
action. Here, an altogether larger claim is being made: consensus can transform
belief into knowledge. This requires us to accept that there is no fundamental difference between belief and knowledge, so long as there is collective agreement on
it by those whose opinions count – the experts who don the mantle of scientific
authority. In this way today's scientists have become modern day alchemists transforming the lead of subjective belief into the gold of objective knowledge.

Such a claim has profound implications for the nature of science itself. It permits
scientists to slip the bounds of what had traditionally been understood as science
into pseudo-science, to adopt Popper's term, and into the realm of futurology. For
sure, eminent scientists can make more of an impact on the popular imagination
with their essays in futurology than with their scientific research. In 2006 the
physicist Stephen Hawking made headlines with his prediction that humans
would have to leave Earth to avoid the risk of the species being wiped out by
an asteroid or a nuclear disaster. In 2003, Britain's Astronomer Royal and future
president of the Royal Society Martin Rees wrote a book, *Our Final Century*
(*Our Final Hour* in the American market), which estimated that mankind had
only a fifty per cent chance of surviving the century thanks to threats even more
dangerous than global warming.

The historian must accept a definition of science as being what scientists do.
However, the historian can make two claims. The first is that the predictive record
of scientists on the future of mankind is rather worse than a random selection
of people off the street. The scientists who put their names to the *Blueprint for
Survival* in 1972 – predicting civilization would come to an end in the lifetimes
of people then living – proved spectacularly poor guides to the future. The actions
of ordinary people who maintained their commitment to the future, the most
important one being the decision to have children, have been vindicated.

Second, the nature and purpose of science has undergone profound change in recent decades. Until around the middle of the last century, science was about making discoveries and solving problems by refining and modifying theories so that they better conformed to what had been found in laboratory experiments and seen through telescopes. The purpose of these activities was the advancement of scientific knowledge.

Bridgman for one rejected the idea of a purpose beyond the pursuit of knowledge. 'To attempt to broaden the concept of science to include social responsibility, as appears to the popular temper at present can result only in confusion,' he wrote in 1959.[14] Three years later, the scientist and philosopher Michael Polanyi said much the same thing in an article extolling the republic of science: 'I appreciate the generous sentiments which actuate the aspiration of [society] guiding the progress of science into socially beneficent channels, but I hold [this aspiration] to be impossible and indeed nonsensical.'[15]

The role of science became like medicine. The point of acquiring knowledge is to diagnose and cure malady. Thus climate science, particularly after 1988, developed as the most important branch of what might be called global therapeutics. The principal justification for climate science is not the pursuit of knowledge for its own sake, but diagnosing the world's ills and defining the parameters of the therapy required by the patient.

With the irruption of global warming into world affairs, scientists mounted a radical extension of Polanyi's republic, in the process turning it upside down. Scientists were now directing science into what they decided were socially and environmentally beneficent channels in a project that required governments to implement the prescription scientists and like-minded experts had devised. The justification for governments to do so was not subject to objective verification, because global warming precluded this. Henceforth, the word of scientists was to be taken on trust.

The rigorous methodology developed by Popper and the verification standards required by a physicist such as Bridgman were now replaced. This did not happen because of the emergence of a superior epistemological standard; that is to say from the development of a new and sharper theory of knowledge. The explanation is quite different. Far from becoming obsolete, they had become inconvenient. Dependence on consensus made it all the more important to ensure that the consensus continued to prevail, especially as the future of the planet was at stake.

This provided strong incentives to sustain the consensus and maintain the world's interest; otherwise, the action which the consensus required would not be taken. For exactly the same reasons, those who expressed their doubts represented a threat. Dissenters needed to be crushed and dissent de-legitimised. They were stooges of oil companies and fossil fuel interests, free market ideologues, or climate change deniers. In the years after 1988, much reputational capital of prestigious scientific bodies and of governments was sunk into global warming, further reducing the incentives for being open to debate and criticism.

For Popper, intolerance and lack of respect for dissenting opinions were antithetical to the precepts of an open society. Dissent is also linked to the success of science in expanding scientific knowledge because criticism is the engine of the growth of knowledge. 'What is called objectivity consists solely in the critical approach,' Popper wrote in 1963.[16] The growth of scientific knowledge came not from the accumulation of observations, Popper argued, but from the repeated overthrow of scientific theories and their replacement by better or more satisfactory ones. This marked science out from virtually all other fields of human endeavor. Since the beginning of the Scientific Revolution in the seventeenth century, the West left far behind all past civilizations – surpassing even the brilliance and originality of the ancient Greeks. Scientific advance represents the supreme intellectual accomplishment of Western civilization.

A precondition for the Scientific Revolution was the freedom to question orthodoxy and the rejection of authoritarianism. Scientists based their claim to progress by pointing to the standards later distilled by Popper. Global warming's inability to meet the verifiability and falsifiability standards set by the Scientific Revolution constitutes a reversion to pre-modern modes of defining what should be accepted as knowledge based on appeals to authority.

The significance of global warming in the history of science is not that it represented a change of paradigm within a branch of science. It was a change in the paradigm of science itself.

13
GREEN WARRIOR

*The core of Tory philosophy and for the case for protecting the environment
are the same. No generation has a freehold on this earth. All we have is a life
tenancy – with a full repairing lease.*

Margaret Thatcher, October 14th 1988

In 1988, climate scientists put global warming onto the international political agenda. In the three and a half years to the Rio Earth summit in June 1992, politicians joined scientists. In signing the UN Framework Convention on Climate Change, they committed themselves to the convention's objective of stabilizing greenhouse gases at a level that would avoid 'dangerous anthropogenic interference in the climate system'. After Rio, debating the science of global warming became superfluous. Politics had settled the science.

The speed with which the rhetoric of alarm was ratcheted up is astonishing. A March 1989 article in the *Financial Times* on the Green Revolution in international relations spoke of heightened concern about the environment, 'rising panic is scarcely too strong a phrase'. A conference the previous month in New Delhi warned of apocalyptic scenarios. 'Global warming is the greatest crisis ever faced collectively by humankind,' its final report claimed. The next month saw an international conference at The Hague organized by the French, Dutch and Norwegian prime ministers. Twenty-four governments signed a declaration suggesting that human life was under imminent threat. 'The right to live [sic] is the right from which all other rights stem,' the Hague Declaration began. 'Today, the very conditions of life on our planet are threatened by the severe attacks to which the earth's atmosphere is subjected.'[1]

In addition to the three sponsoring governments, signatories included Canada, West Germany, Australia, New Zealand, Brazil, India and Zimbabwe, whose oppressed citizens had more reason to fear their ruler than the composition of the atmosphere. News of global warming reached Buckingham Palace. In her 1989 Commonwealth Day message, the Queen spoke of threats to the environment so far-reaching it was difficult to grasp them. 'We hear, for example, of the possibility of radical changes in our climate leading, among other things, to a rise in the sea level, with all that would mean for small islands and low-lying regions.'

For some, the second environmental wave came just too late. A rival accused Al Gore of running for national scientist in his 1988 presidential campaign. 'I started to wonder whether the issues I knew to be important were peripheral after all,' Gore wrote in *Earth in the Balance*, 'I began to doubt my own political judgment.'[2] Global warming barely registered in the presidential election that year. True, the Democratic party platform called for regular world environmental summits to address global threats such as the 'greenhouse effect', but its presidential candidate Michael Dukakis framed the environment as a law enforcement issue. 'We're going to have an Environmental Protection Agency that is more interested in stopping pollution than in protecting the polluters,' he told the Democratic National Convention in Atlanta.[3]

Not to be outdone, Vice President Bush told the Republican Convention a month later that he was going to have the FBI trace medical wastes and infected needles dumped into America's lakes and rivers.[4] At a campaign stop in Michigan, Bush said he would use the 'White House effect' to tackle the greenhouse effect and pledged to convene a global conference on the environment at the White House during his first year as president.

James Baker, the new secretary of state, was quicker off the mark. His first speech was to the IPCC's Working Group III on developing response strategies, ten days after taking office. 'We face the prospect of being trapped on a boat that we have irreparably damaged, not by the cataclysm of war, but by the slow neglect of a vessel we believed impervious to our abuse,' Baker told representatives from forty countries.[5]

The World Resources Institute, one of Washington's leading environmental pressure groups, praised Baker's speech. It was 'a positive shift in commitment' compared to the Reagan administration. Bolin thought otherwise. Baker had not realized the scope of oncoming climate change or the scale of the response required.

Things weren't so positive in a closed door session at which the US officials proposed a plan to collect more data before acting. Some observers blamed the call for delay on mid-level staff held over from the Reagan administration. They couldn't have been more wrong. Opposition came not from remnants from the previous administration, but from the most senior White House staff.

Responding to international pressure, in May the Bush administration conceded that it would support the negotiation of a framework convention. At the G7 summit in Paris, host President Mitterrand observed that environmental issues had never before been the subject of as many conversations and so many decisions. The G7 leaders agreed that 'decisive action' was urgently needed to understand and protect the Earth's ecological balance. This included 'common efforts to limit emissions of carbon dioxide and other greenhouse gases, which threaten to induce climate change, endangering the environment and ultimately the economy', adding that they strongly supported the work of the IPCC. Protecting the environment required 'early adoption, worldwide, of policies based on sustainable development'.

Risking international isolation, the Bush administration maintained a clear line: it would not agree to anything that legally bound the United States to targets or to reduce its emissions of carbon dioxide by a specific time. This led to fierce criticism. William Nitze, a deputy assistant secretary of state in the Reagan and Bush administrations until he resigned in 1990, charged that the eventual outcome had been a success for American diplomacy, but a failure of presidential leadership. (He went on to serve in the Clinton administration at the Environmental Protection Agency.) Al Gore compared Bush's policy to the Senate's refusal to ratify the Versailles Treaty that prepared the way for the Second World War. By not providing the world with leadership in the face of what Gore called the 'assault by civilization on the global environment', Bush was 'inviting a descent toward chaos'.[6]

That the Bush administration's position was not a function of some hard-line ideology can be seen with the Reagan administration's response to fears about depletion of the ozone layer. In 1974 two American chemists alerted the world that stratospheric ozone could be destroyed by chlorofluorocarbons (CFCs) being broken down to release chlorine, increasing the amount of ultra-violet radiation reaching the surface of the planet. In the 1970s, countries began unilaterally to cut their consumption of CFCs. In 1985, two months after the Vienna Convention for the Protection of the Ozone Layer had been agreed, the British Antarctic Survey found a hole in the ozone layer (which had thinned by forty per cent in eight years). The Reagan administration played a key role in negotiating the 1987 Montreal Protocol binding signatories to steep cuts in CFCs, which came into force at the beginning of 1989. President Reagan and George Shultz hailed it as a magnificent achievement. Former UN secretary-general Kofi Annan described it as perhaps the single most successful international agreement to date.[7]

US support was a product of hard-headed calculation. According to Richard Benedick, the US lead negotiator, a major breakthrough came with a cost-benefit analysis by the President's Council of Economic Advisers. Despite the scientific and economic uncertainties, the monetary benefits of preventing future deaths from skin cancer far outweighed the costs of CFC control as estimated either by the industry or by the Environmental Protection Agency. However, if the US acted alone, there would be little long-term benefit. It was better for the US to get as many countries to join in as possible. At Reagan's insistence, US negotiators lowered the participation threshold at which the agreement would come into force.

The discovery of the ozone hole provided the world dramatic narrative of industrial pollutants exposing people to increased risk of skin cancers, premature skin ageing and eye cataracts. Finding substitutes for CFCs as aerosol propellants and refrigerants was straightforward and relatively inexpensive. As a result, politicians could and did act quickly and decisively. In the words of the political scientist Scott Barrett, from whose book this account is largely drawn, 'the achievements of the Montreal controls are truly outstanding'.[8] By providing governments with a template as to how they could tackle global warming, the Montreal Protocol misled most of them – the US being the most important exception – into believing that global warming might be amenable to similarly straightforward treatment. Prime among them was the world leader who first raised the alarm.

'Mrs Thatcher, looking back over your life,' the BBC's Michael Buerk asked, 'are you really a friend of the earth?' *The Greening of Mrs Thatcher*, broadcast on March 2nd 1989, drew from Thatcher some of the most surprising things she ever said.

There was more than a hint of green in her final years at Number 10. Unlike Nixon's green phase, hers was a product of conviction, not political calculation. She was changing the political weather, Nixon was reacting to it. On the environment, Thatcher had profound differences with her ideological soul mate Ronald Reagan. She supported action against acid rain, Reagan blocked it; she believed in the dangers of resource depletion, he thought they were baloney. Even when they agreed on the ozone layer, it was for different reasons, hers being environmental, his the health and economic wellbeing of Americans.

Environmental policies used to be mostly about cleaner rivers and smog-free cities, Thatcher answered Buerk. The problems had been localized, there hadn't been the realization of a global dimension, 'there was no greenhouse effect up there somewhere. There was no ozone layer'. She said how she'd overruled scientific advice and saved the British Antarctic Survey. 'I have always been interested in Antarctica. There is some marvellous wildlife there, there is probably a good deal of mineral deposits.' It was a fantastic, icy place that wasn't a wasteland, and recounted a meeting with members of the survey. 'They came into the next room and gave me a marvellous account of everything they are doing just a few weeks ago.'

She was worried about the greenhouse effect: 'We still do not fully understand the greenhouse gases or how they are going to operate, but we do know that we have to do something.' At that stage, emissions cuts were not on that 'to do' list. Her priority was trees. 'We are giving very considerable help on research into forestry and into the planting of tropical forests and into the preservation of tropical forests.'

According to her policy aide, George Guise, Thatcher's thinking had been particularly influenced by her conversations with the billionaire financier, Sir James Goldsmith, who owned an eighteen-thousand-acre estate on the Pacific coast of Mexico and whose brother, Edward, had helped put together the 1972 *Blueprint for Survival*. Thatcher told Buerk that planting trees would help solve Bangladesh's perennial flooding.

> When President Ershad was here recently, I said: 'Look this is quite absurd. You are getting floods year after year. Really, we want the silt out of your rivers, put back on to the hills into the country behind you, into Nepal and India, and planted with trees again.'

The real difficulty with the developing world, she said, was that they wanted higher standards of living and to get out of poverty. 'That is why we have the concept under Mrs Brundtland of sustainable development with which we firmly agree.'

She also favored nuclear energy. 'I would prefer more nuclear power because it is not fundamentally interfering with the world's eco-systems.'

'Finally Mrs Thatcher,' Buerk asked, 'if and when you finish being Prime Minister, would you want to be remembered as somebody who had helped to save the world in this environmental sense?' 'Enormously so, enormously so,' she replied. 'My whole sort of political philosophy is that what you have inherited from your forefathers, it is your duty to add to it … The problems science has created, science can in fact solve and we are setting about it.'

At the end of 1988, the Maltese government sponsored a resolution of the General Assembly of the United Nations on the conservation of the climate as mankind's common heritage. The resolution said rising greenhouse gases could produce global warming which 'could be disastrous for mankind' and encouraged the convening of global, regional and national conferences to raise awareness of global warming. From 1989 there was a marked intensification in the rhythm of international conferences on global warming.

The most important were the ministerial meeting that produced the Hague Declaration in March, the Noordwijk ministerial conference in November, the Bergen Conference on sustainable development in May 1990, and the Second World Climate Conference in Geneva in November. Shortly before the Geneva conference, European Community (EC) environment ministers meeting in Luxembourg agreed an EC-wide goal of stabilizing carbon dioxide emissions at current levels by 2000 (a target that would only be met because German emissions fell with the incorporation of East Germany and Britain's dash-for-gas). The resolution also requested the IPCC to produce its report 'as soon as possible' and in time for the Geneva conference less than two years away.

At the UN, Crispin Tickell regularly held informal gatherings of the ambassadors of the five permanent Security Council members in his New York apartment and briefed the secretary-general afterwards. At one of his meetings, Tickell floated the idea of an environmental conference to mark the twentieth anniversary of the Stockholm conference. The idea was taken up. Initially it was to be at Stockholm again. The Brazilian government wanted the conference in Rio de Janeiro. The Swedish government quickly agreed, so Rio it was.

The conference momentum soon put Thatcher onto the defensive. Asked in the House of Commons why she wasn't going to the twenty-four-nation conference at The Hague, Thatcher gave no quarter: 'The conference is to set up yet another organization, which is not necessary, and it proposes that compensation should be paid – without saying how – and that sanctions should be applied if rules are not complied with – again without saying how.'[9] To the horror of civil servants, Britain did not even send an observer. According to a senior Dutch official, the new world ecological institutions could grow in the same way as the European Commission, Parliament and Court had grown out of an embryonic European coal and steel community. No wonder Mrs Thatcher stayed away.

Seven months later at the Noordwijk conference in November, with more than three times the number of countries, the negotiations went in a different direction – emissions cuts. Britain brokered an agreement between the ambitions of the Europeans for firm commitments to stabilize emissions and the refusal of the US, joined by the Soviet Union and Japan, to do so. The declaration fudged the issue by recognizing the need to stabilize greenhouse gas emissions 'as soon as possible,' while recording the view of many industrialized nations that this should be achieved at the latest by 2000. No one could have foreseen that the disintegration of the Soviet Union and the implosion of its economy would enable the Russian Federation to record a forty-one per cent fall in carbon dioxide emissions between 1990 and 2000, a reminder of the inscrutability of even the near future.[10]

The Dutch government said it would implement previously announced plans for unilateral cuts in greenhouse emissions of eight per cent by 1994, a proposal which had led to the fall of the Dutch government the previous May. In fact, Dutch carbon dioxide emissions rose by 4.6 per cent over the period. While the Dutch set a pattern other countries would follow, there was a one-off event. The delegation from Saudi Arabia told the conference that the world's largest oil producer considered global warming 'a life or death issue for considerable areas of the earth'. There was 'no argument' that carbon dioxide was the main culprit or question about the need to move to non-greenhouse gas energy generation.

The same month, Thatcher addressed the UN General Assembly. The threat to the global environment was an insidious danger threatening irretrievable damage to the atmosphere, to the oceans and to the earth itself, Thatcher declared, comparing it to the risk of global annihilation from warfare.

The speech is significant in two respects. Although the subject of global warming occupies much of the text, its core message of undiluted Malthusianism could have been the same without it.

> More than anything, our environment is threatened by the sheer numbers
> of people and the plants and the animals which go with them ... Put in
> its bluntest form: the main threat to our environment is more and more
> people.[11]

The tragedy of Easter Island was a warning of what might befall the rest of the world. Cutting down a primeval forest had led to warfare over the scarce remaining resources, the population fell and there wasn't enough wood to make boats to sail to another island. 'We must have continued economic growth in order to generate the wealth required to pay for the protection of the environment,' she told the General Assembly. 'But it must be growth which does not plunder the planet today and leave our children to deal with the consequences tomorrow.' In the past, growth happened. Now it had to be the right sort.

The other notable feature of the Mrs Thatcher's speech was its impact on American politics. It generated far more media coverage in the US than the

Noordwijk meeting and more in the US than in the UK. The National Governors Association Task Force on Global Warming was holding its first meeting at the United Nations. Its chair, Republican Jim Thompson of Illinois, called her address 'elegant, straightforward and full of common sense and leadership'. New York's Mario Cuomo used Thatcher's speech to criticize President Bush's refusal to set a timetable for emissions cuts at the Noordwijk conference earlier in the week (a position shared by the British government at that point), 'I would have preferred the President to manifest the leadership of the nation on that occasion.'[12] A copy of the speech was circulated to senior staff in the White House by Robert Grady, an associate director at the Office of Management and Budget (OMB). 'The President could easily have given this speech and taken credit for an aggressive posture with very little change in current policy required,' Grady commented. The three names at the top of the circulation list, John Sununu, the president's chief of staff, Dick Darman, head of the OMB and chief science adviser Allan Bromley were doing all they could to ensure the president did not make such a speech and that US policy did not change.

In May 1990, Gro Harlem Brundtland hosted the Bergen conference on sustainable development. 'I am nervous that time is running out,' Brundtland warned. But she was more rattled by journalists' questions on whether she supported a resumption of commercial whaling, Norway being accused of hypocrisy for hosting an environmental conference while hunting whales. The Canadian government's attempt to line up with the US in opposing emissions targets was stymied by a leak to the Friends of the Earth of a State Department cable which said that Canada would be joining Britain and the US in opposing any move on emissions cuts that went beyond the position agreed at Noordwijk. The outcry led environment minister Lucien Bouchard to pledge that Canada's emissions would be no higher in 2000 than in 1990. Like the Dutch pledge six months earlier, it was an empty promise. Canada's carbon dioxide emissions increased by twenty-three per cent between 1990 and 2000, faster than the seventeen per cent for the US.

British opposition to emissions caps also began to crack. A junior environment minister (Chris Patten, the environment secretary, decided to stay at home) signalled that Britain would have a target for stabilizing or even reducing emissions. It marked a watershed in Britain's approach to global warming, according to the *Financial Times*.

Abandonment by two of its most reliable allies did not change America's position. Timothy Atkeson, assistant administrator of the EPA, said the expected costs were 'as big as you can get.' 'We're talking about costs in excess of the gross national product of the United States,' he told reporters.[13] Although participants declared their willingness to 'assume a major responsibility to limit or reduce greenhouse gases', the outcome of the Bergen conference marked no movement from the position agreed at Noordwijk. Political progress seemed stalled as the prospect of a transatlantic rift opened up.

14
RUSH TO JUDGMENT

There are three roads to ruin; women, gambling and technicians. The most pleasant is with women, the quickest is with gambling, but the surest is with technicians.

President Georges Pompidou

Any remaining doubts must not be allowed to dissuade us from action.

Michel Rocard, November 6th 1990

While governments were engaged in an accelerating round of conferences, scientists were working to meet the deadline for the IPCC's First Assessment Report ahead of the World Climate Conference in November 1990. As chair, Bolin had to overcome a major disagreement between Working Group I, tasked with summarizing scientific knowledge under Sir John Houghton, and Working Group II, led by Soviet academician Yuri Izrael to examine potential impacts of global warming.

One of the Soviet scientists in Working Group II argued that warming might be beneficial at northerly latitudes (an opinion that had been held by Svante Arrhenius and Guy Stewart Callendar in the first half of the twentieth century). Preventative action was therefore not justified. A meeting was held in Leningrad but, according to Bolin, there was still tension. Even within Working Group I, agreement was not always easy.

Bolin had wanted to document and explain reasons for disagreement. 'I had repeatedly pointed out to the working groups that the goal was not necessarily always to reach an agreement, but rather to point out different views when necessary and to clarify reasons for disagreements when possible, but this was still seldom tried,' he recalled.[1] Despite Bolin's encouragement, public airing of disagreements was counter-cultural.

Nonetheless, the First Assessment Report went further in meeting this objective than subsequent ones. It was also more open in highlighting not only uncertainties in the science (one can be uncertain about the particular outcome of tossing a coin, but the parameters of the outcome can be defined with certainty), but also ignorance, especially in the section dealing with past climate variations and change (Section Seven of the Working Group I report). As the IPCC developed, admissions of ignorance gave way to more qualified expressions of uncertainty, often couched in spurious degrees of confidence.

According to the Working Group I Summary for Policy Makers, the observed global mean surface air temperature had increased by between 0.3°C and 0.6°C over the previous hundred years.

The size of this warming is broadly consistent with predictions of climate models, but it is also of the same magnitude as natural climate variability. Thus the observed increase could be largely due to this natural variability; alternatively this variability and other human factors could have offset a still larger human-induced greenhouse warming.[2]

The recent warming was neither unique nor could natural variability be ruled out. Section Seven of Working Group I on past climate variations noted that although the evidence pointed to a real but irregular warming over the last century, 'a global warming of larger size has almost certainly occurred at least once since the end of the last glaciations without any appreciable increase in greenhouse gases. Because we do not understand the reasons for these past warming events it is not yet possible to attribute a specific proportion of the recent, smaller warming to an increase of greenhouse gases'.[3]

Putting the twentieth-century warming in the context of past climatic changes, the report pointed out that the Little Ice Age involved global climate changes of comparable magnitude to the warming observed over the previous one hundred years to 1990, part of which could reflect cessation of Little Ice Age conditions. 'The rather rapid changes in global temperature around 1920–1940 are very likely to have had a natural origin,' the report went on. 'Thus a better understanding of past variations is essential if we are to estimate reliably the extent to which warming over the last century, and future warming, is the result of greenhouse gases,' a need that would give rise to the Hockey Stick and its prominence in the 2001 Third Assessment Report.[4]

The crucial section of the entire report was on the detection of the enhanced greenhouse effect (Section Eight of the Working Group I report). Its authors acknowledged that the 0.3°C to 0.6°C observed rise in average global temperatures permitted a number of explanations. If it had been caused by a 'man-induced' greenhouse effect, then the implied climate sensitivity of carbon dioxide (the warming resulting from a doubling of carbon dioxide in the atmosphere) would be at the bottom of the range. If a significant fraction of the warming had been due to natural variability, the implied value for the climate sensitivity would be even lower than model predictions. If a larger greenhouse warming had been offset by natural variability or other factors, then the climate sensitivity of carbon dioxide could be at the 'high end' of model predictions. Scientists' inability to reliably detect predicted signals should not be taken to mean that 'the greenhouse theory is wrong, or that it will not be a serious problem for mankind in the decades ahead'.[5]

Section Eight also carried an implicit rebuke to James Hansen and his claim two years earlier to have detected the greenhouse effect.

Because of the many significant uncertainties and inadequacies in the observational climate record, in our knowledge of the causes of natural climatic variability and in current computer models, scientists working in this field cannot at this point in time make the definitive statement: 'Yes, we have now seen an enhanced greenhouse effect.'[6]

With the possible exception of an increase in the number of intense showers, there was, the Summary for Policy Makers said, no clear evidence that weather variability would change in the future. Neither was there evidence that tropical storms had increased or any consistent indication that they would be likely to increase in a warmer world.

Nonetheless, belief in anthropogenic global warming shines through the pages of the Working Group I report, especially the twenty-eight-page Summary for Policy Makers. 'We calculate with confidence,' the summary claimed, 'that carbon dioxide has been responsible for over half the enhanced greenhouse effect in the past'.[7] At first glance, it might appear that here the IPCC had nailed its culprit. In actual fact, the claim teeters on the brink of meaninglessness. As the report admitted, scientists were unable to attribute the contribution of the enhanced greenhouse effect to the observed rise in global temperatures over the previous one hundred years. To have said that extra carbon dioxide was responsible for over half of an effect that couldn't be measured or even detected was the modern equivalent of medieval monks counting angels on pinheads.

Distilled to its essence, the message of the First Assessment Report was that global warming is happening, even though the evidence remained equivocal. In one of the most significant passages in the report, the IPCC stated: 'The un-equivocal detection of the enhanced greenhouse effect from observations is not likely for a decade or more, when the commitment to future climate change will then be considerably larger than it is today.'[8] Thus for the IPCC, detection of the greenhouse effect was a question of 'when' not 'if', as Sir John Houghton wrote in the *Financial Times*. 'We are sure that human activities are leading to climate change – although we do not claim yet to have detected it.'[9]

The most loaded claim was the report's call for action. 'If there are concentration levels that should not be exceeded,' the Summary for Policy Makers warned, 'then the earlier emission reductions are made the more effective they are,' hinting at the possibility of undefined tipping points and climate catastrophism.[10] Yet buried in the body of the report, the IPCC noted growing evidence that worldwide temperatures had been higher than at present, at least in summer, around five to six thousand years ago, although carbon dioxide levels were thought to have been quite similar to the pre-industrial era.

Alarm about global warming depended on a critical but unverifiable assumption: increased levels of carbon dioxide trigger positive feedbacks amplifying the direct warming effect from carbon dioxide. Water vapor plays the central role in assumptions about positive feedbacks. It is a more abundant and powerful greenhouse gas than carbon dioxide (the IPCC estimated that if water vapor were the only greenhouse gas present in the atmosphere, the greenhouse effect would be sixty to seventy per cent of the value of all gases included, but if carbon dioxide alone were present, the corresponding value would be about twenty-five per cent). Without positive feedback from water vapor and other sources, the IPCC estimated a value of +1.2°C for the climate sensitivity of carbon dioxide.

However, the IPCC assumed that the impact of this initial warming is amplified by increasing the concentration of water vapor, raising the initial 1.2°C to 1.9°C. There are many other feedbacks; overall, they might be positive and amplify the direct warming effect of added carbon dioxide, or they might be negative and dampen it. More water vapor might lead to more clouds being formed. 'Feedback mechanisms related to clouds are extremely complex,' Working Group I stated. 'There is no *a priori* means of determining the sign of cloud feedback.'[11]

Results from computer climate models cited in the 1990 report gave a range of 1.9°C to 5.2°C for the climate sensitivity of carbon dioxide. Taking into account these results together with observational evidence, the IPCC chose a value of 2.5°C as its best estimate, implying that over half the warming effect of carbon dioxide assumed by the IPCC came from positive feedbacks.

The IPCC's handling of the role of clouds was criticized by Richard Lindzen, professor of meteorology at the Massachusetts Institute of Technology. Without the assumption of an overall positive feedback effect, the IPCC would have assumed a rise closer to 1°C from a doubling of carbon dioxide. If the IPCC and scientific orthodoxy had followed Lindzen, it is safe to say that the post-1988 era would not be an age defined by belief in global warming.

In addition to uncertainty and ignorance over the size and direction of feedbacks, the IPCC made an even more fundamental assumption based – literally – on hope, that, in the words of the IPCC, 'The climate system is in equilibrium with its forcing.'[12] Making any forecasts of long-term temperature depends on the extent to which this assumption holds. While the assumption is necessary and desirable from the perspective of climate modellers, it does not follow that nature will accommodate it. Although some elements of the climate system were chaotic on a century to millennium timescale, others were stable. 'That stability,' the report went on, 'gives us hope that the response of the atmospheric climate (including the statistics of the chaotic weather systems) to greenhouse forcing will itself be stable and that the interactions between the atmosphere and the other elements of the climate system will also be stable.'[13]

Large-scale features of the world ocean circulation system were deemed to be non-chaotic by the IPCC. 'The question,' according to the Working Group I report, 'is whether the existence of predictability in the ocean component of the Earth's climate system makes the system predictable as a whole. However this seems to be a reasonable working hypothesis,' the authors thought.[14] Absence of computing capacity meant that scientists were unable to model the response of the oceans, even though, as the 1990 report said, 'this is crucial for climate prediction'.[15] It is easy to see why. The entire heat capacity of the atmosphere is equivalent to less than three-meters-depth of water.[16] In the North Atlantic, the heat input carried by the ocean circulation is of similar magnitude to that reaching the ocean surface from the sun.

The sheer length of time for the oceans to respond to a warming atmosphere opened up a huge disconnect between the timescales over which scientists were anticipating global warming and the 'children and grandchildren' timescale deployed by politicians to justify action against global warming. According to Working Group I, the global oceans need millennia to reach a new equilibrium, making a couple of generations a rounding error.[17] Even if it were possible to stop all man-made emissions of carbon dioxide, atmospheric concentrations would decline very slowly and would not approach their pre-industrial levels for 'many hundreds of years. Thus any reductions in emissions will only become fully effective after a time of the order of a century or more'.[18]

With positive feedbacks assumed in the climate models, the IPCC's business-as-usual scenario of emissions growth predicted a temperature rise of 0.3°C per decade (i.e., in the subsequent one to two decades, the world would get the whole of the 0.7°C temperature rise observed in the twentieth century), resulting in a 'likely' rise of 1°C by 2025. By 2009, well past the halfway mark of this thirty-five-year forecast, the temperature had risen by 0.18°C since 1990, or an average of just under 0.1°C per decade. To hit the predicted one-degree rise by 2025, the average global temperature would have to rise at an average rate of 0.5°C per decade for the rest of the period.

Throughout the time when the IPCC report was being prepared, one political leader took a particularly keen interest in the scientists' progress. Mrs Thatcher asked Houghton for regular updates. The week of May 21st 1990 was particularly intense. On the Monday afternoon, Houghton presented the conclusions of Working Group I report to the prime minister and other ministers in the Cabinet room at Number 10. The next three days, Houghton chaired the plenary of the working group at a hotel in Windsor, on Tuesday and Wednesday returning to Downing Street to help Thatcher with her speech for the opening on Friday of the Hadley Centre to coincide with publication of the IPCC's assessment report. Houghton was amazed at the prime minister, pencil and erasure in hand, determined to get the science right. She put a lot of her own time into it, Charles Powell, her chief aide, told him.

Your task, she told the Hadley Centre's director and staff, was no less than 'to help us safeguard the future of the planet'.[19] Describing the IPCC's report as of historic significance, she said governments and organizations around the world were 'going to have to sit up and take notice'. She announced that Britain would set itself the target of cutting projected levels of carbon dioxide emissions by thirty per cent by 2005. Because the projected increases were so high, this meant returning to 1990 levels by that date. Unlike the Dutch and Canadians, the UK beat this target by more than five percentage points, largely thanks to privatizing and liberalizing the energy market, enabling the market to respond with a dash-to-gas for generating electricity.*

In October 1990, the second World Climate Conference met in Geneva. Unlike the all but forgotten first conference in 1979, it was addressed by six government leaders. Scientists and academics also came to spread the alarm that was not to be found in the IPCC's equivocation. German meteorologist Harmuth Grassl, a contributor to the Working Group I report, said that while scientists were 'basically here as ruminants, it is very important to get out the main message: that change on earth is now so fast there is no analogy during the last 10,000 years'.[20] It was a claim without any basis in the IPCC report.

Martin Parry, a British geographer and future chair of Working Group II, produced a report claiming that the world could suffer mass starvation and soaring food prices in just forty years. True, there were sharp rises in food prices in the first decade of the twenty-first century. They weren't caused by rising temperatures but by climate change policies diverting agricultural resources to the production of biofuels, contributing to a wave of food riots in 2008.

The political response was divided between the minority of governments that took seriously the IPCC's equivocal verdict and those arguing that the need for action overrode it. Despite pressure from the European Community – 'I hope that Europe's example will help the task of securing world-wide agreement,' Thatcher told the conference – the US and the Soviet Union, the world's two largest carbon dioxide emitters, weren't budging.[21] Yuri Izrael for the Soviet delegation emphasized the doubt and uncertainty of climate change. More scientific research was needed, Izrael concluded. John Knauss, director of the National Oceanic and Atmospheric Administration, who led the US delegation, said that Washington had refused to set targets because 'it does not believe in them', adding, 'it's as simple as that. We are not prepared to guarantee our projections'. American officials thought European targets lacked credibility, which they characterized as political goals not easily put into effect.[22]

Maurice Strong, recently appointed secretary-general of the Rio Earth Summit, described the evidence of global warming as 'compelling, if not yet definitive'. It posed 'the greatest threat ever to global security'. Action had to be taken before the scientific evidence was definitive, Strong told delegates. 'It is not feasible to wait for the post mortem on planet earth to confirm our diagnosis. If there is ever an instance in which we must act in accordance with the precautionary principle, this surely is it.'[23]

Michel Rocard, the French prime minister, declared that the time for words was over. 'What we need now is action,' Rocard told the conference. 'The race

---

* With coal, the ratio of hydrogen to carbon is 0.5 to one; with natural gas, it is four to one, i.e., the proportion of carbon in natural gas is one-eighth that of coal.

against time is on. The very survival of our planet is at stake.'[24] Despite the appeal for action made at The Hague and the summits at Noordwijk and Bergen, 'nothing decisive has yet been accomplished'. Beyond the problem of global warming, the world community faced a fundamental question on the enforcement of international environmental law. 'What point is there in holding meetings and conducting research if there is no certainty that the standards we adopt and the commitments we enter into will actually be respected?' Rocard asked.[25]

Acknowledging the need for more research, Mrs Thatcher argued it should not be used as an excuse to delay 'much needed' action. 'There is already a clear case for precautionary action at an international level,' she said. 'We must not waste time and energy disputing the IPCC's report.'

Her argument did not impress *The Times* science correspondent Nigel Hawkes. 'Computer models predicting temperature rises very much smaller than their proven margins of error are being used by a prime minister who claims to be a scientist as grounds for imposing economic sacrifices on the entire world,' Hawkes wrote. 'A couple of cold winters will take the froth off the debate, and allow us the time we need to discover whether or not the earth is really warming up.'[26]

By this stage of her premiership, it says much for Thatcher's commitment to global warming that she was dedicating so much time to it. The week before, Sir Geoffrey Howe, her nominal deputy, had resigned. Before the month was out, she was no longer prime minister. In retirement, her views on global warming appear to have evolved. Her last book *Statecraft*, published in 2002, speaks of her concern about anti-capitalist arguments deployed by campaigners against global warming and the lack of scientific advice available to leaders from experts who were doubtful of the global warming thesis.

These were not the sentiments of her Geneva speech, her last pronouncement on the subject as prime minister. Instead she used language straight out of the green lexicon, talking of the growing imbalance between 'our species and other species, between population and resources, between humankind and the natural order.'[27] She called for as many countries as possible to negotiate a framework convention on climate change for agreement in 1992 with binding emissions cuts, following the lead taken by European governments. In this endeavor,

> the International [sic] Panel's work should be taken as our sign post: and the
> United Nations Environment Programme and the World Meteorological
> Organisation as the principal vehicles for reaching our destination.[28]

The IPCC did play the role Thatcher envisaged, but the WMO and UNEP – neither of which can be remotely described as purveyors of doubt about global warming that Thatcher said in her 2002 book was needed – were sidelined as negotiating vehicles. The G77 group of developing countries have more leverage within UN fora than inter-governmental arrangements, in which it is easier for OECD countries to hold sway.*

The previous year, Brazil and Mexico had led an initiative to get increased representation with the formation of a special committee on the participation of developing countries. Both countries had objected to Bolin's draft of the synthesis report. Frankly admitting that IPCC was a political bargaining process, the report was changed: 'Recognizing the poverty that prevails among the populations of developing countries, it is natural that they give priority to achieving economic growth,' it now reads. 'These sentences are of course politically inspired, but they are basically factual,' Bolin explained.[29] Bolin was right. Since the 1972 Stockholm conference, this has been the consistent position of the G77 on environmental issues, right the way through the negotiations to hammer out UN Framework Convention on Climate Change, the Kyoto Protocol in 1997 to the Copenhagen climate conference in 2009.

In December 1990, UN General Assembly adopted a resolution establishing the Intergovernmental Negotiating Committee for a Framework Convention on Climate Change (INC). Four intensive rounds of negotiations in the course of 1991 were marked by procedural wrangles and failure to produce a draft negotiating text. A fifth round in February 1991 was marked by deadlock. According to Daniel Bodansky's account of the INC negotiations, 'States still seemed engaged in a battle of nerves, hoping that, with the Rio Summit fast approaching, the other side would blink first.'[30]

In April – less than two months to go before the summit – INC chairman Jean Ripert held extended meetings with key participants in Paris. No one blinked.

'Don't go wobbly on me, George,' Thatcher is reported to have told President Bush on facing down Saddam Hussein in 1990. On climate change, the Europeans wanted Bush to wobble. The prospect of the Rio Earth Summit without a convention on climate change was being taken right down to the wire.

---

*   Mostafa Tolba, UNEP executive director at the time, complained that governments took the preparation of the framework convention away from UNEP. 'It has been speculated that the developed countries were not yet ready for the positive action and concrete action advocated by the UNEP executive director.' Other than his testimony, there is little evidence to support this view. Mostafa K Tolba with Iwona Rummel-Bulska, *Global Environmental Diplomacy: Negotiating Environmental Agreements for the World, 1973–1992* (1998), p. 95.

15
A HOUSE DIVIDED

I say this on climate change: we're not going to enter into commitments we don't keep.

President Bush, June 7th 1992

Negotiations on the climate change framework convention were marked by what the head of India's delegation described as a 'fundamental and irresolvable difference' between the US and European Community.[1] Apart from the INC's first session, US negotiators maintained the line that they would not accept any text which bound them to timetables and targets for carbon dioxide emissions. Diluted formulations calling for countries to commit to measures 'aimed at' stabilizing emissions were rejected.

The negotiations stalled.

Less than two months before the summit, British and American negotiators hammered out a compromise. Britain's environment secretary, Michael Howard, was sent to Washington to persuade key members of the administration that the President could sign the convention.

If history judges George H.W. Bush's supreme achievement to bring the Cold War to a peaceful conclusion, then global warming could have been designed to wrong foot him. In the post-Cold War New World Order, America found itself isolated. Fundamentally an internationalist, it would have been hard for Bush to stay away from the most numerous gathering of world leaders ever assembled. But there was no upside for him in going to Rio. Election year made the politics harder still and Bush did not have the political skills to pull it around. Bush lacked Nixon's cynicism in his handling of the Stockholm conference twenty years earlier or Bill Clinton's deftness in negotiating the Kyoto Protocol but not putting it to a vote in the Senate.

In the 1988 election, the Bush campaign had decided that the environment was critical to winning key battleground states. To win suburban voters in these states, the strategy was to go to the right on tax, defence and crime and to the center on childcare and the environment. On election night, exit polls gave Bush and Dukakis a dead heat on the environment and Bush wound up winning New Jersey, Illinois and California, the last Republican candidate to do so.

This was more than 'read my lips' campaign rhetoric. William Reilly, president of the Conservation Foundation, provided briefings to both presidential campaigns. Mostly ignored by the Dukakis camp, they were devoured by the Republicans. Bush hosted a dinner with leading environmental campaigners, the candidate sitting between Russell Train, who led the US delegation at the Stockholm conference, and Reilly. After the election, William Ruckelshaus, the Environmental Protection Agency's first administrator, suggested to Bush he appoint Reilly to lead the EPA.

Reilly became the first environmentalist to head the EPA. He would not have many allies in the administration, except the most important one. On one occasion, budget director Richard Darman commented, 'The problem is we have an environmentalist running the EPA and,' pointing to the Oval Office, 'the bigger problem is we've got an environmentalist sitting in there.'[2] According to Bush campaign manager Lee Atwater, Bush regarded people like Atwater as necessary tools, but had come into politics to work with people like Reilly. Bush's interest in environmental policy was genuine, a conservationist in the mould of Teddy Roosevelt averse to command-and-control solutions to environmental problems.[3]

The priority of the senior White House staff was different: improving America's economic performance. With Bush mostly focused on foreign policy, Reilly's was an isolated voice. Nonetheless, the Bush administration notched up a number of achievements, notably the 1990 Clean Air Act Amendments, which pioneered the use of tradable emissions permits.

The prospect of the administration's differences being played out under the floodlights of the Rio Earth Summit meant that, one way or another, the divergence between the environment candidate and the priority accorded the economy needed to be resolved. That posed a particular challenge for an administration held together by loyalty to the president and, when John Sununu was chief of staff, the discipline he exerted, rather than a shared sense of mission. Reilly recalls remarking to Barbara Bush at a social event on how well everyone got on with each other. 'They're all old friends, except for you Bill,' the First Lady replied. 'You're the only one we didn't know.' When loyalty gave way to other agendas, as happened during the Rio conference, the result was spectacularly damaging.

This vulnerability was accentuated by another. As a communicator, Bush did not use speeches to build a case, but to convey sentiments ('a kinder, gentler nation', most famously) or state policy positions ('the day of the open check book is over', with respect to the Earth Summit). This mattered when the global warming policy adopted by the Bush administration ran counter to where the rest of the world was heading. It needed advocacy, but the administration had no advocate.

Technically, the Bush White House was unusually well equipped to appraise the science and economics of global warming. Sununu held a Ph.D. in mechanical engineering from MIT and had worked on thermal transfer problems. Darman held a Harvard Ph.D. Robert Watson, who later worked in the Clinton White House and succeeded Bolin as chairman of the IPCC, described Sununu and Darman as 'incredibly bright'.[4] Michael Boskin, chairman of the Council of Economic Advisers, had devoted much of his career as a Stanford professor to studying the interface between technology and the economy. Allan Bromley, the White House science adviser, was a nuclear physicist. (During his confirmation hearings, Al Gore tried to convince Bromley that Bromley did not understand the greenhouse effect.) Bromley wrote a memoir of his time in the White House. There was no question that Sununu and Darman considered many of the claims of environmental activists to be seriously overblown, 'but they were prepared to listen to any reasonable argument. What they refused utterly to do was to stipulate that we were *already* in a crisis, as many of the activists demanded as a basis for discussion'.[5]

Of them, Darman was the most skeptical. He first brought Hansen's claims to Sununu's attention. Darman's approach to micro-economic policy was much more conservative than on fiscal and size of government issues. He deployed a cost-benefit approach that a particular environmental measure would cost X thousand dollars per fish saved, a mode of analysis he'd learnt from David Stockman, Reagan's first budget director.

Early in the Bush presidency, some of the leading global warming advocates had a meeting with Sununu, Darman and Bromley. Sununu probed the computer models. Did they couple the atmosphere and the ocean? No, came the reply. Sununu pointed out that the thermal capacity of the oceans could not be ignored. Computer models that only took account of the atmosphere were meaningless. Although his interlocutors were not happy that the White House chief of staff was unsupportive, Sununu authorized a large increase in funding for climate modelling.[6] In two years, spending more than doubled to $1.03bn.

Reilly and Sununu often clashed. Shortly after James Baker's speech on global warming, Reilly made a similar speech on the subject. The next day, Sununu rang to tell him it wasn't administration policy. Reilly replied that he has taken the same line as Baker. 'He has?' asked Sununu, 'I'll talk to Jim.' Baker subsequently announced he was rescuing himself from involvement in the area to avoid potential conflicts of interest over his personal investments. 'You'll never win against the White House,' Baker's deputy, Bob Zoellick, told Reilly.[7]

In October 1989, Bush asked Bromley to chair the White House climate change working group. The next month he and Reilly led the US delegation to the Noordwijk ministerial conference. 'Neither we nor anyone known to us had any detailed economic or technical understanding of what would be involved in achieving this level of emissions [reductions],' Bromley recorded.

> The lack of economic analysis was astonishing … I asked the head of one of the major European delegations how exactly his country intended to achieve the projected emissions goals and was told, 'Who knows – after all it's only a piece of paper and they don't put you in jail if you don't actually do it.'[8]

At the Malta summit with President Gorbachev in December 1989, when the two announced the end of the Cold War, Bush said he would host a White House conference on the environment and global warming the following spring. The Council of Economic Advisers studied the economics of global warming, its conclusions forming part of the Council's 1990 annual report published in February that year. Its chairman, Michael Boskin, was concerned at reliance on primitive attempts to model the climate and the failure of other governments to analyze the economics. Global warming champions were environment ministries. Their lexicographical preference, to use Boskin's term, was to treat the impact on economic performance as ancillary. Boskin alerted the economic ministries of other governments, but was frustrated that they continued to take a back seat to environment ministries.

The Council of Economic Advisers did some preliminary cost/benefit and risk analysis. Boskin brought in outside expertise, notably Yale economist William Nordhaus, a former Carter administration official and the leading economist in the field. The Council noted that there was 'an extremely high level of uncertainty' regarding possible future climate change. Unlike policies to combat depletion of the ozone layer, there were no low-cost substitutes for fossil fuels. The cost of gradually reducing US carbon dioxide emissions by twenty per cent over the course of the next one hundred and ten years was estimated at between $800bn under optimistic scenarios to $3.6 trillion under pessimistic ones, between thirty-five to one hundred and fifty times greater than the EPA's estimates of the costs of completely phasing out CFCs by the end of the twentieth century.[9]

The study estimated the impact on the economy of carbon dioxide reduction policies by reference to the 1973 and 1979 oil price shocks and their impact in reducing the energy intensity of economic activity. With no growth in energy consumption between 1973 and 1985, carbon dioxide emissions were flat. But these were also years of economic weakness. Although caused by many factors, 'higher energy prices clearly played an important role', the report said. Policies designed to stabilize emissions could halve the growth rate of the global economy.

Reducing emissions was even harder for the US because of its dependence on coal-fired power stations, which contributed fifty-six per cent of America's electricity in 1986. Canada, France and Sweden generated more than eighty per cent of their electricity from nuclear, hydroelectric or geothermal sources. Germany had a similar reliance on coal, but had an ace up its sleeve.

With the costs of substantially slowing carbon dioxide emissions likely to reach trillions of dollars, what, the report asked, might be the benefits? Most sectors of industrialized economies were not climate sensitive. Estimates of the impact on world agriculture of a doubling of carbon dioxide ranged from $35–70 billion a year on pessimistic scenarios, with the US losing $1 billion annually, to small net gains.[10] By comparison, trade-distorting agricultural policies were reckoned to cost $35 billion a year for the world and $10 billion a year for the US.

The report concluded that without improved understanding of the impacts and likelihood of global warming, there was no justification for imposing large costs on the American economy. The adoption of many small programs, each of which failed a standard cost-benefit analysis, could significantly slow economic growth and eliminate jobs, the Council warned. Any strategy to limit aggregate emissions without worldwide participation was likely to fail, the report stated.

The Bush administration was the only Western government to seriously analyze the economics of global warming, widening the rift between it and the rest of the West. This became particularly evident in April at the White House conference on the environment. Held in a Marriot hotel, the president gave the opening and closing addresses. 'Two scientists, two diametrically opposed points of view – now where does that leave us?' Bush asked in his first address, pointing to a couple of scientists who had been arguing about the science on a TV talk show. Unimpressed, was the verdict of many participants. After a round of polite applause, European delegates quickly headed for the lobby with critical comments for reporters. Germany's environmental minister, Klaus Töpfer, criticised the US. 'Gaps in information should not be used as an excuse for worldwide inaction,' Töpfer told the *Washington Post*.[11] Lucien Bouchard, Canada's environment minister, chimed in: 'The price of inaction is too high.'[12] Bert Bolin, who managed to get an invitation although he had not been on the original guest list, criticized those hiding behind 'the concept of uncertainty'. Because of the inertia of the climate system and the energy stored in the oceans, Bolin told the conference, 'We are [therefore] committed to a further change [of the climate] of perhaps [an additional] fifty per cent.'[13] Fifty per cent of what? Bolin didn't say.

Administration officials tried to push back. 'Up until now, the conferences I've been to haven't focused at all on economics,' Reilly told a journalist. 'I don't trust a commitment that is made without some knowledge of the cost.' But Allen Meyer of the Union of Concerned Scientists was dismissive. 'This conference is yet another example of the yawning chasm between George Bush's campaign rhetoric on global warming and the reality of his administration's policy of inaction.'[14]

Not everyone was hostile. 'President Bush is absolutely correct,' said Soviet Deputy Prime Minister Nikolay Laverov on Bush's comments linking environmental wellbeing and economic welfare. 'The two are interwoven, and that is what differentiates this conference from other conferences of this kind.'[15]

Negative reactions to the president's speech prompted a change of heart. Reilly suggested to Bush and Darman's deputy, Robert Grady, that it was possible to cool it down and join the ranks of the concerned without abandoning the administration's opposition to targets and timetables. Grady produced a draft. 'Above all,' the president said in his closing address, 'the climate change debate is not about research versus action, for we've never considered research a substitute for action.'

Traveling in the president's car, Bush asked Reilly if the speech had gone all right. Reilly began to answer but Bush cut him off: 'I know; I showed I give a shit.'[16]

'Bush does about-face at warming conference,' the headline said in *USA Today*.[17] Bolin was pleased. 'I take the president's speech to be a clear signal to proceed very vigorously with what we are trying to do with the IPCC,' he told journalists.[18] The president was inching closer to Rio.

Canada proposed that Maurice Strong be the conference secretary-general. Bush had known Strong from his days as American ambassador at the UN. Despite being seen as a Democrat, Bush, in 'typically gentlemanly fashion indicated that I was OK'. The view was not reciprocated. 'A phoney' and a 'horse's rump' was how administration insiders came to view him.

Rio was to be the culmination of a two-decade long effort to bring environmental issues from the side-lines to the centre of international politics. As Strong told the 1990 Geneva climate conference:

> It will focus on the need for fundamental changes in our economic behavior and in international economic relations, particularly between North and South, to bring about a new, sustainable and equitable balance between the economic and environmental needs and aspirations of the world community.[19]

For the developing world, the Earth Summit promised to be a bonanza. 'This is about sharing power,' said Rizali Ismail, Malaysia's UN ambassador.

> When it was East vs. West, our development needs were ignored unless you were a marionette of the Soviet Union or the US. Now with the environment seriously frightening many people in comfortable paradise areas, for the first time people are taking us seriously.[20]

Pakistan's Mahbubul Haq estimated that the industrial countries would have a 'peace dividend' of $1,200 billion to distribute over the next ten years. India's lead climate negotiator, Chandrashekhar Dasgupta, contrasted the split between the US and the rest, and the cohesiveness of the developing world, with China speaking on behalf of the G77 plus China, enhancing the effectiveness of the South in the negotiations.[21]

The potential for a feeding frenzy set alarm bells ringing in Brazil. José Goldemberg, the Brazilian minister in charge of preparing for the summit, became convinced that the summit was heading for disaster. Brazil wanted to be seen making amends for its previous negative attitudes towards. In the 1970s, Brazil had been wary of environmentalism as international concern about preservation of the Amazon rainforest was regarded by Brazil's military rulers as interference in the country's national sovereignty.

That changed when President Sarney came to power in 1985, ending military rule. Sarney pressed for Brazil to host the 1992 summit in response to an international outcry about forest fires in the Amazon. Brazilian scientists had concluded that deforestation was not as extensive as the international press had claimed, but that it could also significantly harm the regional climate of parts of the country.

Fernando Collor de Mello, who succeeded Sarney in 1990, had an added incentive. Under a cloud of corruption charges, Collor hoped that hosting the world's largest summit might win him a reprieve (he was forced to resign at the end of 1992). To prepare for the summit Goldemberg, a physics professor, met Sununu, the engineering Ph.D. Although Sununu had, Goldemberg thought, a reasonable grasp of the science, he felt Sununu was typically American in thinking if temperatures did rise then technology and air conditioning were the answer.[22]

In April 1991, reinforcements landed in Brazil. Sailing up the Amazon, the Royal Yacht Britannia berthed at Belém. It was Prince Charles's idea to bring together politicians, businessmen and NGO representatives. The most important invitee to the seminar was Collor. When he got wind that Collor might not come, he sent a hand-written letter. 'The Royal Yacht is sailing all the way out to Brazil especially for this seminar so as to provide what I hoped to be a reasonably neutral and relaxed setting for such a gathering,' the prince wrote. If Collor didn't come, the prince continued, 'I think it would give the wrong signals to many people who are looking forward towards the importance of the 1992 United Nations Conference.'[23]

It was just as well Collor came, as the prince had also invited Reilly. Brazilian protocol meant that the venue for the meeting between Collor, Goldemberg and two other ministers in Collor's Cabinet was held on a vessel of the Brazilian navy. Sitting around a table on the boat's deck, there was a lively discussion on prospects for the conference. The planning looked chaotic, Reilly pitched in. The summit was heading for disaster. As things stood, he would recommend that the president should not go. Collor said that the presence of sixty-five heads of state was worth less than the president. If Bush comes, we will not allow him to be embarrassed, Collor promised. On his return, Reilly switched his recommendation. The Brazilians kept their side of the deal, Reilly believes. 'They put on a first rate show.'[24]

Sununu's departure from the White House at the end of 1991 raised hopes that the administration would soften its opposition to targets and timetables. They were quickly disappointed. On coming to the White House to head the Policy Coordinating Group, Clayton Yeutter was aghast. 'Why in the world is this summit meeting being held and, for heaven's sake, why in our presidential election year?' he asked.[25] Rio was putting the administration in an impossible position as the US was bound to be criticised whatever it did. Yeutter tried to find out who in the administration had agreed to the summit in the first place, but got no satisfactory answer. At the same time, the White House became increasingly aware that the other developed nations were playing by different rules and approaching Rio in a different spirit. Yeutter spoke to a senior European diplomat. Would his country be prepared to accept the pledges and commitments expected of them? Of course, came the reply. Would his country be able to carry out those pledges and commitments? Of course not.

Hypocrisy has always played a role in international relations. Why couldn't the US play by the same rules as everyone else? When other governments sign a treaty, the hard work is over. For an American president, it is just the beginning. There is more risk in treaty ratification than in negotiating it in the first place. Treaties, especially ones with potentially enormous implications for domestic policy, present institutional challenges almost entirely absent in countries with parliamentary systems. In those, the government controls the legislature and writes the legislation that turns the provisions of a treaty into domestic law. If the courts start to interpret in ways not envisaged by the executive, the government can always amend the original legislation.

In the US, obtaining a two-thirds majority in the Senate requires the president to commit his time and prestige. Deals might have to be cut with reluctant senators. The Senate can attach 'reservations' or 'understandings', changing what the president proposes. Further uncertainty comes from how the courts might interpret the treaty. A global warming treaty with emissions cuts hardwired in the text risked being a blank check for Congress, with the courts determining the terms of payment.

Domestic and foreign policy advisers in the Bush White House disagreed whether it mattered if the US failed to meet its targets. Yeutter stated that the administration should not agree to binding targets. Brent Scowcroft, national security adviser, argued that in international conventions, every day, the US made commitments it wasn't sure it could keep.

A meeting at the White House in April 1992 brought home to Goldemberg and the Brazilians just how precarious the prospects for American participation in the summit had become. Despite Goldemberg's lowly status, Scowcroft invited him to meet the president where he explained what the summit meant to Brazil.

The White House insisted that all language about guilt and crimes of the developed world be removed from the Earth Charter. Collor despatched Goldemberg to visit China and India to reconcile the G77 stance with the American position and tone down the language. It didn't prevent the April meeting of the G77 in Kuala Lumpur from reaffirming the developed world's crimes and guilt.

If Bush was going to Rio, there had to be an agreement that he could sign. All the indications were that the US would not sign the biodiversity convention (President Clinton did in June 1993, but the Senate did not ratify it). That meant the administration had to agree the climate change convention or the president staying away. At crunch meetings in Paris, the American negotiating team tested the waters to see if there might be a way to bridge the gap with the Europeans. They knew there was no way that the US or, for that matter, most of the other developed nations could scale their carbon dioxide emissions back to 1990 levels. Staring into the abyss of failure, the British side approached the Americans to see if there might be language to finesse the issue. The Americans returned to Washington and worked up a text.[26]

At the end of April, Michael Howard went to Washington with a version of the text in his pocket. On April 29th he visited five different federal departments and agencies. At his 9am meeting at the EPA, Reilly told Howard that it was essential for the US to sign. The outcome depended on Howard's meeting later that day at the State Department. At the Energy Department, its secretary, Admiral Watkins, also told Howard that his meeting at the State Department was crucial, but that the US should have nothing to do with the convention. Howard had lunch at the British embassy with two senior White House aides who expressed diametrically opposite views. The only thing they could agree on was that the outcome depended entirely on Howard's meeting at the State Department.

At his meeting with Zoellick, Howard went through the draft text of the convention line by line. At the end of it, Zoellick declared that it was a document that the US could sign. Then on to the White House and a meeting with Scowcroft. Howard reported the outcome of the meeting with Zoellick. 'If it's good enough for Bob, it's good enough for the president,' Scowcroft said.[27] The next day, he phoned Howard to say that the president wanted a minor change. Howard replied that the Europeans wouldn't wear it. Zoellick replied that he knew Howard was going to say that and that they would go ahead and sign anyway. The outcome was a coup for British diplomacy. It is doubtful any other country could have pulled it off. Zoellick's secretary told Howard that they had more calls to and from the British embassy than all the rest put together.

Bush called John Major to confirm the deal, and followed up with calls to President Mitterrand of France (who kept his side of the deal) and Chancellor Kohl of Germany (who didn't). The text agreed by Howard and Zoellick formed Article 4 2 (a) and (b) of the convention. It requires developed countries to adopt policies and measures to limit greenhouse gas emissions, thereby demonstrating that they are taking the lead in modifying the long-term trend in emissions. The Europeans got their favored emissions path into the text. Returning to the 1990 emissions level by 2000, the convention states, would contribute to such a modification. The US got a non-binding formulation and recognition of the need to maintain strong and stable economic growth. The way Darman put it, the US was making a moral, but not legal, commitment to cut carbon dioxide emissions.

'There is nothing in any of the language which constitutes a commitment to a specific level of emissions at any time,' Yeutter wrote to conservative Republicans, describing the outcome as masterfully vague.[28]

16
PRESIDENT BUSH GOES TO RIO

The time of the finite world has come.

UN secretary-general Boutros Boutros Ghali, June 3rd 1992

I did not come here to apologize.

President Bush addressing the Earth Summit, Rio de Janeiro on
June 12th 1992

Having agreed that the president could sign the convention, the next question was whether he should go to Rio. If anything this debate provoked even more discussion than the convention itself. The White House was split down the middle. Darman was the most vocal opponent, but the president wanted to go, even if it had a political cost back home. After finding out that virtually all the leaders of important countries were going, Bush felt he should be there.

There was a logistical consequence of the lateness of President Bush's decision to attend. The White House staff and press corps were booked into the VIP's Motel, one of Rio's famed guesthouses, offering comfort, privacy and an impressive array of mirrors.[1]

Because of his relationships with fellow environment ministers, the White House decided that Reilly should lead the American delegation to Rio. The caveat was that he would not make any binding agreements without first clearing them with Yeutter. According to Yeutter, Reilly did a splendid job. Reilly continued to press his agenda, but as a team player. Even after agreement of the compromise text on the climate change convention, Reilly kept up the pressure for targets and timetables.

In early June, he arranged for Gro Harlem Brundtland to be seated next to the president at a dinner to press the case. The next day, Bush called Reilly from his running machine. Could he provide the president with a memo on whaling? At the dinner, the chairman of the Brundtland Commission had spent two hours lobbying the president on the need for 'scientific' whaling.

Publicly, Brundtland had a different message. Speaking on the White House driveway, she expressed her disappointment at American unwillingness to sign the biodiversity convention. 'I believe you can combine environmental concern with an increase in jobs,' although she didn't specifically mention those of Norwegian whalers.[2]

Splits within the Bush administration and its differences with the rest of world served to shift the spotlight from splits between other nations. Europe's environment commissioner Carlo Ripa di Meana threatened to boycott the summit after failing to secure support for a European carbon tax and to express his disgust at the compromise with the US. It meant that Europe was being forced to accept a treaty with lower standards than it was adopting for its own members.

There were also splits among the nations of the South. Saudi Arabia wanted forests to be seen as carbon sinks that remove carbon dioxide. Malaysia argued that the industrial countries had caused the climate problem, so should reduce their emissions rather than trying to solve the problem by locking up forests in the developing world.

Once he got to Rio, Reilly met Goldemberg to explore possible changes to the biodiversity convention that would enable Bush to sign it. Reilly cabled Yeutter his suggestions. Shortly after, while being interviewed on live television, Reilly was handed a copy of the cable that had been leaked to the *New York Times*. He abruptly ended the interview. 'I was personally embarrassed just to be handed the goddamn thing,' he said later.[3]

The leak had a devastating impact on the administration's ability to contain what anyway was going to be a difficult situation. Two days later, an exasperated Bush told a joint press conference with the British prime minister, 'I'd like to find the leaker, and I'd liked to see the leaker filed – fired.' Once found, the culprit would be 'gainfully unemployed'. All this was greeted with glee by Democrats in Rio. Al Gore, leading the Senate delegation to the conference, said it had set off a firestorm of criticism. 'Once again, the president has overruled his EPA chief. This time, the whole world is watching.' Reilly countered that his resignation was not on the cards, 'I do not want to give that satisfaction to my enemies.'[4]

Up till then, Gore had been more measured than some of his colleagues. Senator Wirth called America's position a disgrace. 'Instead of a commie under every bed, it's now an eco-terrorist behind every tree.' Jerry Brown, the once and future governor of California, accused the president of being in the pocket of special interests. Greed and corruption would always win the day. To applause from an audience of mostly American environmentalists, Brown added, 'Bush and the administration's position on the environment are completely crackpot.'[5] This kind of attack on foreign soil was too much even for Wirth. 'Our side is getting hammered,' he complained. 'It's the complete inability of the White House to explain what the US has been doing the last twenty years.'[6]

The *Washington Post* described the US as virtually under siege at the conference. The Bush team's headaches didn't come from the Third World. Collor had been as good as his word. True, Fidel Castro received the loudest ovation with his claim that consumer societies were fundamentally responsible for environmental destruction, but at least kept to the allotted five minutes speaking time. Austria, Switzerland and the Netherlands launched an initiative to get like-minded countries to sign up to the targets that had been taken out of the climate change convention. The administration made a clumsy counter-attack. 'They treated us like we are some kind of colony,' an Austrian diplomat complained, while a Swiss diplomat said the US was 'shooting sparrows with a cannon'.

Soon the sparrows were joined by a German eagle. No American ally had benefited more from the Bush administration than Germany. Now it was payback time. Going back on the agreement between Bush and Kohl, Klaus Töpfer said Germany would seek to have all members of the EC sign a separate declaration at Rio reinstating specific emissions targets. Zoellick, who had joined Reilly in Rio, counter-briefed that Germany and Japan were engaged in a guilt-induced attempt to be politically correct. 'All this chaos … the circus and the rhetoric' at the summit were laid at the door of 'the guilty developed-world logic' in which the wealthiest feel they 'owe the rest of the world.'[7]

American difficulties didn't restrain the German government from playing its ace. It announced that it would cut its carbon dioxide emissions by between twenty-five to thirty per cent by 2005. On July 1st 1990, the economies of the two Germanys were unified. Soon after, the economy of the former East Germany went into a deep slump. Industrial production fell by more than half and heavily polluting power stations were closed. Reunification enabled Germany to proclaim its virtuousness at Rio and Germany's emissions fell by nearly eighteen per cent between 1990 and 2005, the steepest achieved by any advanced economy. At a conservatively estimated cost of DM750 billion ($523 billion) for the first five years, viewed as a policy to cut carbon emissions, reunification is the world's most costly global warming policy to date – equivalent to more than twenty per cent of Germany's 1995 Gross Domestic Product.

President Bush spent as little time at Rio as he decently could. He gave a speech, signed the climate change convention, and had his photo taken with the one hundred and seven other world leaders at the summit. Bush also had a private meeting with the ecologist and film maker Jacques Cousteau, who had given a lecture claiming that population growth would lead to a world where people could only survive like rats. 'Even if we found a way to feed this human tidal wave, it would be impossible to provide this multitude with decent living conditions,' Cousteau told an audience that included Collor and the King and Queen of Sweden.[8]

In reality, the summit was part ceremonial and part soap box, its substantive business having been concluded beforehand. The docking of a replica Viking ship, the Gaia, marked the opening of the Global Forum, an alternative summit for NGOs. Not everybody welcomed the Vikings. 'Go home Gaia. $5 million rich men show off,' said one banner. Compared to the four hundred or so NGOs at Stockholm's Hog Farm, the number of NGOs accredited at Rio was some one thousand four hundred and fifty. James Bond actor Roger Moore, who had been chased around the Sugarloaf by Richard Kiel's Jaws in *Moonraker*, proclaimed June 3rd to be 'the first day of the rest of the world' and Gro Harlem Brundtland argued that whales could be hunted if it was done on a sustainable basis.

Four days later, the Beach Boys were doing a gig and pledging their support for the Global Forum. Its sound system had been turned off after failing to pay its electricity bills and run up a $2 million deficit amid allegations of corruption. In another part of Rio, tenor Plácido Domingo was singing for the great and the good and those who could pay $100. The Rio Refuse Collection Authority complained that people were ignoring recycling signs on the one hundred and sixty special bins placed throughout the conference centre. 'I wouldn't trust him to save the planet,' grumbled one refuse collector as an Australian ecologist threw a large piece of pizza into a bin for recyclable material.

Writing in the *New York Times* on the conference's first day, Czechoslovakia's president Vaclav Havel argued that two years after the collapse of Communism, a new polarization was developing, this time between the rich countries of the North and the poor ones of the South. 'The states of the South find it difficult to overcome their mistrust of the North,' Havel wrote.

> They believe that the northern countries should finally understand that today's patterns of production and consumption, besides not being sustainable, are the principal cause of the threat facing the global eco-system, and that the northern states therefore have to accept substantial blame for environmental degradation in the poorer countries.[9]

On the conference's fourth day, Greenpeace and three other NGOs organized a press conference at Rocinha, home to a quarter of a million of Rio's poorest people. The assistant editor of the British Medical Journal, Fiona Godlee, went with them. At a school, the NGOs outlined their plans to save the summit and the planet, with demands for legally binding cuts in greenhouse gas emissions and a reduction in the standard of living of the North.

Sabarina Uega, who ran the Rocinha residents' association, guided the visitors around over heaps of rubbish and open drains and answered journalists' questions. What did the people of Rocinha want? Lack of clean water and decent sanitation was the biggest problem. Many children died; perhaps as many as one in five before the age of one. What did Sabarina think of all the money that had been spent on prettifying Rio for the conference? She smiled. 'Just because we are poor it doesn't mean we don't want the city to be beautiful.'[10]

Opposition from Third World countries meant the proposed convention on forests became a statement of principles. Similarly the legal status of the Earth Charter, envisaged by Strong as the keystone of the summit's architecture, was downgraded to a non-binding 'Rio Declaration'. Everyone had thought that the charter would be a namby-pamby, platitudinous statement, said Barbara Bramble of the National Wildlife Federation, the largest US environmental group. Instead it turned into a knockdown, drag-out fight between North and South. 'They took a draft about ecologic philosophy and turned it into economic power politics.'[11]

The Declaration pronounced a 'new and equitable global partnership', in which developed countries acknowledged their responsibility for the pressure their societies placed on the global environment (principle seven). Sustainable development required states to eliminate unsustainable patterns of production and consumption and 'promote appropriate demographic policies', in a coy reference to belief in the benefits of population control (principle eight). In the absence of scientific certainty, the precautionary principle should be applied to protect the environment (principle fifteen). Signatories declared themselves in favor of eradicating poverty – 'an indispensable requirement for sustainable development' (principle five) – and against war – 'inherently destructive of sustainable development' (principle twenty-four). The roles of women, youth, and indigenous people in sustainable development were all highlighted and the rights of oppressed peoples (the Palestinians) to have their environment recognized (principles twenty to twenty-three) asserted.

Without a concrete plan of action, Strong believed the high sounding rhetoric in UN declarations would be just that. The purpose of Agenda Twenty-One was to avoid the Rio Declaration following the same fate. 'Humanity stands at a defining moment in history,' the preamble to Agenda Twenty-One claimed. The world was confronted with a perpetuation of disparities within and between nations; a worsening of poverty, hunger, ill health and the continuing deterioration of ecosystems. In reality, the world had already crossed a threshold into an era that saw the largest numbers of people lifted out of poverty in human history, an accomplishment that the doctrine of sustainable development denied was or could ever happen within an internationally liberal economic order. Described by Goldemberg as a naïve attempt by NGOs to reorganize the basis of the world economy, Agenda Twenty-One's estimated cost of $600 billion a year, $125 billion of which was meant to come from developed countries, ensured its irrelevance.

At the time, climate change was not the all-consuming issue at Rio. Twenty years later, Rio's importance is defined by it. Without the summit, the convention would not have been brought into being so quickly – just four years after the alarm had been raised in 1988 – and possibly not at all. For European governments, the Convention's teeth had been pulled with the removal of legally binding targets. Strong expressed extreme disappointment over this, but wrote that the convention marked an historic milestone in the development of international law. 'It was clear from inception that it would involve some very fundamental changes in industrial civilisation.'[12]

The convention did not change the basis of industrial civilization. Its significance lay elsewhere. In Article Two, the international community accepted the central proposition of global warming by committing themselves to stabilizing greenhouse gas concentrations to prevent dangerous anthropogenic interference in the climate system.

Rio therefore was a decisive event in the history of global warming. Even though the IPCC in its First Assessment Report two years earlier had returned an open verdict, the science had been settled by environment ministries and diplomats. Thus President Bush's signing of the convention was more important than the convention's omission of legally binding targets for emissions reductions. One hundred and ninety-three nations of the world, including the most powerful one, remain formally bound by international treaty to accept the view that man-made global warming is dangerous.

Should President Bush have gone to Rio?

As Clayton Yeutter describes it, the White House was in a no-win position. It was a matter of choosing the least damaging option. On the other hand, John Sununu believes the decision was the single biggest mistake of Bush's presidency. Until Sununu's departure from the White House, Bush was all set to have given it a miss.

Suppose the conference had been held four years earlier, what might Bush's predecessor have done? Ed Meese, who served Ronald Reagan in the White House and as chief of staff when he was governor of California, thinks that Reagan would have gone to Rio to explain why he wouldn't sign the climate change convention. His record as governor showed he was keen on protecting the environment, but as president, he was wary of international conventions. He would have been very suspicious of the science on which the climate change convention was based, Meese told the author, and was opposed to 'environmental extremism' being used as a way of advancing the grasp of government.[13] If Reagan had been watching ABC's *This Week* on the eve of Rio, he would have had confirmation of that fear. 'The task of saving the Earth's environment is going to become the central organising principle in the post-Cold War world,' Al Gore told David Brinkley.

In negotiating and signing a climate change convention without legally binding commitments to cut carbon dioxide emissions, the Bush administration ended up where the American political system and public wanted to be. The convention was swiftly and overwhelmingly ratified by the Senate that autumn. If Bush had not signed it, his successor would have. If the Bush administration had not negotiated in good faith, the convention would have ended up with similar provisions to the Kyoto Protocol that Clinton signed but couldn't get through the Senate. In retrospect, Bill Reilly, the most vocal supporter of targets and timetables within the administration, believes their removal had been the right call. The strong economy of the 1990s would have blown the emissions caps and any practical chance to stabilize carbon dioxide emissions.

Not that Bush got any credit from voters. A CBS News poll, released towards the end of the conference, found that only nine per cent of those surveyed expected the summit to produce substantial results to help solve the world's environmental problems. Seventy per cent said the president had been insincere in his expression of support for environmental issues. Asked if protecting the environment was so important that requirements and standards should be set regardless of cost, sixty-seven per cent replied in the affirmative, an indicator of public sentiment at the time, but a meaningless guide to what voters would actually accept in the absence of a context or consequences to the question.[14]

There was a further irony, perhaps the biggest of all. In Rio, Bush administration officials were deeply frustrated that they were playing defence, as if America was the dirty man of the world. Since the presidency of Teddy Roosevelt, when America became the first nation to pass laws to preserve its natural heritage, no country had done more to preserve its wildernesses, canyons and coastline. Post-war environmentalism was born in America. In the 1970s, its government imposed the most comprehensive environmental standards of any major economy. Yet America was put in the dock at Rio, as was President Bush, accused by Bill Clinton of being Rio's lone holdout against environmental progress. Demonstrating the political dexterity that Bush lacked, Clinton flipped the position that he wouldn't sign an agreement that took risks with jobs and the economy, and turned it against him. 'When you're weak at home, it weakens you abroad,' Clinton told a newspaper.

Clinton would bring those skills to the politics of global warming, matchlessly outclassing his predecessor in his political handling of the issue, but ending up precisely where Bush had left off in 1992.

17
TWO PROTOCOLS

If there is one person in the world who has the history, understanding and reputation to bring this all together, it is Al Gore.

Michael Oppenheimer, Environmental Defence Fund,
Kyoto, December 1997

Rio begat Kyoto.

The Kyoto Protocol has not begotten a son, only a string of COPs – Conferences of the Parties – and MOPs – Meetings of the Parties. 'Since 1991, legions of well-meaning diplomats, scientists, and environmentalists have undertaken excruciatingly complicated negotiations in what is essentially a political exercise that creates the illusion of mitigating climate change while actually accomplishing little more than raising public consciousness,' Richard Benedick, who led the US team that negotiated the Montreal Protocol, has written.[1]

What accounts for the success of the Montreal Protocol in eliminating CFC emissions and the failure of the Kyoto Protocol ten years later? Writing in 2003, Scott Barrett argued that the Kyoto Protocol was likely to fail because it did not solve the enforcement problem. Appeals to a state's sense of its responsibilities, offers of assistance or threats of naming and shaming were, Barrett thought, inadequate substitutes for having a credible enforcement mechanism. Barrett would press negotiators on their hesitation to address the issue. 'Always I received the same unsatisfying response: enforcement was something that was best addressed later.'[2] Although the Kyoto Protocol specified that the parties should agree a compliance regime at the first COP after Kyoto, it only came up for discussion at The Hague COP6 in November 2000. By then, it was too late.

Unlike Kyoto, the Montreal Protocol has strong incentives for countries to join the Montreal regime and not to leave it. It provides the threat of trade sanctions on any goods produced using CFCs and other controlled substances by countries that had not ratified the protocol. Countries can be suspended if they have not complied with their obligations, freezing their rights and privileges.

Why, then, didn't the international community incorporate similar mechanisms being used to reverse stratospheric ozone depletion to the global warming treaties? Put another way, how was it that President Reagan, allegedly a unilateralist who once said trees cause more pollution than automobiles, led the international community to agree an environmental treaty with teeth, whereas President Clinton and a vice president who had compared global warming to an ecological *Kristallnacht* signed a toothless protocol which he did not ask the Senate to ratify?

Robert Reinstein was alternate head alongside Benedick leading the US team that negotiated the Montreal Protocol and subsequently led the negotiations for the US on the climate change convention. In a 1992 paper, Reinstein highlighted differences in the nature and uses of the gases themselves and in the status of the science between ozone depletion and global warming. The initial international response to ozone depletion focused on eight synthetic gases that were manufactured for a limited range of specific applications, such as refrigerants and foam-blowing. Greenhouse gases such as carbon dioxide, methane and nitrous oxide occur naturally and are emitted mostly as by-products in processes that are basic to human survival. Whereas there were a limited number of chemical plants producing CFCs, every person who burns fossil fuels is a 'producer' of carbon dioxide and every farmer raising cattle or sheep is a 'producer' of methane.

Then there is the respective status of the scientific knowledge of ozone-layer depletion and global warming. Although the causes of stratospheric ozone depletion had not been proven, certain chemicals had been demonstrated to destroy ozone in laboratory tests under conditions comparable to those in the stratosphere. By comparison, Reinstein observes, the science concerning global climate is extremely complex. Many different layers of the atmosphere must be analyzed, many different physical and chemical reactions; the role of clouds, oceans, land masses, vegetation must all be taken into account, as well as complex interactions with radiation from the sun and other parts of space.

The 1992 Climate Change Convention was adopted on the basis of the precautionary principle, Article 3.3 speaking of the need for the parties to take 'precautionary measures'. Subsequently the official scientific consensus on global warming hardened with the IPCC's 1995 Second Assessment Report. 'The balance of evidence suggests that there is a discernible human influence on global climate,' although the IPCC acknowledged scientists' limited ability to quantify the effect because the expected 'signal' was still emerging from the noise of natural variability (another way of saying the signal hadn't unambiguously emerged).[3]

Even if the scientific uncertainties could be resolved, there was a gulf separating the respective economic cases for the Montreal Protocol and the Kyoto Protocol. In 2007 Cass Sunstein, a law professor at the University of Chicago who subsequently became President Obama's head of the White House office of information and regulatory affairs, compared the two. Sunstein's analysis focused on the differences between the ratios of perceived benefits to costs flowing from policies to reverse the depletion of the ozone layer and to slow down global warming, especially for the United States as the world's largest emitter of CFCs and carbon dioxide. These differences helped explain and drove the differing assessments of national interest, the response of consumers and the role of powerful private actors, Sunstein argued.

The conclusion was stark. Of all the countries in the world, the US was expected to gain the most from the Montreal Protocol and lose the most from the Kyoto Protocol. The perceived costs of complying with Kyoto were $313 billion higher than the costs of Montreal while the perceived benefits of Kyoto were some $3,562 billion lower than the perceived benefits of Montreal.[4] Put another way, each $1 billion spent complying with Kyoto Protocol was estimated to yield $37 million in benefits. By contrast, each \$1 billion spent on complying with the Montreal Protocol was anticipated to yield \$170 billion in benefits. For the world as a whole, Montreal was estimated to produce net benefits greater than \$900 billion. Kyoto, on the other hand, was expected to generate *negative* net benefits of \$119 billion to \$242 billion (the band reflects different treatments of the money spent by the US buying other countries' excess emissions allowances).

According to Sunstein, the difference in the cost benefit assessments of the two was corroborated by differences in consumer behavior. These in turn drove political and business incentives. Vivid warnings about the consequences for human health of CFCs and the trivial costs to consumers of mitigating the risk led Americans to cut purchases of aerosol sprays by more than half. Politicians responded quickly. Who would want to run for election in the Sunbelt on a platform in effect favoring skin cancer for light-skinned people? In 1978, Congress banned the use of CFCs as an aerosol propellant. After DuPont developed HCFCs as viable (and profitable) substitutes for CFCs, businesses followed, pledging to phase out CFCs and lobbying for international controls.*

Given their comparative advantage over foreign competitors, American businesses could expect to benefit from global moves that generated higher demand for replacements for CFCs. 'If environmentally unfriendly products are unpopular in the market, industry is likely to respond with safer substitutes.'[5] Again, the contrast with global warming is telling. Despite the enormous media coverage of global warming, if judged by their actions as consumers, Americans do not rate climate change as a serious risk compared to the benefits they derive from the burning of fossil fuels. 'Contrary to their behavior in the context of ozone layer depletion, American consumers and voters are now putting little pressure on either markets or officials,' Sunstein observed.[6]

To Sunstein, US leadership in obtaining international agreement to cut CFC emissions conforms to the model of a global hegemon providing public goods because it benefited from doing so.

How does Kyoto fit this model?

Not very well. Insofar as there were any benefits accruing to the US from complying with Kyoto, they were a fraction of the costs. Although the Clinton administration pledged to reduce emissions by seven per cent compared to eight per cent for the European Union, it was vastly more challenging for the US. The 1990 base year chosen for the Kyoto Protocol was the trough of recession in America, which occurred in 1992 for Europe. In that period, US carbon dioxide emissions rose

---

*   Reinstein argues that the success of the Montreal Protocol was achieved at an acceptable cost because negotiators used a 'bottom-up' approach, working closely with businesses on what was technically and economically feasible, in contrast to the 'top-down' negotiation of emissions caps under Kyoto which depend on 'technological forcing' of low cost substitutes for fossil fuels that do not exist at present (Robert Reinstein, 'Ozone Protection and Global Climate Change: Is the Montreal Protocol a Good Model for Responding to Climate Change?', 1996).

by 1.3 per cent. In Europe, thanks to recession and the implosion of the former communist economies, emissions fell by nearly five per cent.[7]

Special factors affected Europe's two largest emitters which had nothing to do with policies to reduce their carbon dioxide emissions. Helmut Kohl's Germany was one of the most strenuous in demanding deep emissions cuts, but its approach was based on extraordinarily shallow analysis. After the 1988 Toronto conference had called for developed countries to cut emissions by twenty per cent by 2005, a German study team concluded that the goal was under-ambitious. In early 1990, Kohl asked the BMU, the federal ministry with responsibility for the environment, to prepare a carbon dioxide reduction target.

After only four weeks looking at the issue, the ministry concluded that a 30.5 per cent reduction was feasible. In June, the government adopted a twenty-five per cent target compared to 1987 levels for West Germany. The target for the former West Germany was reaffirmed in November, the government stating that it expected larger reductions from the former East Germany. It proved wildly optimistic. Although German carbon dioxide emissions fell in twelve of the fifteen years from 1990 to 2005, at the end of the period they were 17.7 per cent below 1990 levels, half the fall coming in the first two years after reunification.

For Britain, the speed of its fall in carbon dioxide emissions was in response to a major policy error in privatizing the electricity industry. The original idea had been to create a power generating duopoly so the larger company could own the country's nuclear power stations. Preparations for privatization revealed what many economists and environmental groups had long argued – after taking into account decommissioning costs, nuclear power was fundamentally uneconomic. The nuclear power stations were dropped from the initial privatization package, but the generating duopoly was preserved. To provide some competitive pressure, the electricity regulator encouraged local electricity distributors to build their own gas-fired power stations by letting them earn temporary super-profits from vertical integration. According to Dieter Helm, the leading authority on the British energy industry, 'The consequence was that gas came on faster than would have been dictated by competitive markets, and the coal industry contracted more quickly.'[8] Together with the effect of a sharp recession at the beginning of the 1990s, the switch from coal to natural gas helped Britain's emissions to fall by nearly seven per cent from 1990 to 1997.

As the negotiations progressed, the EU proposed their members share the British and German reductions under an EU 'bubble', even though they were simultaneously arguing that other countries should be held to flat rate reductions. Yet the high cost of meeting Kyoto's seven per cent cut for America was not the decisive factor behind America's non-ratification of Kyoto. The fundamental reason lay in the architecture of the Protocol, which followed the ground plan of the convention. This divided the world into two, with developed nations listed in Annex I of the convention. Non-Annex I nations therefore comprised the rest of the world – the least developed, oil-rich exporters, successfully industrialized and the world's fastest growing economies.

At the convention's first COP in Berlin in the spring of 1995, the parties agreed that the commitments of the Annex I parties needed to be strengthened. As a condition for allowing the process to proceed, the G77 plus China stipulated that no new commitments should be introduced for non-Annex I parties. This agreement was incorporated into the Berlin Mandate, which defined the objectives and parameters for the negotiations that resulted in the Kyoto Protocol.

At COP2 in Geneva the following July, Tim Wirth, now serving as undersecretary of state for global affairs, announced that the Clinton administration would be urging Annex I countries to negotiate 'realistic, verifiable and binding targets' to reverse the trend of rising greenhouse gas emissions. 'This is a big deal,' Wirth told the *New York Times*.

It certainly was. A year later, the US Senate adopted the Byrd-Hagel resolution by ninety-five to zero; America should not sign any protocol which imposed limits on Annex I parties unless it also imposed specific, timetabled commitments on non-Annex I countries within the same compliance period. In some respects, the resolution represented the Senate taking a second look at the convention it had ratified less than five years earlier by a similar margin as it had adopted Byrd-Hagel. The principle of 'common but differentiated responsibilities' – the first of the convention's five principles – and the bifurcation between Annex I parties and the rest of the world were central to the convention's ground plan.

There was always a high risk of the Annex I bifurcation becoming unbridgeable, as the convention did not provide an automatic mechanism by which non-Annex I countries could or should graduate to Annex I. However it contained two provisions that might have served as a basis for bridging the divide. Article 4.2(f) stipulated that by the end of 1998, a future COP should have reviewed available information 'with a view to taking decisions' on the matter, so long as any move into Annex I was with the approval of the party concerned. The provision was essentially stillborn, only being used to remove Turkey, considered a developed country for the purposes of global warming, from the Annex II donor countries, which mirrors Annex I, so Turkey remained in Annex I but does not have Annex II donor obligations. The second provision, Article 4.2(d), required the periodic review of adequacy of the commitments of Annex I and non-Annex I parties. As we shall see, the fate of this provision provides one of the most telling pieces of evidence as to the attitude of the developing world to the global warming negotiations.

The effect of the Berlin Mandate was not to change the structure of the convention, but to build a wall around the existing Annex I parties and thereby institutionalize the division between North and South. To date, only Malta has crossed this new Berlin Wall, and that was because it had joined the EU (Cyprus, the only other non-Annex I country to become a member of the EU, so far has not done so). If there was any issue that risked triggering the collapse of negotiations at Kyoto, it was in response to any attempt to fragment or erode the solidarity of the G77 plus China on this issue.

By contrast, the Montreal Protocol avoided this by having a unified ground plan based on objective criteria. While bearing in mind the developmental needs of developing countries, to use the words of the Montreal Protocol's preamble, there is no list of countries subject to quantitative limits; the annexes simply list the various categories of controlled gases. Developing countries are subject to the full control regime while capping their per capita consumption of ozone depleting gases. In the case of CFCs, developing countries which consume less than 0.3 kg per capita of CFCs a year are granted a grace period of ten years 'in order to meet its basic domestic needs' before having to comply with the full rigour of the Protocol's controls, as long as they keep their per capita consumption below 0.3 kg a year.

Thus the Montreal Protocol provides a tough, universal regime with powerful sanctions for compliance. The bifurcated regime adopted in the 1992 climate change convention and developed further by the Kyoto Protocol can only be judged superior to the Montreal Protocol if the principal objective is assumed not to be the reduction of greenhouse gas emissions.

History helps explain why such divergent approaches were taken. The Montreal Protocol was negotiated between 1985 and 1987, in the brief period after the rise of Mikhail Gorbachev and before the formal adoption of the doctrine of sustainable development in 1988. Agreement on the final text was secured five months after publication of the Brundtland Report in April 1987. By contrast, the climate change regime is an offspring of the doctrine of sustainable development, a term that does not appear anywhere in the Montreal Protocol, and its fusion of First World environmentalism and the Third World's demand for the New International Economic Order.

Of the world's top ten per capita carbon dioxide emitters in 2000, six are outside Annex I, including the top three, Qatar (with per capita emissions three times those of the US and nearly six times the average of the developed world), the United Arab Emirates and Kuwait. Neither is per capita GDP a criterion for inclusion. By 2009 for example, non-Annex I South Korea had per capita GDP of $23,407, just $278 less than the EU's at $23,685.

Instead of being based on objective criteria, the Annex I dividing line closely follows the 1980 Brandt Line, purportedly delineating the unbridgeable chasm dividing the rich North from the poor South. Thus Annex I countries comprise the old OECD (i.e., excluding Mexico, Chile, Israel and South Korea, which were not OECD members at the time) plus the countries of the former Soviet bloc, but excluding the Soviet Union's Asian republics, such as Turkmenistan and Kazakhstan. The resulting North-South demarcation, as carved in stone in global warming agreements, is a product of history. It has a political explanation, but no objective economic justification.

With the Montreal Protocol, developing countries had not engaged early in the process or *en bloc*. After it came into force in 1989, developing countries asserted their need for new and additional transfers, agreement being reached at the London ozone conference in July 1990, when the Chinese and Indian delegations announced that they were recommending that their countries ratify the protocol. Coming late to the party had a cost, because the developed nations, led by the US, had already settled the Protocol's main terms. They weren't going to make the same mistake twice. In 1989, Brazil and Mexico pressed for increased developing country representation in negotiations to tackle global warming, creating a Special Committee on the Participation of Developing Countries. The following year, the UN resolution establishing the INC specified that it should be open to all member states.

Putting developing countries on an equal footing had profound implications for the conduct of the negotiations as well as their outcome. The Montreal Protocol was negotiated by around thirty countries. Typically the climate change COPs have been attended by one hundred and fifty or more countries and more than one thousand, three hundred delegates. Progress is achieved through UN-style consensus rather than majority voting, meaning that there must be no stated or formal objections to a decision.

Given all these constraints, to have produced a treaty signed by over one hundred and ninety nations was the result of a diplomatic *tour de force*. Michael Oppenheimer, an astrophysicist subsequently specializing in atmospheric physics and chemistry, was in Kyoto as science adviser to the Environmental Defence Fund. Al Gore, Oppenheimer said, was the only person with the history, understanding and reputation to bring it together. 'If this fails, a large part of the blame falls on the administration.'[9] He was wrong on both counts.

There was one person who could and did bring the negotiations to a successful conclusion and that was Raúl Estrada-Oyuela, the Argentine diplomat who chaired the negotiations over a period of thirty-two months. The resulting agreement was, Estrada wrote two years later, the 'best compromise the international community was able to reach at that time'. It was a diplomatic accomplishment of the highest order. 'I thought Raúl did a strong, credible job in a most challenging negotiation,' Tim Wirth told the author.[10]

To have blamed Gore if the conference had broken up without an agreement or for its failure to produce a treaty which met the requirements of the Byrd-Hagel resolution would also have been unfair. But it wasn't in the interests of NGOs like the Environmental Defence Fund to point their finger in the direction of the South, but at America and its most prominent environmentalist.

At a symbolic level, though, Gore invited the charge because he had made it himself: Western civilization was the root cause of the environmental crisis. Nature was in crisis because Western man was sick. 'Ecology and the human spirit' is the somewhat Germanic sounding subtitle of *Earth in the Balance*, in which Gore set out these and other thoughts. At other times, Gore conceded that Western man might not be wholly to blame. Thinking about human beings and the environment led him to pose the biggest environmental question of all. Had God made a mistake when giving mankind dominion over the earth?

Gore's ecological philosophy brought together an American tradition of environmentalism, extending from Thoreau and Muir, with crankier elements imported from Europe, notably Schumacher. What made Gore unique was that no ecologist before him had attained high political office, let alone being a heartbeat away from the presidency of the United States. The harsh criticism Gore directed at the preceding and succeeding administrations contrasts with the Clinton administration's performance with respect to the Kyoto Protocol, the subject of the next two chapters.

18
CHINA SYNDROME

It's the economy, stupid.

Clinton-Gore campaign war room, 1992

*Ours are survival emissions. Theirs are luxury emissions. They have
two people to a car and yet they don't want us to ride buses.*

Shukong Zhong, China's chief negotiator, Kyoto,
December 1997

As a teenager, the future vice president and his sister read and talked about *Silent
Spring*. A happy and vivid memory, Al Gore recalled. Rachel Carson's picture
hangs in his office and her example inspired Gore to write *Earth in the Balance*.
It is one of the most extraordinary books by any democratic politician seeking
high elective office, for it constitutes an attack on Western civilisation and a
fundamental rejection of two of its greatest accomplishments – the Industrial and
Scientific Revolutions.

Searching for a better understanding of his own life and how he was going
to rescue the global environment, Gore concluded that modern civilisation was
suffering from a spiritual crisis. Although one of a number of environmental crises, global warming symbolised the collision between civilisation and the Earth's
ecological system. Global warming turned the internal combustion engine into
a more deadly threat than any military foe America was ever likely to face. The
current generation might even experience a year without a winter, Gore warned.

Western man had only escaped the Malthusian trap by making a Faustian
pact. Men were to blame, for Western civilisation had emphasised a 'distinctly
male' way of relating to the world. A solution might be found by 'leavening the
dominant male perspective with a healthier respect for female ways of experiencing
the world'.[1] Western civilisation was a dysfunctional family, impelled by addiction
to inauthentic substitutes for direct experience of real life, leading to the frenzied
destruction of the natural world. It was well known, Gore observed, 'that the vast
majority of child abusers were themselves abused as children'.[2]

The chain of abuse went back to the two philosophers who anticipated the
Scientific Revolution. 'The unwritten rules that govern our relationship to the environment have been passed down from one generation to the next since the time
of Descartes, Bacon, and the other pioneers of the Scientific Revolution.'[3] Gore
reinterpreted medieval metaphysics as an ecological philosophy connecting man
to nature in a web of life, matter and meaning, now lost to the modern world. By
breaking with Aristotleanism, Bacon and Descartes had separated man from nature and science from religion. For them, facts derived from science had no moral
significance. 'As a consequence,' Gore wrote, 'the scientific method changed our

relationship to nature and is now, perhaps irrevocably, changing nature itself.'[4] If science had kept its link with religion, Gore thought humans might not be threatening the earth's climate balance.

Bacon was morally confused, because he had argued that science was about the advancement of knowledge and making scientific discoveries without reference to any moral purpose. The divorce of facts from values and morality had terrible consequences in the twentieth century, Gore argued, Bacon and the scientific method thereby contributing to the extreme evils perpetrated by Hitler and Stalin.[5] Gore's accusation against science shows an extraordinary misreading of history. The Nazis did not commit their crimes because they lacked values, but because their values were evil. Moreover the Nazis enacted the most environmentally friendly laws in Europe. They passed anti-vivisection laws (Gore criticised Bacon for dissecting animals for the sake of knowledge) but used humans instead. Nazi ideologists rejected the proposition that science is morally neutral, most horribly in their racial theories. Similarly, Stalin supported Trofim Lysenko's genetic theories, not because of their superior explanatory power, but because they conformed to Marxist-Leninist ideology.

Yet Gore's assault on the Scientific Revolution met with silence from leading academies and societies. Collectively scientists tolerated an extraordinary attack on the integrity and morality of their discipline because they were united by a common enemy – global warming and fossil fuel interests.

The book was well timed. Bill Clinton's Arkansas had one of the worst environmental records in America. Environmental policy was not, as Clinton admitted, his strong suit. Gore gave Clinton a signed copy. 'I read it, learned a lot, and agreed with his argument.'

With Gore as vice president, there was no debate within the new administration on the science of global warming. Gore regularly hosted breakfast seminars with leading scientists, exposing agency heads to what scientists were thinking. Bob Watson, who worked in the White House during President Clinton's first term, recalls Gore being an avid reader of *Nature*. He would often telephone, 'Bob, what do you think of this paper?' In preparing scientists for the seminars, Watson would tell them to speak for a maximum of seven minutes, as Gore will interrupt and ask questions. Gore's biggest strength was in synthesising and connecting issues and Watson had no hesitation in putting him in front of a pure science audience.

The economics presented a greater challenge. In his book, Gore strongly criticised the Bush administration for threatening to torpedo the Rio summit because of its refusal to sign up to targets and timetables. Ratcheting back emissions to their 1990 level was a target the US could 'easily' meet. Here debates within the Clinton administration in the run-up to Kyoto were similar to those in the Bush administration before Rio. 'Al has discovered it's a lot easier to write a book about the subject than to grapple with the economic costs,' one of Clinton's top aides said, 'but he's getting the hang of it.'[6]

From his position in the State Department, Tim Wirth advocated an aggressive plan for significant cutbacks in greenhouse gas emissions. The plan was scaled back as Clinton listened to warnings of senior economic advisers, notably Larry Summers and Janet Yellen of the Council of Economic Advisers. 'The spin is that we won,' an economic adviser told the *New York Times*. 'We agreed there needed to be goals, even aggressive goals. But there also needed to be escape hatches, in case the economic effects turned out to be a lot more damaging than we thought.'[7]

Rescuing the environment was not the central organising principle of the Clinton administration. Within four weeks of being sworn in, Clinton announced what was – deficit reduction to induce a bond market rally, encourage private investment to spur productivity, job creation and growth. He proposed cutting spending and raising taxes. Gore argued for a broad-based tax on energy. Clinton called the BTU energy tax his toughest tax call, particularly after he had dropped his election pledge of a middle-class tax cut. Lloyd Bentsen, the treasury secretary, joined Gore in pressing him. 'Finally, I gave in.' A few days later, Clinton announced the BTU tax in his State of the Union speech in February.

Environmentalists were thrilled. A retrospective paper by one called it 'brilliantly conceived in every way. It was simple, clean, easy to administer and raised significant revenue'.[8] The BTU tax quickly became the most controversial part of Clinton's deficit reduction package. Manufacturers claimed it would cost more than half a million jobs. When the Senate deleted the BTU tax two months later and substituted a 4.3 cents a gallon gas tax, Clinton's relief was palpable. 'The bad news was that the gas tax would promote less energy conservation than the BTU tax; the good news was that it would cost middle-class Americans less.'[9]

The loss of the BTU tax was enormously consequential. Even when Democrats controlled Congress, taxation as a policy response to global warming was not politically feasible. So the Clinton administration turned away from energy taxes to champion the major environmental policy innovation of the Bush administration – tradable pollution permits. Emissions trading had solved a politically intractable problem that had stalled progress on tackling acid rain and led to a market-based way that enabled the Bush administration to propose the most ambitious target of a fifty per cent emissions cut with the most creative means. Although Gore had been a bystander in the Clean Air Act debates, emissions trading and flexible market mechanisms became the central plank of the Clinton administration's negotiating strategy.

Transposing a mechanism designed to cut emissions from the chimneys of a few dozen power stations in one country to creating a market for the right to emit a gas used in processes too numerous to count, a gas, moreover, that is part of a naturally occurring cycle and therefore influenced by land use changes, then extending the market to cover developed countries and, through the Clean Development Mechanism, to embrace virtually every country in the world, posed technical, legal, verification and compliance challenges several orders of magnitude more complex.

Carbon taxes would have been simpler, easier and cleaner. Focusing negotiators on setting quantities, in the form of emissions caps, rather than setting prices, in the form of taxes, also created irresistible incentives for gaming. It incentivised countries to target emissions reductions that would have happened anyway (essentially the position of the European Union) or to negotiate trading mechanisms to take advantage of other nations' emission reductions that were happening anyway, principally those caused by the collapse of the Soviet bloc (the American goal).

The outcome was very different from the Montreal Protocol. All the reductions of CFCs and other substances controlled by the Montreal Protocol occurred as a direct result of regulatory actions designed to reverse the depletion of the ozone layer. On the other hand, the interest of the Annex I parties in negotiating the Kyoto Protocol was to free ride to the greatest extent possible reductions that would have happened anyway or existed only on paper.

The Clinton administration's journey along this road began in October 1993 with its climate change action plan. The forty-nine-page document listed forty-four actions designed to meet the president's personal commitment to reduce emissions to 1990 levels by the end of the decade (they actually increased by 16.1 per cent compared to the 3.4 per cent fall needed to return to 1990 levels). It was thin stuff. All the actions were voluntary and the plan involved only $1.9 billion in new and redirected spending between 1994 and 2000.

Some environmental NGOs disguised their disappointment. Others couldn't. The Sierra Club said the administration had looked global warming in the eye and blinked. The National Wildlife Federation compared it to date rape. And there was an endorsement the administration could have done without. Fred Singer, a leading scientist opposed to the scientific consensus, wrote that the voluntary measures made 'a certain amount of sense'.

Losing both houses of Congress in the November 1994 mid-term elections put the Clinton administration's climate policies on the defensive at home when its presence was required on the international stage. In March 1995, Helmut Kohl addressed the first COP in Berlin. There was a certain irony as Kohl urged delegates to remember the lesson of Berlin. Never again should walls of enmity be erected between peoples and nations, Kohl declared, for the Berlin Mandate institutionalised a new division across the world between North and South.

The demarcation line had been agreed at a late stage in the negotiations on the Climate Change Convention three years earlier at an INC session in Paris during Holy Week. There were various attempts to define developed and developing countries. GDP per capita was felt not to be a good measure, as there were lots of countries in between. Because there were more developing countries, it was easier to define the developed countries, which was done on the basis of membership of the OECD (a definition that put Turkey in an anomalous position as a developing country which was also an OECD member) plus Eastern Europe.

Angela Merkel, the German environment minister, had to fashion a compromise to reconcile the contradictory demands of North and South. Growing up in East Germany, Merkel was versed in the uselessness of inflexibility. She was helped by Britain's environment secretary, John Gummer, one of the most pro-European members of John Major's government who had replaced the Atlanticist Michael Howard two years earlier. Gummer's views were closely aligned with mainstream European attitudes and was an early admirer of Merkel. 'Very, very able,' Gummer found her.[10] China had to be kept onside, recognising Chinese sensitivities to anything that appeared to them to infringe their sovereignty (at times, Gummer recalls negotiators spent longer debating the rights and wrongs of the Boxer Rebellion than climate change). In Gummer's view, the difficulties America had over China were not fundamentally about climate change, but sprang from fears about China's rivalry with the US.

Apart from the EU and the G77 plus China, the other main negotiating bloc was the US-led Umbrella Group, loosely comprising Australia, Canada, Iceland, Japan, New Zealand, Norway, the Russian Federation, and the Ukraine as a counterweight to the EU. Merkel's strategy was to target Canada's environment minister, Sheila Copps, the leading left-winger in Jean Chrétien's Liberal government. Copps was peeled away from the Umbrella Group, which helped convince the Chinese delegation that they should have confidence that Annex I parties genuinely accepted overwhelming responsibility for taking action to tackle global warming. Shuttling between two rooms, one with developed countries, the other with developing countries, she produced a compromise text at six in the morning and declared the Mandate adopted despite protests from OPEC members.

Throughout the rest of the negotiations, OPEC was carefully managed. At COP2, OPEC members were bought off with a fund to compensate oil producers for the loss of income resulting from policies to cut carbon dioxide emissions. There is evidence that Japan offered Saudi Arabia a side deal in return for its cooperation at Kyoto.

No threats or inducements could subsequently shift the G77 plus China from what they had wrung from COP1. The text of the Berlin Mandate states that the process will not introduce any new commitments for Parties not included in Annex I. Its implications went further than a literal reading might suggest, which neither added nor subtracted anything to what was already in the text of the convention, as it precluded even specifying what the obligations of non-Annex I parties in the original convention might be.

Other than the State Department, the Berlin Mandate did not receive high level attention in the Clinton administration. Although American negotiators in Berlin recognised that it might go down badly on the Hill, agreeing it was seen as a 'tactical step to keep the process moving'. It proved a grave miscalculation.

A decade and a half on, the man who discharged the Berlin Mandate believes it was the only basis on which the process could have been taken forward. As host, Japan would by custom hold the presidency of the conference. Internal splits meant it was unable to field a candidate of sufficient standing. Before taking up his new post as Argentina's ambassador to Beijing, Raúl Estrada-Oyuela was elected chair of the Ad Hoc Group of the Berlin Mandate.

As an Argentine, Estrada brought to the job the economic perspective of his fellow countryman, Raúl Prebisch, the intellectual grandfather of the development half of sustainable development. The two got to know each other when they were living in Washington in the early 1970s where their sons went to the same kindergarten. A decade later, Estrada attended Prebisch's funeral in Santiago; a complicated affair, with Prebisch's two concurrent wives each taking a share of his ashes. Some of Estrada's most penetrating economic analyses of the climate change issues owe much to Prebisch's economics.

Estrada had been involved in climate change negotiations since attending the Geneva World Climate Conference in 1990. He consciously applied the approaches used by his predecessors to forge consensus: Ripert in consulting privately with delegations on every issue to understand their thinking; Tommy Koh of Singapore, who chaired the preparatory committee for the Rio summit and addressed delegations by their first names, making emotional appeals at plenary sessions; and Merkel's devotion to constructive compromise. The chair needed to possess an instinct for the sense of the room; to know when to press forward and when to wait. And he needed determination. 'You have to be optimistic. Particularly in Kyoto, I was paid to be optimistic ... the Chairman has to be perseverant, persistent, otherwise you are lost.'[11]

Estrada was supported by the convention secretariat, headed by the Maltese Michael Zammit Cutajar for its first eleven years. Zammit Cutajar began his career in UNCTAD, providing a further link with Prebisch, who was his first boss. When questioned about the position of the climate change secretariat in the negotiations, Zammit Cutajar would recall what Prebisch used to say:

> 'As a secretariat we are objective, but we cannot be indifferent to development. We cannot be neutral. We are fighting for development.' So when people try to block the [climate change] process, we can admire their negotiating skill, but we cannot be indifferent ... We have a commitment.[12]

Although the commitment to tackle global warming was seen through the prism of Prebisch's economics and the development needs of the South, the chair and secretariat did not permit their commitment to be about promoting the interests of the South. At one point, in Kyoto, Estrada accused Brazil of coming to the conference 'with an open hand', triggering a walk out by the Brazilian ambassador. He opposed the Clean Development Mechanism, which was designed to pump money from North to South. When Saudi Arabia and Kuwait challenged one of his decisions, he rounded on them. 'From the very beginning ... a group of countries was trying to stop the process ... I will do everything to overcome those countries ... I am not going to be [held] hostage.'[13] OPEC wouldn't risk a walk out and Estrada knew he had the votes.

This was not the case for the G77 plus China on the issue that bound them together: individually and collectively its members would be held totally immune from the costs of policies designed to tackle global warming. President Clinton tried; Estrada had more success – but nothing could break the solidarity of the South.

After the 1994 mid-term elections, the Clinton administration requested that the negotiations be slowed down. At COP2 in July 1996, with Clinton cruising to re-election, the Clinton administration called for an international agreement on binding targets to be met with the maximum flexibility. It also put its name to a ministerial declaration led by Canada, which stated the continued rise of greenhouse gas concentrations 'will lead to dangerous interference with the climate system'. It was an important moment. The convention had been adopted on the precautionary principle. The ministerial declaration replaced the modal auxiliary 'might' with the future tense 'will'. The convention's objective is to avoid dangerous anthropogenic interference in the climate system, but it did not define what should be considered dangerous. Now the politicians had. Because delegations from Australia, New Zealand, Saudi Arabia and Venezuela disagreed (and Russia's Yuri Izrael continued to dispute that global warming would be harmful), the COP noted, but did not adopt, the declaration.

The Kyoto conference had to resolve three sets of issues. The first was the level of emissions cuts for Annex I countries. This pitted the US and the EU against each other, a nervous Japan on the sidelines, worried about the burden on its economy, but as host not wanting to cause the conference to fail. The second set revolved around the flexibility of those targets and timetables and the means by which they should be achieved. This was another battle between the US and the EU, joined by a deeply suspicious South, that saw flexibility as a means by which Annex I countries could evade their responsibilities by cutting emissions of poorer countries rather than their own. The third was the extent to which non-Annex I parties should indicate willingness to assume some form of obligation to limit carbon dioxide emissions at some point in the future.

The stage for the confrontation on the first of these was set in June 1997 at the Denver G8 summit, the only one hosted by Bill Clinton. On his return to Washington, Clinton told Taylor Branch, his oral historian, how the Europeans, including Tony Blair and Helmut Kohl, had ganged up on him. 'They upbraided him, said the president, even though they had no idea how they would meet their standards,' Branch recorded.[14] Blair had been elected less than two months earlier on a manifesto which included the pledge to cut carbon dioxide emissions by twenty per cent by 2010 (a target the Labour government missed by fifty per cent).

More painful was Kohl – 'almost like a blood brother', according to Branch. It had been eighteen years since a German chancellor had attacked an American president, but their respective positions had changed one hundred and eighty degrees since Helmut Schmidt had taken on Jimmy Carter in Tokyo in 1979. Then it was a realist German expressing his scorn for an American president's crusade against imported oil. In the intervening years, German politics had been transformed by the rise of the Greens.

Schmidt was a hard-headed pragmatist. His definition of a successful leader was to prevent his country being overrun by war or by need and hunger. Schmidt's call for a NATO response to the Soviet Union stationing of SS-20 nuclear missiles helped fuel the rise of anti-nuclear sentiment in Germany. In 1978 and 1979, massive, sometimes violent, demonstrations against nuclear power and nuclear weapons made the extra-parliamentary Left the voice of radical environmentalism that until then had mainly been the province of old and neo-Nazis. In October 1980, the Green Party was formed to bring the radicals into the parliamentary system for the first time in post-war Germany. The rise of the Greens was a gift to Kohl and the CDU, who used the Greens against the SPD to split the centre-left. At the same time, the CDU had to develop pro-environmental positions to prevent losing conservative voters to the Greens.

In July, the US Senate fired its warning shot across Clinton's bows on the third strategic issue by passing the Byrd-Hagel resolution. According to co-sponsor Chuck Hagel, there was not a single senator who had not been concerned about what might come out of Kyoto and worried that the president was going too far.[15]

Three months later, Clinton announced the instructions he was giving American negotiators in a speech at the National Geographic Society. The US would commit to stabilising emissions at 1990 levels between 2008 and 2012.* Although he talked of achieving meaningful reductions 'here in America', that goal was contradicted by Clinton's second goal – flexible mechanisms including emissions trading and 'joint implementation', whereby firms could invest in projects in other countries and receive credits for those reductions at home. On bringing non-Annex I countries in to the net, Clinton could hardly have been clearer: 'The United States will not assume binding obligations unless key developing nations meaningfully participate in this effort.'

Less than a month before the Kyoto conference was due to begin, Tim Wirth announced he was quitting. His place was taken by Stuart Eizenstat, who, at short notice, flew in from Switzerland where he had been negotiating the restitution of Nazi gold. Where Wirth's time in Congress had made him a somewhat divisive figure, Eizenstat was highly regarded by Democrats and Republicans. Clinton could hardly have made a better received appointment.

If the Kyoto conference had only been about flexibility mechanisms, the outcome would have been an unqualified success for the Clinton administration. US

---

* The original draft of the speech called for a five per cent cut, but a member of Clinton's economic team changed the text. Amy Royden, *U.S. Climate Change Policy Under President Clinton: A Look Back* (2002), fn. 119.

negotiators gained every one of their main objectives. America wouldn't have to cut its own emissions if it could buy other countries' excess emissions – the Russian Federation having enough 'hot air' to satisfy American needs, or so the thinking went. (It was never likely that the US Congress would approve a law which resulted in American businesses and consumers sending billions of dollars to Russia for its hot air.) It was on the third of Clinton's three objectives that no ground was given nor taken at Kyoto.

Rio had been a signing ceremony. The Kyoto COP3, which started on December 1st 1997, was a real time negotiating conference with cameras present at the final climatic session. NGOs came, but their purpose was to influence the negotiators and raise the stakes for failure. Scientists also played their customary Cassandra role. 1997 was expected to be the Earth's warmest year on record, they said. 'We are beginning to see the fingerprint of man's impact on the climate,' a scientist from Britain's Hadley Centre told the media.

Not everyone was impressed. *Washington Post* columnist Charles Krauthammer reminded his readers of a leading climate scientist, Stephen Schneider, who was now arguing that it was 'journalistically irresponsible' to present both sides of the debate. Twenty-five years previously, Schneider had been arguing that the real threat was global cooling. Then, he had dismissed fears about global warming by claiming that a doubling of carbon dioxide would produce a temperature change of less than one degree centigrade.

A congressional delegation was in Kyoto to provide oversight. 'This is not a conference about environment,' Hagel told a local newspaper. 'This is a conference about economics,' one with an underlying agenda of wealth transfer. 'I've never seen so many silly people,' the senator added. Senator Liebermann was challenged by a Chinese delegate: 'Do you expect us to keep our people poor? Is that what you want?' To which the senator, a future sponsor of cap-and-trade legislation, replied, 'We can't ask our people in the US to accept the burdens associated with reducing greenhouse gas emissions, if at the same time, the developing nations accept no responsibility.'

Senator Kerry took a more emollient approach. America must tell the Third World: 'The mistakes we made should not be repeated and we're willing to help you grow in thoughtful ways. So please, when you get cars, think about unleaded gasoline and emissions controls, about the virtues of mass transit and trains.'[16]

There had been an end of June deadline for countries to submit proposals for consideration. Out of more than two hundred pages, Estrada distilled a twenty-five-page draft protocol, closing minor issues but leaving open the large contentious ones. He then challenged the Annex I countries to put their cards on the table.

The EU's opening position had been a flat fifteen per cent cut. In March the EU re-allocated this among themselves under the 'EU bubble', in the process weakening their credibility in pressing for a flat rate for everyone else, although

John Gummer believes it was perfectly reasonable for the world to view the EU as a single economy. Indeed he describes his support of the EU bubble as his proudest moment as a minister of the British crown. It meant Britain would have to do more so Ireland could do less, as some recompense for eight hundred years of oppression.

On the third day of the COP, the US indicated that it might introduce a small symbolic cut to take its target below 1990 levels. Three days later, Estrada proposed a ten per cent cut for the EU, five per cent for the US and two and a half per cent for Japan. The EU objected, saying the three should have the same target.

On December 8th, Gore arrived. 'A one day cosmetic trip that would not make up for years of neglect by the Clinton administration,' Philip Clapp, a former aide to Tim Wirth and president of the National Environmental Trust, called it. Environmental groups chanted 'read your book' outside rooms where Gore was meeting.

Gore's decision to go had been made only two weeks before. Although he had spent months wavering, he insisted that political calculation had not entered into his reasons for making the trip. Representative John Dingell, a Michigan Democrat and long-time opponent of tough emissions curbs, thought otherwise. Dingell warned that Gore's performance at Kyoto 'could affect many things, including his nomination and election'. Harlan Watson, a senior congressional aide, had first known Gore in the early 1980s, when he had been a congressman. He had always been struck by how bright, quick and extraordinarily articulate Gore was. A different man was on stage in Kyoto, hesitant and looking around for prompts. Perhaps it was jet lag.

Then Gore dropped his bombshell. In remarks added thirty minutes beforehand, the vice president said that after meeting the US negotiating team and speaking with President Clinton by phone, 'I am instructing our delegation right now to show increased negotiating flexibility if a comprehensive plan can be put in place.'[17]

The message was clear. The Clinton administration would do whatever it took to avoid being isolated. 'We were taken aback,' recalled Watson.[18] The congressional delegation had been completely frozen out; it hadn't been helpful not to know what its own government was doing, Hagel later recounted. Perhaps the American negotiating team didn't know what was happening either. Watson remembers Eizenstat looking as surprised as everyone else.

Estrada made a new proposal of eight, five, and four and a half per cent reductions for the EU, the US and Japan respectively. Two days later on December 10th, at around midnight in Washington, Eizenstat called the White House to say they were close to agreement. Eight, seven, six would clinch the deal. At 2am Washington time, Gore telephoned Japanese Prime Minister Ryutaro Hashimoto. After reminding him how far the EU, the US and Japan had come, he laid it on the line. The last thing anyone wanted was for people to say that the thing that prevented a deal was the host country not moving a final percentage point. And that's where the three main Annex I parties ended up, although during the conference's closing moments Australia got its cap raised from plus five to plus ten per cent.*

On flexibility, the US scored a string of wins; a five-year commitment period from 2008 to 2012, joint implementation and generous accounting for forest sinks. Emissions trading was strongly opposed by the G77 plus China. From their perspective (and to some in the EU), it was wrong to have a mechanism to enable America to cut other countries' emissions to avoid cutting its own. It provoked a confrontation between China and the US, drawing a rebuke from Estrada. 'It might be better if we have no agreement,' he said. 'I invite you to reflect.'

During the pause, Chinese and American negotiators stood nose-to-nose, snarling at each other in a pre-dawn showdown.

The conference had been due to end on December 10th, but at four in the morning the following day Estrada announced he was deleting an OECD text on emissions trading, putting in its place a new article permitting trading but stating it must be supplemented by domestic action. Estrada banged his gavel. Emissions trading was in the Protocol.

The US also got a bonus that wasn't in its original list. Three days before the June deadline for proposals for Kyoto, Brazil tabled a complex plan to allocate greenhouse gas emission commitments based on countries' historic contributions to the increase in global temperatures. Although none of it was adopted, the G77 plus China used the hook of a Clean Development Fund to be financed by fines levied on Annex I countries for non-compliance. In November, Brazilian and American negotiators realised that paying a fine for non-compliance was functionally identical to buying a licence to remain compliant without the stigma. The US had found another flexibility mechanism. Even better, it could in principle extend emissions trading to the whole world.

The Clean Development Mechanism (CDM) turned out to be one of the most controversial parts of the Kyoto Protocol. After the conference, the Clinton administration argued that the CDM represented a 'down payment' on developing countries' future efforts to reduce greenhouse gas emissions. In reality, it was nothing of the sort. 'Though I facilitated approval of this proposal, I did not like it,' Estrada wrote in 1998. 'I do not understand how commitments can be implemented jointly if only one of the parties involved is committed to limit or reduce emissions and the other party is free from the quantitative [restrictions].'[19]

---

*   In post-conference briefings, Clinton administration officials claimed that -7 per cent was equivalent 'at most' to -3 per cent on their original proposal of stabilisation because of differences in accounting for forests and land management (sinks) and inclusion of extra greenhouse gases with a 1995 base year. See for example Stuart Eizenstat's testimony to the Senate Foreign Relations Committee, February 11th 1998 http://www.iitap.iastate.edu/gcp/kyoto/protocol.html

The hypothesis that mitigation costs are lower in developing countries is true only if market distortions are adjusted, Estrada argued, otherwise everything is cheaper in developing countries – a disparity that has been the root cause of every colonisation since the time of the Greeks.

The US only won emissions trading because it had comprehensively lost on the third of Clinton's three objectives. The response to Gore's instruction of flexibility on the part of the US was met by total inflexibility on the part of the G77 plus China. Early in the conference, a US negotiator indicated that all they were looking was for some movement on the issue of developing country participation, while Hagel said not all one hundred and forty developing nations need sign on. Those that did need only agree to a general commitment to limit emissions. The G77's response was swift. 'We have said categorically *no.*'

Estrada tabled a draft article to enable non-Annex I countries to make voluntary commitments. A number of non-Annex I countries gave it qualified support, including the Association of Small Island States, Argentina, South Korea and the Philippines. The majority didn't. OPEC members, perhaps recognising that the article might increase the chance of Senate ratification, argued that the article be deleted. They were joined by the host of the Earth summit, along with India and China. Recognising there was no consensus, Estrada said the article should be deleted.

Then New Zealand launched an initiative for future commitments from non-Annex I countries based on Annex I countries delivering theirs, with talks beginning in 1998. The G77 plus China said that it would not participate in them as a matter of principle. In his speech to the conference, the spokesman for the G77 plus China concluded with one word: 'No.' The proposal was not discussed again.

Some delegations had already left. Contracts for the conference translators had expired, leaving the Russian and Chinese delegations without interpreters. Having worked through the night, at around 1pm on December 11th, Estrada said he was happy to submit a Kyoto Protocol and declared that it had been unanimously recommended.

At a wrap-up briefing for the congressional delegation, Watson was sitting next to Dingell and a member of his staff. Eizenstat came over to them. 'Well, Mr Chairman, I did the best I could,' Eizenstat said. 'Don't worry Stu,' Dingell replied. 'You can't make chicken salad when you're handed chicken shit.'[20]

Nine hundred and seventy-nine days after COP1, Estrada had discharged the Berlin Mandate. After it was formally adopted, Estrada took a nap. Then he went to see the city with his wife.

19
THE MORNING AFTER

Our most fateful new challenge is the threat of global warming; 1998 was the warmest year ever recorded. Last year's heat waves, floods, and storms are but a hint of what future generations may endure if we do not act now.

President Clinton, January 19th 1999

During his twenty-four hours in Kyoto, Al Gore told reporters he welcomed the prospect of a 'knock-down, drag-out' fight to ratify the Kyoto Protocol. 'It would be high stakes and a lot of fun.'[1]

His bravado scarcely lasted the return flight. The Protocol faced 'bleak prospects', Trent Lott, the Senate majority leader, warned. 'I have made clear to the President personally that the Senate will not ratify a flawed climate change treaty,' Lott reassured Chuck Hagel.

A tug of war ensued between the Senate and the White House. The day after the Protocol's adoption, the *Washington Times* reported that the Clinton administration had decided to delay submitting it to Congress. 'As we said from the very beginning, we will not submit this agreement for ratification until key developing nations participate in this effort,' an administration spokesman said. Lott argued that Clinton should not withhold the treaty from the Senate for 'cynical, political reasons.'[2]

Clinton nominated Frank Loy to succeed Tim Wirth as undersecretary of state for global affairs. On taking up his post, Loy immediately recognized that even with a friendly Senate, the Protocol was not in a state to be ratified. The text was an outline, being totally silent or sparse in setting out how the Protocol's emissions goals should be met. While Loy and his colleagues recognized they were never going to get absolute quantified emissions targets from even the richest of the G77, getting something from some of them was high on the administration's agenda both in terms of meeting the convention's objective of avoiding dangerous interference and for its political importance.

In doing what they could to improve the prospects for ratification, the Clinton administration got little help from the Europeans, in particular Germany, France and the Scandinavians, which had Green environment ministers. Far and away the most thoughtful and realistic was the UK and John Prescott, environment secretary and deputy PM. Prescott was 'extremely helpful', according to Loy.

In October 1997, Clinton had visited Argentina. Speaking in the magnificence of the Nahuel Huapi National Park, Clinton invoked Theodore Roosevelt and Perito Moreno, who had visited Patagonia together in 1912. He promised $1 billion to help developing countries find alternative energy sources and praised his host, President Menem, for stating that developing countries should have emissions targets. Argentina's backing was a coup. Historically Argentina had provided the intellectual leadership of the developing country movement and the first COP after Kyoto was being held in Buenos Aires.

Meanwhile a fierce debate raged between the Clinton administration and Congress over the economic implications for the US of adopting Kyoto. In July, the administration produced its analysis. Emissions trading, joint implementation and the Clean Development Mechanism would enable the US to buy its way out of the problem at minimal cost. Trading among industrialized countries would more than halve the costs of climate change policies. Supplemented by the Clean Development Mechanism, trading might reduce costs by up to eighty-seven per cent of a domestic-only approach. Overall the report estimated annual costs to the US of $7–12 billion, equivalent to 0.07–0.11 per cent of GDP, a fleabite on the back of the booming US economy.

The assessment was received with considerable skepticism on Capitol Hill. Jim Sensenbrenner, chairman of the Committee on Science, held hearings and asked the Energy Information Administration, an independent agency within the Department of Energy, to offer a second opinion. In October, the EIA produced a more comprehensive and detailed analysis which suggested that the costs would be an order of magnitude higher than the administration claimed. It projected a reduction in 2010 GDP of $61–183 billion (if revenue from auctioning emissions permits was used to reduce social security taxes) and a range of $92–397 billion (if permit revenues were returned to taxpayers in a lump sum), implying a range of 0.65–4.2 per cent of GDP for the two approaches.*

Having a debate on the economics of global warming made the US exceptional. The UK, which had taken a pragmatic attitude to the climate change negotiations, did not examine the economic consequences of Kyoto. Few people were better placed to see what went on than Andrew Turnbull. Cabinet secretary and Britain's top civil servant under Tony Blair between 2002 and 2005, before that, Turnbull had been a Treasury highflyer, working in Number 10 for Margaret Thatcher in her last two years as prime minister, permanent secretary at the Department of the Environment from 1994 before returning to the Treasury in 1998 as permanent secretary. According to Turnbull, at no stage was anyone inside the British government prepared to step back and reappraise the issue. Thatcher's championing of global warming had settled the issue. Her reputation as a politician not afraid

---

* In 2004, the General Accounting Office analyzed the factors that led to these dramatically different cost ranges. It found the Council of Economic Advisers used a model that generally assumed that the economy adjusts smoothly to new policies over the longer-term. By contrast, the Energy Information Agency model used a more comprehensive cost measure and was thus able to capture certain costs that the Council of Economic Advisers' model did not. Other differences included assumptions about international trading and the proportion of reductions that would be achieved domestically. General Accounting Office, 'Estimated Costs of the Kyoto Protocol' GAO-04-144R, January 30th 2004, p. 3.

to challenge orthodoxy and her scientific mind continued to have a huge impact long after she'd gone. Rising global temperatures through the 1990s made carbon dioxide appear the villain of the piece and policies to deal with it looked like a good idea. But the Treasury never did any serious work on the economics.

As a policy disaster in the making, global warming reminds Turnbull of the poll tax. At the beginning, people went along with it because they thought it a small-scale, incremental policy. After it went wrong and helped bring about Thatcher's fall, they would say, 'It's not my fault it blew up, I wouldn't have done it if I'd known what happened subsequently.' The European dimension of global warming reinforced this tendency. Here was a policy where Britain wasn't being a foot-dragger. Since Thatcher's Royal Society speech in 1988, the UK had been in a lead position. The dash-to-gas allowed Britain to show off at little apparent cost. Only later would the costs emerge in terms of closed steel mills, distorted tax policies and unattainable targets for renewable energy.[3]

The Buenos Aires COP4 was held in the first two weeks of November 1998. It adopted the Buenos Aires Plan of Action to put flesh on the bones of the Protocol by the end of 2000. More importantly, it turned out to be the high point of the Clinton administration's campaign to get some meaningful participation from developing countries.

It began promisingly. On the conference's first day Maria Julia Alsogaray, Argentina's secretary of natural resources and sustainable development, told the conference that while Argentina did not bear historic responsibility for the climate change problem, it wished to belong to the group of countries which had responsibility for finding a solution. Developing countries too had some responsibility for climate change and an ethical duty to ensure sustainable development.

Nine days later, Carlos Menem told delegates that at the next COP Argentina would make a commitment to cap its emissions for the period 2008 to 2012. 'This is a major, major move,' Stuart Eizenstat said in Buenos Aires, 'truly historic.' Environmentalists were also ecstatic. 'It is a major breakthrough,' Michael Oppenheimer of the Environmental Defence Fund said, describing Argentina's move as 'a significant first step' in satisfying the Senate's requirements.[4]

The next day, Kazakhstan announced that it wanted to join Argentina and voluntarily assume Annex I obligations, which would enable it to sell its surplus emissions as 'hot air' to America. It looked like the Americans were on a roll. Speculation had been mounting that the administration would build on the momentum of Argentina's announcement by signing the Protocol. Senator Byrd, co-sponsor of the Byrd-Hagel resolution, warned Clinton against 'making empty gestures that will only make the potential future approval of the Protocol by the Senate more difficult.' From Buenos Aires, Senator Liebermann urged Clinton to sign. 'If we are not at the table, we cannot cajole or convince the developing nations to become part of the solution,' Liebermann said.[5]

The day after Menem's speech, Peter Burleigh, America's acting representative to the United Nations, signed the Kyoto Protocol in New York. Publicly Clinton said nothing. Instead a statement was put out on behalf of Vice President Gore. Signing Kyoto imposed no obligations on the US, the statement said. 'We will not submit the Protocol for ratification without the meaningful participation of key developing countries in efforts to address climate change.'

The downbeat spin in Washington contrasted with the stir it created in Buenos Aires. 'I am not gilding the lily when I say there was near euphoria among the delegates here,' Eizenstat told the *New York Times*. Lieberman said it gave America the credibility to be at the table. 'That means we can not only make sure it happens, but that it happens in the way that we prefer.' Hagel dared Clinton to invite the Senate to ratify it. 'If this treaty is good enough to sign, it's good enough to be submitted to the Senate for an open, honest debate.'

Clinton avoided battle on the Senate floor. Instead Buenos Aires marked the furthest extent of the Clinton administration's global warming diplomacy. Like Napoleon's defeat at Borodino, it was the start of a two-year retreat. Argentina and Kazakhstan could announce their intention to assume Annex I obligations, but there was no mechanism in the Protocol for them formally to do so. Agreeing a mechanism required consensus. Consensus, or rather the lack of it, was like General Winter to the retreating French; not an outcome decided in pitched battle, but worn down through steady attrition. By the time the administration had signed the treaty, the battle was already lost, defeat being confirmed at the conference closing plenary two days later.

At the start of COP4, there was a battle on whether voluntary commitments should be on the conference agenda. Speaking on behalf of the G77 plus China, Indonesia said the issue had been deliberated at length, but no consensus had been reached and proposed the agenda be adopted without it. India recalled the debate at Kyoto which had rejected the idea of voluntary commitments. A number of OPEC members warned that such a discussion was bound to be divisive and could lead to the imposition of voluntary commitments. China said developed countries' 'luxury' emissions were rising and that voluntary commitments would create a new category of parties under the convention.

Speaking in favor were Australia, Japan, Canada, New Zealand and the EU, which recognized that the question of broadening commitments in the long-term was necessary and unavoidable. Of the non-Annex I nations, only Chile spoke in support. The agenda was adopted without the proposed item.

Two days later, the conference discussed the adequacy of commitments of both Annex I and non-Annex I parties to attain the convention's objective. The first review had been three years earlier and had resulted in the Berlin Mandate. Article 4.2(d) of the convention required a second review not later than December 31st 1998 (and thereafter at regular intervals). The article, suggested by American negotiators, was meant to be a periodic spur; without extending commitments beyond the Annex I parties, it was numerically impossible to stabilize greenhouse gas concentrations and thus achieve the convention's objective.

The G77 plus China said that there was consensus that existing commitments were inadequate, as developed countries were shirking their responsibilities. The issue was passed to a contact group which met over the weekend. This failed to produce agreement other than to remit four different drafts to another body, this time the Subsidiary Body on Implementation. The US and Australia noted that the IPCC's evaluation confirmed that developed country actions by themselves would be insufficient, while China interpreted this as an attempt to extract commitments from developing countries. On the Tuesday (the COP was due to finish at the end of the week), the Canadian co-chair of the contact group reported that they agreed that commitments were inadequate, but not on the reasons or on any actions that might be required.

At the COP final plenary two days after America had signed the Protocol, Alsogaray reported that the parties had not been able to review the adequacy of commitments as required by the convention.

The issue was left for COP5 in October 1999 in Bonn.

As with COP4, settling the agenda was the first issue for COP5. The draft agenda included the Article 4.2(d) adequacy review. Again, the G77 plus China objected. This time it wanted the wording of the item changed from 'adequacy of commitments' to 'adequacy of their implementation', changing the sense and purpose of the review requirement in the convention. After what many had felt to be a difficult COP in Buenos Aires, delegations experienced, in the words of convention secretary Zammit Cutajar, an unexpected mood of optimism, a mood bought at the cost of not attempting to resolve divisive issues. On the final day of the COP, the conference president, Poland's Jan Szyszko, said no agreement had been reached to resolve the adequacy review, recorded China's amendment, and gavelled the decision, saying the item would be taken up at COP6.

A similar fate befell Kazakhstan's attempt to join Annex I. While Argentina announced its adoption of voluntary greenhouse gas growth targets, it had backed down from trying to change its non-Annex I status. In response to the EU's suggestion of agreeing to increasing global participation after the first 2008-2012 commitment period, China said it would not undertake commitments until it achieved 'medium development level'. (By 2008, China's per capita carbon dioxide emissions were above the world average. Reinstein recalls a comment by China in one of the last sessions of the INC negotiating the convention: 'China will always be a developing country,' which he interpreted not as a statement about China's economic aspirations but as a firewall against China being dragged into OECD-like commitments.)

Although Kazakhstan's proposal to join Annex I was supported by a number of Annex I parties, several non-Annex I countries said they did not have enough information on whether Kazakhstan could fulfil its obligations. There was no consensus and the COP decided that the issue should be taken up by COP6.

So on it rolled. The US, supported by Canada, Australia and New Zealand, pressed for COP6 to be held in early 2001, after the American elections. The G77 plus China pushed for November; the decision going their way.

The Hague COP6 was held six days after the disputed presidential election and Florida's hanging chads. The American team at The Hague did not know whether Al Gore or George W. Bush, who had spoken against Kyoto in the election, would be taking office in January. The attempt to get some evidence of future commitments from non-Annex I countries was now all but over. The draft agenda was adopted except for the item on the second review of the adequacy of commitments. No consensus on the matter had been found by the end of the COP, the G77 plus China saying the topic was sensitive so it would be better not to discuss it further.

Instead the American side had to contain a counter-attack from the EU which in American eyes amounted to an attempt to reopen the basis of the deal struck at Kyoto. They also had to contend with a noisy NGO participation that stormed a meeting with locked arms and refused to leave. Something was thrown at Loy. Wiping cake from his face, Loy reminded everyone that the day was the anniversary of the assassination of President Kennedy, who had urged Americans not to be swayed by those 'confusing rhetoric with reality'.

The air of unreality at The Hague started with the reading of some poems by the Dutch poet laureate.

> *He lost his way within a maze*
> *In search of silver and of gold –*
> *He searched a lifetime and he found*
> *He was where he'd been from of old*

Presumably it sounded better in Dutch. Much discussion revolved around sinks. Jan Pronk, the Dutch president of COP6, recalled lengthy debates about the definition of a sink, the definition of a forest and even the definition of a tree. When is a tree a tree? 'All this went on year after year, month after month, seminar after seminar, workshop after workshop, conference after conference. And during the conferences and negotiations themselves, day after day, hour after hour, night after night,' Pronk recounted.

Since Kyoto, European politics had turned deep green. Helmut Kohl lost the 1998 federal elections and was replaced by a Red-Green coalition led by Gerhard Schröder, Merkel being succeeded by the Green Jürgen Trittin. He joined Dominique Voynet, the environment minister in Lionel Jospin's government and one of the founders of the Greens in France. Trittin and Voynet constituted a Green motor at the heart of Europe. Instead of defining the precise rules and mechanisms needed to implement Kyoto, COP6 was to be Europe's hour when it cleansed Kyoto of its loopholes and forced rich countries (i.e., America) to face up to their responsibilities and cut their own emissions, rather than buy up poorer countries' emissions allowances.

France held the EU's rotating presidency. President Chirac's speech to the conference was a call to arms against the common enemy – America. There was no doubt that global warning had set in, he said. Without action, there would be dreadful consequences – rising sea levels, floods, extinctions of plants and animals, storms, typhoons, cyclones, hurricanes, the spread of deserts and the emergence of environmental refugees. 'That is why, I can confirm to you here, Europe is resolved to act and has mobilized to fight the greenhouse effect.'

While acknowledging President Clinton's personal commitment, Chirac reminded the conference that each American emits three times more greenhouse gases than a Frenchman. 'It is in the Americans, in the first place, that we place our hopes of effectively limiting greenhouse gas emissions on a global scale.' The bulk of efforts in meeting the Kyoto targets should be through efforts to curb domestic and regional emissions. Flexibility mechanisms were a complement: 'They should definitely not be seen as a means of escape.' For the first time, he declared, humanity was instituting a genuine instrument of global governance. 'If the South lacks the capacity to act, the North all too often lacks the will.'[6]

Battle was joined over sinks such as forests and how they should be counted in reducing countries' emissions. Doing so reopened the basis on which the emissions numbers had been agreed. 'We went to Kyoto intending to accept a target of no reduction from 1990 levels, but we ended up with a seven per cent reduction,' Loy explained a couple of months after the conference. 'One of the ways we were able to justify that to ourselves was that there was a provision for sinks.' (It's also worth recalling that, at Kyoto, the EU moved from its opening position of a fifteen per cent cut to eight per cent, but on more generous accounting.) The same went for America's ability to use the Protocol's flexibility mechanisms to meet its target. 'We would not have signed it if they hadn't been in there,' Loy commented.[7]

The treatment of sinks deadlocked the conference. John Prescott, the most thoughtful and realistic of the Europeans according to Loy, tried to broker a deal. The deal fell apart after Voynet took it to other EU environment ministers. It led to a spectacular falling out between her and Prescott. Voynet accused the former seaman of behaving like an 'inveterate macho' after Prescott had complained that Voynet had scuppered a deal because she had got cold feet and was tired and exhausted. 'I did not say the lady was tired,' Prescott told the House of Commons. 'She constantly said it herself. She was too tired to take in all the complexities. I quoted her words.'[8]

Less than three weeks after COP6 flamed out, the Supreme Court declared George W. Bush the next president.

During the last three years of his presidency, Clinton was deeply engaged on global warming. His final State of the Union message in January 2000 described global warming as the greatest environmental challenge of the new century. 'If we fail to reduce the emission of greenhouse gases, deadly heat waves and droughts will become more frequent, coastal areas will flood, and economies will be disrupted. That is going to happen, unless we act,' Clinton warned.

He seems to have believed it. In a conversation with President Jiang Zemin two years before Kyoto, Clinton told the Chinese leader that he didn't want to contain China:

> 'The biggest security threat China presents the United States is that you will insist on getting rich the same way we did.' And he looked at me, and I could tell he had never thought of that. And I said, 'You have to choose a different future, and we have to help. We have to support you. And that does not in any way let us off the hook. But it just means that we have to do this together.'[9]

At the White House global warming wonkfest in October 1997, where he recalled the conversation, Clinton showed an easy mastery of the policy implications across all its dimensions, ranging from the apparent paradox of more droughts and more floods, to policy mistakes in the 1970s that stopped gas-fired power stations (the federal government had grossly underestimated domestic natural gas reserves – his Democrat predecessor, Jimmy Carter, had a big hand in that), to it being better to recover waste heat from electricity generation and to encourage consumer conservation than to force electricity companies to change their power plants.

Of course, Clinton understood the politics better than anyone else in the room: 'Number one, we can't get to the green line unless there is a global agreement that involves both the developing and the developed countries.' In his post-Kyoto testimony, Eizenstat told the Senate Foreign Relations Committee that the problem of global warming could not be solved unless developing countries got on board. By around 2015, China would be the largest overall emitter of greenhouse gases and by 2025 the developing world would be emitting more greenhouse gases in total than the developed world.[10] He was right, but underestimated the speed at which it was happening. China overtook the US in 2007, eight years before Eizenstat predicted, and non-Annex I emissions overtook Annex I emissions in 2008, seventeen years before Eizenstat thought they would.

Administration officials believed if they could come back with a reasonable regime that included the larger developing country emitters, they would be able to get the Senate to go along with it. To succeed, they needed to breach the new Berlin wall. In retrospect, Loy believes that agreeing the Berlin Mandate was a serious mistake. Rio had got it more right, but the Berlin Mandate had hardened the structure into a bifurcated regime.[11]

Although global warming was a priority towards the end of Clinton's presidency, it wasn't at the beginning. National prosperity, social progress and the longest period of economic growth in America's history topped the list of administration accomplishments in his final State of the Union message. Like Thatcher a decade before, who had championed the environment in her last two years in Downing Street, fixing the economy was what Clinton was elected to do. If Clinton had emulated Carter, more likely than not, he would also have been a one-term president.

A *New York Times* poll published on the first day of the Kyoto conference found that sixty-five per cent of those surveyed agreed that the US should take steps to cut its emissions regardless of what other countries did. Although fifty-seven per cent of respondents said environmental improvements must be made regardless of costs, when asked of the most important problem facing America only one per cent answered the environment. Voters wanted symbolism and that's what Clinton gave them.

Yet the whole policy was based on a massive illusion. Through thick and thin, developing countries stated that the developed world had to take the lead in cutting carbon dioxide emissions. If America had met its Kyoto commitments through using the Kyoto flexibility mechanisms to the hilt, all it would have shown is that the super-rich America lives by its own rules and could buy other countries' emission reductions so it didn't have to cut its own. It is hard to believe that developing countries would have viewed that as a realistic basis for their participation in a regime to cap their emissions. The Europeans too had a point, except they were sharing the consequences of the collapse of communism within the EU bubble.

Ultimately Clinton's global warming policy failed. For sure, he managed the politics superbly and pulled out every stop to secure some measure of developing country involvement. At Buenos Aires, even the experienced Eizenstat got carried away when in reality the battle was over. But the longer-term legacy of his efforts was to perpetuate a myth that Kyoto would have been viable but for five hundred and thirty-seven votes in Florida.

In this respect Clinton's predecessor left a more durable legacy by risking isolation to exclude emissions targets and timetables from the 1992 climate change convention. As Clayton Yeutter sees it, the strong economy of the 1990s would have made it impossible for the US to have complied with any probable targets and timetables that might have emerged at Rio. Reasonable Americans might have argued about this in the early 1990s, but twenty years on President Bush's decision was, in Yeutter's view, clearly the right one.[12] What might someone who served at a similar level in the Clinton White House say? The final negotiating instructions to the American delegation in Kyoto were relayed over the speaker phone in the chief of staff's office. Erskine Bowles, who held the position at the time, told me he was unable to recall his impressions of the evolution of the policy with sufficient accuracy from that long ago.[13] Silence speaks volumes.

On March 6th 2001 Chuck Hagel and three other senators wrote to Clinton's successor to clarify the new administration's position on Kyoto. A week later, President Bush wrote back: 'I oppose the Kyoto Protocol because it exempts eighty per cent of the world, including major population centers such as China and India.'[14] Only later in the month did the implications finally sink in: Bush was not submitting Kyoto to the Senate. 'The President has been unequivocal,' a White House spokesman said. 'It is not in the United States' economic best interest.'

It ignited a firestorm. 'The world is tottering on the brink of climate disaster,' Friends of the Earth raged. The State Department was inundated with cables reporting the dismay of foreign governments. A more measured analysis was provided by MIT economics professor Henry Jacoby, who explained that Kyoto was two different things. 'It's the current text and numbers, and it's a process,' Jacoby said. 'The numbers are no longer going to work, but the process is going to go on.'[15] Viewed like that, Bush had little choice, as Senate ratification was about giving its consent to the text and numbers in the Protocol.

The White House was taken aback by the international reaction. On the eve of his first European trip in June, Bush had a second go and began to map out the beginnings of an alternative approach. Somewhat naively, chief of staff Andy Card told reporters that other nations would come to appreciate Bush's decision. 'The emperor of Kyoto was running around the stage for a long time naked,' he said, 'and it took President Bush to say, "He doesn't have any clothes on."'[16] Only in the fable does the boy get thanked for saying so.

'Breaking up is so very hard to do if you really love him,' the Walker Brothers sang in their sixties' hit. 'Oh baby, it's so hard to do.' It wasn't for Bush. His no to Kyoto was how the new president introduced himself to the world. Breaking up wasn't hard at all. That was what probably hurt the Europeans most of all.

The Europeans should have seen it coming. They had done little to assist the Clinton administration in securing developing world participation. The result was the outcome of the logic of the decision made by COP1 in Berlin six years earlier. To keep the G77 plus China in the process, the Berlin COP made a trade-off that increased the probability of losing the US further down the track. But the view the parties took then was if they hadn't agreed to the developing world's demands, there would have been no process at all.

Columbia University's Scott Barrett has criticized the Bush rejection for lessening the chances of adoption of a more viable approach than Kyoto. 'In rejecting the treaty in the way that he did – and, crucially, in doing so without offering an alternative – President Bush only reinforced the view that Kyoto had to be the only way forward; and he only made other signatories, especially members of the European Union, *more* determined to conclude the negotiations and bring the treaty into force.'[17] No one gets criticized for blaming Bush, but the idea that the world was willing to entertain an alternative to Kyoto was for the birds.

Four months later, the suspended COP6 resumed in Bonn. When it concluded, the EU environment commissioner, Sweden's Margot Wallström, commented, 'I think something has changed today in the balance of power between the US and the EU.' What actually happened was rather different.

The threshold at which the Protocol came into force had been set to accommodate American non-ratification. America's exit gave the remaining members of the Umbrella Group, especially Japan, Canada, Australia and Russia, much more leverage because they could decide whether the treaty ever came into force. Ironically the EU conceded far more to them than they had refused the US the previous November at The Hague. COP7 at Marrakesh in November 2001 doubled Russia's sink allowances from seventeen to thirty-three mega tonnes of carbon dioxide. The Japanese delegation, over eighty strong, gained a reputation for intransigence. At one point, a delegate speaking for the G77 plus China responded to Japan's request with 'you must be joking'. They weren't. They too gained concessions on sinks and significantly weakened the compliance regime, a negotiating objective they shared with the Canadians and Australians.

In 2002, Canada announced that it would unilaterally claim a further thirty per cent for its exports of hydro-electricity to the US, prompting the *Globe and Mail* to comment: 'If a country like Canada can claim credits in violation of the agreement and get away with it, more deceitful ways of breaking with the agreement can easily be found by other countries.'[18] Canada, a midwife of sustainable development in the 1980s, had travelled a long way in the fourteen years since the Toronto conference had called for twenty per cent emissions cuts below 1990 levels.

Kyoto ended up with a compliance regime that gave Annex I countries a free pass if they left the regime altogether (as Canada would do) and no financial or other economic penalties for staying outside it. By carrying forward excess emissions and adding an extra thirty per cent of the over-emission to the next commitment period, the compliance regime created incentives for countries to exit and incentives for everyone to forgive the overshoot to keep them inside it. Indeed, at COP16 in Cancún in December 2010, Japan confirmed it would not participate in a second commitment period of the Kyoto Protocol. For the EU and the environmental NGOs, Kyoto had become too big to fail, enabling other large Annex I countries to punch holes in Kyoto with near impunity.

Kyoto's only rationale was as a proof of concept in which the developed countries in Annex I took the first step in reducing their greenhouse gas emissions, with the rest of the world following at some unspecified point in an unspecified manner. The numbers tell their own story. In 1998, Tom Wigley, a leading climate scientist, published a paper which attempted to estimate the implications of the Protocol for the climate at the end of the twenty-first century. The analysis used the IPCC's value for climate sensitivity of carbon dioxide of 2.5°C in its Second Assessment Report. Adhering to Kyoto would, Wigley estimated, lead to a reduction in global mean temperatures of 0.08–0.28°C, depending on what happened to emissions after 2010.[19] This reduction compares to a projected 2°C rise by 2100 according to the IPCC's IS92a scenario. If fully implemented, Kyoto was estimated to delay global warming by between four and fourteen years over the course of the twenty-first century. The impact of the Kyoto reductions accrued even more slowly on sea level rise. 'The prospects for stabilizing sea level over coming centuries are remote, so it is not surprising that the Protocol has such minor effects,' Wigley wrote.

At one level, Kyoto could be counted a success. In 2008, reported Annex I emissions of the Protocol's six greenhouse gases were 9.6 per cent below their level in 1990.[20] However, including methane and four other greenhouse gases, as required by the Protocol, leads to a distorted picture because different gases remain in the atmosphere for different lengths of time. Wigley found that there was no single scaling factor to convert between carbon dioxide and methane emissions. A further distortion is that reported emissions also include estimates from changes in land use and forestry. In his February 1998 testimony, Eizenstat explained that changes in accounting rules meant that for the US, reducing greenhouse gas emissions by seven per cent 'is quite close to the President's original proposal to return emissions to 1990 levels by 2008–12'.

Stripping out reductions from land use changes and emissions of other greenhouse gases to focus on core carbon dioxide emissions results in a quite different picture. The 9.6 per cent Annex I reduction shrinks to 2.2 per cent. If the US is excluded, reported carbon dioxide emissions fell by 1,142.7 million tonnes, or 11.6 per cent. But this reduction masks two opposing trends. The 1,142.7 million-tonne reduction comprises a fall of 1,590.1 million tonnes from the former Communist economies of Central and Eastern Europe (Economies in Transition, or EIT parties, in Protocol parlance) partially offset by a 447.4 million-tonne rise in carbon dioxide emissions from the other Annex I countries. Other than the US, the largest absolute increases came from Turkey, Australia and Canada, which all increased their emissions at a faster rate than the US (110.1 per cent, 44.1 per cent and 25.9 per cent respectively compared to 16.1 per cent for the US). Japan notched up a 6.2 per cent increase. Even Norway, which in 2008 announced its intention to become carbon neutral by 2030, increased its reported emissions by 27.0 per cent.

Disaggregating the numbers produces a similar picture for the EU. The reunified Germany reported a 203.6 million-tonne fall in carbon dioxide emissions, but ninety-nine per cent of this fall was offset by increases from the four EU members that increased emissions the most: Spain, despite its colossal investments in wind and solar farms (up 109.3 million tonnes, or 47.9 per cent); Italy (up 32.3 million tonnes, 7.4 per cent); Greece (up 26.9 million tonnes, 32.4 per cent) and the Netherlands (up 16.4m tonnes, 10.3 per cent).

Sweden, Denmark and France recorded small reductions (11.0 per cent, 3.9 per cent and 0.7 per cent respectively, amounting to 10.9 million tonnes in all). Overall, the 42.3 million-tonne reduction by the non-EIT members of the EU, including Germany, was more than accounted for by the UK's 54.4 million-tonne fall, stemming principally from its dash-to-gas. The EU's reputation had been saved by UK electricity privatization.

Overall, the collapse of communism has been by far and away the single biggest factor in delivering the Kyoto emissions reductions – a truly one-off, epochal event. But these reductions have been swamped by the dramatic economic growth of the developing world. Wigley's study indicated that Kyoto would shave 0.08–0.28°C off a two-degree rise over the current century. A joint IEA / OECD 2009 study found that global carbon dioxide emissions were rising much faster than Wigley had assumed. Such an emissions growth trend would be in line with the IPCC's worst case scenario in its 2007 Fourth Assessment Report, which projects average world temperatures rising by between 2.4°C to 6.4°C by 2100.[21] Taking the midpoint of this rise would imply Kyoto delaying global warming by between twenty-two months and a little over six years over the course of this century.

Not that miracles don't happen. COP7 at Marrakesh agreed to permit Kazakhstan to become an Annex I party to the Protocol, but not to the convention. Because it had made no declaration when the Protocol was adopted, Kazakhstan does not have an emissions target listed – and no hot air to sell. Like the soul of an unbaptized infant in Catholic theology, Kazakhstan exists in a special state of climate change limbo all on its own.

20
TURNING UP THE HEAT

The scientists tell us the 1990s were the hottest decade of the entire millennium.

President Clinton, Address before a Joint Session of the
Congress on the State of the Union, January 27th 2000

*In holding scientific research and discovery in respect, as we should, we must
also be alert to the equal and opposite danger that public policy could itself
become the captive of a scientific-technological elite.*

President Eisenhower, Farewell Address, January 17th 1961

Climatically, Kyoto was nigh on perfectly timed. According to one widely used global temperature series, February 1998 recorded the highest ever average global temperature in the instrumental record. 1998 was the warmest year so recorded.

The revised HadCRUT3 temperature series records a rise of slightly less than half a degree centigrade (0.496°C) from 1900 to 2000. Average observed temperature rose by 0.346°C from 1900 to 1944. Then, for almost a third of century, recorded average temperatures fell; the HadCRUT3 value for 1976 is actually lower than for 1900. It then rose by 0.526°C to 2000.

Put another way, over one hundred per cent of the recorded increase in average global temperature in the twentieth century occurred in its last quarter.

So perhaps it's not surprising that alarm about global warming rose through the 1990s. Nonetheless, the multi-decadal fall in temperatures before 1976 required some kind of explanation, particularly to bolster the credibility of computer predictions of future temperature increase.

The IPCC sought to provide an answer to this conundrum in its Second Assessment Report. Doing so was important. If the climate models cited by the IPCC were unable to reproduce the broad outlines of the changes in global average temperatures, why should governments place much confidence in their ability to predict the future? Just because computer models could reproduce passable versions of the twentieth-century temperature record does not mean they will accurately foretell the future. But if they couldn't reproduce the past, why should anyone have faith in them?

The Second Assessment Report claimed to have solved the puzzle. Including an assumption about the effects of sulphate aerosols enabled computer models to account for the thirty-year decline in observed temperatures from around 1944. It also implied a higher value for the climate sensitivity of carbon dioxide. Computer simulations without aerosol forcing produced a best fit of a 1.5°C rise from a doubling of carbon dioxide in the atmosphere. Including sulphate aerosols resulted in the 4.5°C assumption providing the best fit.

How can we be sure that the assumption about sulphate aerosols wasn't a case of adding epicycles? The Working Group I report recognized some shortcomings in knowledge. While there was a body of observations on aerosols, 'The sensitivity of the optical properties to the size distribution of the particles, as well as to their chemical composition, made it difficult to relate the aerosol forcing in a simple manner.' Additionally the global impact of aerosols on the thickness and extent of cloud cover was 'recognized but could not be quantified'.

There was a larger issue the report was silent on – the validity of relying on computer models. From the Second Assessment Report onwards, the IPCC used the same language to describe computer model results as for empirical evidence derived from experiments on nature – 'experiments with GCMs' (general circulation models); 'model experiments'; 'recent multi-century model experiments'; 'this hypothesis [dismissing the suppression of the greenhouse gas signature being masked by sulphate aerosols] is not supported by recent GCM experiments'. Computer models thus attained an independent reality in the minds of climate scientists.

The Second Assessment Report noted that GCMs had to be calibrated by introducing systematic 'flux adjustments' to compensate for model errors. In a post-Kyoto analysis, three distinguished MIT professors wrote that computer models were based on incomplete knowledge about the key factors that influence climate. 'Today's climate models cannot reproduce the succession of ice ages and warm periods over the last two hundred and fifty thousand years, let alone the smaller climatic fluctuations observed over the last century.'[1] Canadian mathematician Christopher Essex has described climate models as intricate pastiches to artfully simulate climatological phenomena in the absence of a comprehensive physical theory. Essex argued that climate models' dependence on *ad hoc* constructions, known as parameterisations, meant that climate models could only represent the current climate regime where these *ad hoc* invariants hold.[2] Their values cannot be empirically validated for climate which had not happened.

A further challenge in trying to detect human impact on the climate was trying to define the envelope of natural variability, a problem that the climatologist Hubert Lamb believed to be practically insoluble. On the basis that natural variability plus anthropogenic forcing equals the instrumental record, the Second Assessment Report recounts naive attempts by climate scientists to calculate a value for natural variability by subtracting the assumed temperature changes in response to greenhouse gas forcing. To work out by how much extra carbon dioxide would warm the atmosphere, you need to know the amount of natural variability. To work out the amount of natural variability, you need to know by how much higher levels of carbon dioxide have been warming the atmosphere. As the IPCC noted, deriving an estimate for natural variability this way 'depends critically' on the accuracy of the models' assumption of the climate sensitivity of carbon dioxide and other greenhouse gases.[3] The exercise risked toppling down the vortex of a circular argument.

In the absence of empirical data to fill critical assumptions, treating computer model results as *de facto* experimental evidence indicated a collective IPCC mindset that risked loss of contact with reality. After the IPCC Fourth Assessment Report, Bert Bolin wrote of new results from a 'much richer set of experiments with global climate models'. Commenting on a critical analysis by dissenting scientists, Bolin wrote: 'The statement in their report that "computer simulations … can never be decisive as supporting evidence" is of course formally correct.' Relying on them was justified because the IPCC's conclusions were 'still very plausible, being based on the wide variety of model experiments'.[4]

This sort of thinking was something that shocked Professor Michael Kelly, a Cambridge University physicist and panellist in one of the 2010 Climategate investigations into the University of East Anglia's Climatic Research Unit. 'I take real exception to having simulation runs described as experiments (without at least the qualification of "computer" experiments),' he emailed a colleague:

> It does a disservice to centuries of real experimentation and allows simulations output to be considered as real data. This last is a very serious matter, as it can lead to the idea that 'real data' might be wrong simply because it disagrees with the models! That is turning centuries of science on its head.[5]

The question the world most wanted the IPCC to answer was also the most difficult to give an unambiguous response: Were humans warming the planet and, if they were, by how much? The issue was covered in Chapter Eight of the Working Group I report. Statements on whether man-made global warming was actually happening are 'inherently probabilistic in nature. They do not have simple "yes-or-no" answers'.

This should matter less for politicians than scientists, the IPCC suggested. A distinction should be drawn, the report argued, between 'practically meaningful' i.e., what policy should be based on, and what was 'statistically unambiguous attribution' to scientists.

> While a scientist might require decades in order to reduce the risk of making an erroneous decision on climate change attribution to an acceptably low level (say one-five per cent), a policymaker must often make decisions without the benefit of waiting decades for near statistical certainty.[6]

The report also sought to pre-empt counter-arguments based on scientific uncertainty.

> The gradual emergence of an anthropogenic climate signal from the background noise of natural variability guarantees that any initial pronouncement that a change in the climate has been detected and attributed to specific causes will be questioned by some scientists.[7]

The answer to the difficult question of when detection and attribution of human-induced climate change was likely to occur 'must be subjective'.

Discussion on the Second Assessment Report's top-line conclusion took a day and a half. Two of the Chapter Eight lead authors, Ben Santer and Tom Wigley, argued for a formulation saying that the balance of evidence pointed to an 'appreciable human influence' on global climate. As chair, Bolin proposed 'discernible' instead of 'appreciable'.

> There were no objections to this proposal which better emphasized the uncertainty. Even though the precise meaning of the word 'discernible' was still somewhat unclear, to my mind it expressed considerable uncertainty as well as the common view that it was impossible to provide a more precise measure.[8]

Interviewed during the Kyoto conference, Bolin went further. Global warming was not something 'which you can prove', going on to say: 'You try to collect evidence and thereby a picture emerges.'[9]

When published in June 1996, the Second Assessment Report ignited a row about the final text of Chapter Eight of the Working Group I report. Frederick Seitz, one of America's foremost physicists and a former president of the National Academy of Sciences, wrote in the *Wall Street Journal* that in more than sixty years as a member of the American scientific community, he had never witnessed a 'more disturbing corruption of the peer-review process'. After an IPCC meeting in Madrid the previous November, Seitz claimed that more than fifteen sections of Chapter Eight had been changed or deleted, including the sentence: 'None of the studies above has shown clear evidence that we can attribute the observed [climate] changes to the specific cause of increases in greenhouse gases.'[10]

Santer, as lead author of the Chapter, replied two weeks later in a letter co-signed with forty other scientists. They denied any procedural impropriety but admitted the draft had been changed at the behest of governments. 'The changes made after the Madrid meeting were in response to written review comments received in October and November 1995 from governments,' as well as from individual scientists and NGOs. Conceding that deletions had been made, the letter implied that they didn't matter: 'The basic content of these particular sentences has not been deleted.' If so, why did the government reviewers insist on changing the text?

It is easy to see why the changes were politically necessary. The IPCC's headline claim to have detected a discernible human influence on global climate would have been hard to square with the statement that no study had shown clear evidence that changes in the climate could be attributed to increases in greenhouse gases. A deputy assistant secretary at the US State Department, Day Mount, wrote to Sir John Houghton, chairman of the IPCC Working Group I, on November 15th: 'It is essential ... that chapter authors be prevailed upon to modify their text in an appropriate manner following discussion in Madrid.'[11]

Santer and his fellow scientists were correct in suggesting that acceding to political pressure conformed to IPCC principles and procedures. In addition to nominating lead authors, governments effectively had the final say on what the IPCC published. Bolin had drafted the principles and procedures adopted by the IPCC in June 1993 for the Second Assessment Report. 'Review is an essential part of the IPCC process,' they stated. 'Since the IPCC is an intergovernmental body, review of IPCC documents should involve both peer review by experts and review by governments.'[12] Advertised as summarising scientific knowledge, these procedures made IPCC reports politico-scientific documents in which scientific integrity took second place to political expediency in global warming's hierarchy of needs.

This was made clear in a 1997 article co-authored by Stephen Schneider, who regularly acted as a spear-carrier for the global warming consensus. The Second Assessment Report, he wrote, 'was fraught with political significance' because it was published shortly before COP2 in Geneva at which the Clinton administration announced its support for binding emissions targets.[13] Rejecting the implication that rules had been broken, Schneider conceded that Seitz's attack had 'demonstrated that a hybrid science/policy organisation like the IPCC needs better, more explicit rules of procedure'.

Addressing the scientific issues underlying Seitz's criticism, Schneider agreed with long-time critics such as Pat Michaels that direct observational evidence of global warming effects were not well-matched to climate predictions of one degree Celsius compared to the half degree observed in the twentieth century. While critics put the rise down to natural variability, doing so without characterizing the probability was, Schneider asserted, 'a scientifically meaningless claim'. This is an odd position, as the 'null hypothesis', one where there is no statistical relationship between two phenomena, does not need to be explained. Odd, too, is Schneider's criticism, given Bolin's choice of 'discernible' to describe human influence on the global climate *because* of its imprecision. Schneider's reasoning implies that the IPCC's conclusion is also scientifically meaningless.

What, Schneider asked, is the probability that a half-degree warming trend was a natural accident? The answer could not be found in the thermometer record alone as it only stretched back a century. Doing so would be like trying to determine the probability of 'heads' by flipping a coin once. Climate scientists had to look at proxy records of climate change such as tree ring widths, deposits left by glaciers and ice cores. The proxy records, Schneider thought, suggested that a half-degree Celsius warming was not unprecedented, but neither was it common. 'In my judgment,' Schneider wrote, 'this circumstantial evidence implies that a global surface warming of half a degree has about an eighty to ninety per cent likelihood of *not* being caused by the natural variability of the system.'[14]

It was a bold claim. At a National Academy of Sciences hearing nine years later, a group of paleo-climatologists were asked by a member of the panel whether the science was such that they could determine the average century-scale temperature one thousand years ago within half a degree Celsius. Every presenter said 'no' — apart from one, who claimed to know the average century-scale temperature one millennia ago to within 0.2°C.

Schneider's probability claim underlines the importance of the pre-thermometer climate record in estimating the probability of whether mankind was causing the planet to warm which, he believed, computer models by themselves could not. It is also important in the history of the evolution of the scientific consensus on global warming. Before the scientific consensus claimed that the pre-thermometer record was unimportant when it had become discredited, it had claimed it was important.

Until the IPCC turned its attention to the pre-thermometer climate record, the widely accepted view was of considerable climatic variability, with a Medieval Warm Period starting around the turn of the first millennium, followed by falling temperatures and a Little Ice Age in the seventeenth century, followed by an erratic warming interspersed by reversions to colder conditions before a warming in the twentieth century.

Hubert Lamb believed that the Medieval Warm Period, which was well documented in Europe, extended across the northern hemisphere. Lamb cited a Viking report of a cousin of Erik the Red swimming two miles across a fjord in Greenland, deducing that water temperatures were at least four degrees Celsius warmer than in the second half of the twentieth century. Tree ring records from California indicated, in Lamb's words, a sharp maximum of warmth, much as in Europe, between 1100 and 1300.

Warmer temperatures at more northerly latitudes coincided with a 'moisture optimum' in the Lake Chad basin, with a maximum occurrence of the pollen of monsoon zone flora. Population estimates for the Indian sub-continent also suggested sequences that roughly paralleled temperature trends in higher latitudes, falling temperatures there being mirrored in population declines in the sub-continent, Lamb thought.

The IPCC's re-writing of the pre-thermometer climate record began in the Second Assessment Report. 'Based on the incomplete evidence available, it is unlikely that global mean temperatures have varied by more than 1°C in a century' since the end of the transition from the last ice age.[15] At that stage, revisionism was sporadic. The report also presented evidence that conflicted with this, supporting the views Lamb had expressed in his last book. Recent evidence from ice cores drilled through the Greenland ice sheet indicated that 'changes in climate may often have been quite rapid and large, and not associated with any known external forcings'.[16] It also contained a warning that turned out to be prescient: 'Because the high quality of much-needed long-time series of observations is often compromised, special care is required in interpretation,' a warning the authors of the Third Assessment Report should have heeded.[17]

In 1995, David Deming of the University of Oklahoma published a study inferring temperature records from boreholes which suggested North America had warmed somewhat since 1850. As Deming later recalled, it gave him credibility within the community of scientists working on climate change. A major person working in the area emailed him: 'We have to get rid of the Medieval Warm Period.'

Fred Pearce, the British environmental journalist and author, discovered in 1996 that Tim Barnett of San Diego's Scripps Institution of Oceanography had teamed up with Phil Jones of University of East Anglia's Climatic Research Unit to form a small group to mine paleo-climate records for signs of global warming and to summarize their research in the IPCC's next assessment report. 'What we hope is that the current patterns of temperature change prove distinctive, quite different from the patterns of natural variability in the past,' Barnett told Pearce.[18]

How could people be persuaded to believe that current temperatures were unnaturally high, when it was it widely believed that temperatures had been higher several centuries before the start of the Industrial Revolution?

There was a way.

In 1912, fragments of the lower jaw of an orangutan, a human skull and a chimpanzee tooth were recovered from a quarry in the south-east of England. The bones, which had been stained with iron solution and acid, were taken to the British Museum. Man's evolutionary missing link had been found. In the teeth of entrenched opposition from the paleontological establishment, an anatomist correctly identified Piltdown Man as a hoax, a finding which took a further thirty years to be definitively accepted.

Tree rings from high altitude bristle cone pines would supply the raw evidence. In 1998, a paper written by a post-doctoral academic, Michael Mann, together with Raymond Bradley and Malcolm Hughes, based on Mann's Ph.D. thesis, was published in *Nature*. It presented a new temperature reconstruction of the Northern Hemisphere going back six hundred years to 1400. The following year Mann (the sole expert who said that he could know the average global temperature to within 0.2°C) and his co-authors published a second paper, extending the series a further four hundred years to 1000.

The picture it presented was of a stable climatic regime for the northern hemisphere, which lasted nine centuries, followed by unprecedented variability in the twentieth century, with the final decade averaging warmer temperatures than any time in the previous nine hundred and ninety years. The Medieval Warm Period had disappeared along with the Little Ice Age. The chart showed a long smooth line of low temperature variability trending downwards – in a 2007 Australian radio interview, Hughes claimed that the late nineteenth century was 'one of the coolest, if not the coolest of the last several thousand years' – followed by a sharp upward movement from around the beginning of the twentieth century.

The Hockey Stick gave the strong impression that the twentieth-century warming was anomalous. Equally the Medieval Warm Period and the Little Ice Age no longer needed to be explained or put within the envelope of natural variability because now they hadn't happened. If Mann's reconstruction was robust, it would help answer Lamb's objection of the impossibility of defining a figure for natural variability and enable the IPCC to move decisively beyond the equivocal conclusion of the First Assessment Report – that the observed warming in the twentieth century had been within the bounds of natural variability.

There was one problem. It was too good to be true.

The series of tree rings which generated the hockey stick shape had been selected by a researcher, Donald Graybill, for an earlier study into the possible effects of carbon dioxide fertilization. Graybill selected samples from bristlecone and closely related foxtail pines in the western United States for 'cambial dieback', where the bark had died around most of the circumference of the tree. Graybill and fellow researcher Sherwood Idso reported anomalously high twentieth-century growth for these compared to full bark trees unrelated to temperature data from nearby weather stations, a fact known to Mann and his co-authors. Self-evidently they would make highly unreliable temperature proxies.

Their inclusion by Mann and his associates in their multi-proxy temperature reconstruction needed an additional step. To put measurements from different proxy series on a common basis to be statistically analyzed, the data are standardized using a mean and standard deviation of the whole timespan of the series. At an early stage in the computational sequence, Mann used an algorithm derived from only the post-1901 portion of the data. The effect was similar to having a Google search algorithm to give hockey stick-shaped series the highest ranking in the analysis.

Did Mann know what he was doing?

He had graduated from Berkeley with a degree in applied mathematics and physics in 1989. Adopting this non-conventional approach to standardizing proxy data would have required a conscious decision. To get the hockey stick from the data, Mann needed both the algorithm and Graybill's tree ring data.

Even if the algorithm had been selected erroneously, Mann had reason to know that producing a hockey stick depended on using the anomalous bristlecone ring series. Inside his directory of North American proxy data, Mann had a folder, BACKTO_1400-CENSORED, containing the North American data excluding all sixteen of Graybill's series from the North American data. When the numbers from the CENSORED folder were run, the hockey stick disappeared. If Mann hadn't realised that his algorithm searched for hockey sticks, removing the Graybill series should have alerted him to the flaw in his method and meant that Mann's public claims of robustness for the Hockey Stick were contradicted by what he'd found but buried in his CENSORED folder.

In 1998, Mann was nominated a lead author of the critical Chapter Two ('Observed Climate Variability and Change') of the Working Group I contribution to the Third Assessment Report, a signal achievement for someone only awarded his Ph.D. that year. The following year, he was appointed assistant professor at the University of Virginia. In 2000, the Hockey Stick got the highest possible political endorsement. In his final State of the Union message, President Clinton told Americans that the 1990s were the hottest decade of the entire millennium. Two years later, *Scientific American* named Mann one of fifty leading visionaries in science and technology.

Mann and his Hockey Stick didn't have the entire field to himself. Competition came from Keith Briffa of the University of East Anglia's Climatic Research Unit, who published a proxy temperature reconstruction within a couple of months of Mann's first effort. Thanks to the Climategate emails released on the internet in November 2009, we now know how much tension this caused and the way the IPCC process resolved the issue in favour of Mann's Hockey Stick.

Briffa wrote a paper for *Science* comparing the two temperature reconstructions. After he'd sent a draft to Mann, in April 1999 Mann contacted the *Science* editor. 'Better that nothing appear, than something unacceptable to us,' he emailed, copying in Raymond Bradley, Mann's co-author. Privately, Bradley immediately dissociated himself from Mann's remarks. 'As though we are the gatekeepers of all that is acceptable in the world of paleoclimatology seems amazingly arrogant,' Bradley emailed Briffa.

After Briffa's paper was published, Mann tried to patch things up. 'Thanks for all the hard work,' he emailed colleagues in May. The sentiment didn't go down well with Bradley. 'Excuse me while I puke,' Bradley emailed Briffa.

John Christy, who was also a Chapter Two lead author, recalled that the Hockey Stick featured prominently in IPCC meetings from 1999, telling the House Science, Space and Technology Committee in March 2011:

> those not familiar with issues regarding reconstructions of this type (and even many who should have been) were truly enamoured by its depiction of temperature and sincerely wanted to believe it was the truth. Skepticism was virtually non-existent.[19]

When the chapter lead authors met in Tanzania at the beginning of September, they were shown Mann's Hockey Stick and Briffa's reconstruction, which showed a sharp cooling trend after 1960. The decline was more than a presentational problem. If tree ring growth declined when temperatures rose, what could be reliably inferred from them? And if the tree rings hadn't responded to warming in the second half of the twentieth century, how could we know that they hadn't done the same thing in response to possible medieval warmth?

In a follow-up email on September 22nd, IPCC coordinating author Chris Folland of the UK government's Met Office said that a proxy diagram of temperature change was a 'clear favourite' for inclusion in the summary for policymakers. However, Briffa's version 'somewhat contradicts' Mann's and 'dilutes the message rather significantly', adding: 'We want the truth. Mike [Mann] thinks it lies nearer his result (which seems in accord with what we know about worldwide mountain glaciers and, less clearly, suspect about solar variations)'[20]

The same day, Briffa emailed acknowledging the pressure to present

> a nice tidy story as regards 'apparent unprecedented warming in a thousand years or more in the proxy data' but in reality the situation is not quite so simple.[21]

Briffa said he did not believe global mean annual temperatures had cooled progressively over thousands of years:

> I contend that there is strong evidence for major changes in climate over the Holocene [the current geological epoch] (not Milankovich [climate changes induced by changes in the Earth's orbit, tilt etc.]) that require explanation and that could represent part of the current or future variability of our climate.[22]

The same day, Mann said he would be happy to add back Briffa's temperature reconstruction. Including it risked giving 'the skeptics' a field day 'casting doubt on our ability to understand the factors that influence these estimates and, thus, can undermine faith in the paleo-estimates', unless there is a comment that 'something else' was responsible for the discrepancies.[23]

Briffa hurriedly re-did his reconstruction. He sent it to Mann at the beginning of October 1999, but the new version produced a larger post-1960 decline.

When the diagram was presented at the next lead author meeting in Auckland, New Zealand in February 2000, Briffa's 'disagreeable curve' was not the same any more. It had been truncated around 1960; as Christy testified in 2011, 'No one seemed to be alarmed (or in my case aware) that this had been done.'[24] And that is how the IPCC decided to present it in the Third Assessment Report, without noting the data deletion.

Star billing went to Mann's Hockey Stick. A picture is worth a thousand words and the Hockey Stick appeared twice in the synthesis report, twice more in a diagram combining past and future temperature change and on the third page of the Working Group I Summary for Policy Makers. The text placed the Hockey Stick into the context of progress since the Second Assessment Report. Working Group I's contribution claimed that additional data from new studies of current and paleo-climates, together with 'more rigorous evaluation of their quality', had led to greater understanding of climate change.[25] Computer models suggested that it was 'very unlikely' (a term indicating a less than ten per cent chance) that the rise in observed temperatures could be explained by internal variability alone and the thousand-year temperature reconstructions indicated that this warming was unusual and unlikely (a ten to thirty-three per cent chance) to be entirely natural.

This opinion was translated to the front of the four-hundred-page synthesis report to answer the question: What is the evidence for, and causes of, changes in the Earth's climate since the pre-industrial era? 'The Earth's climate system has demonstrably changed on both global and regional scales since the pre-industrial era, with some of these changes attributable to human activities,' it stated in big, bold, blue letters.

The Hockey Stick had given the IPCC more than the single coin toss Schneider said was necessary to establish the probability of whether man-made global warming was real.

21
QUIS CUSTODIET?

New analyses of proxy data for the Northern Hemisphere indicate that the increase in temperature in the twentieth century is likely to have been the largest of any century during the past one thousand years. It is also likely that, in the Northern Hemisphere, the 1990s was the warmest decade and 1998 the warmest year.

IPCC – Third Assessment Report (2001)

Obfuscation, denial and a cover-up that blew spectacularly on the eve of the most important climate change negotiations since Kyoto – the events following publication of the Third Assessment Report are the most astonishing in the science of global warming. Leading scientists and science academies and societies had to decide: were they to be guardians of scientific orthodoxy or upholders of scientific standards? They could not be both.

It took four decades for the scientific establishment to accept the truth about the bones known as Piltdown Man. During that time, hundreds of scientific papers were devoted to the concoction – nearly as many as to all legitimate specimens in the fossil record put together.

That episode is a corrective to what the American physicist and philosopher Thomas Kuhn described as the persistent tendency of scientists to make the history of science appear as a linear, cumulative growth of knowledge. Combined with the generally unhistorical air of science writing, the impression was created that 'science has reached its present state by a series of individual discoveries and inventions that, when gathered together, constitute the modern body of technical knowledge'.[1]

Kuhn's 1962 classic *The Structure of Scientific Revolutions* rejected this. 'Cumulative acquisition of unanticipated novelties proves to be an almost non-existent exception to the rule of scientific development,' Kuhn argued.[2] He proposed instead a succession of incompatible paradigms. Scientists develop specialized vocabulary and skills and a refinement of concepts increasingly removed from commonsense prototypes. Increased professionalization, Kuhn thought, leads to 'an immense restriction of the scientists' vision and to a considerable resistance to paradigm change'.[3]

An established paradigm gives scientists the rules and the tools so they can focus on solving problems, elucidating and extending the paradigm, and provides a textbook framework to initiate younger scientists into the field. Scientists were not particularly adept at the kind of thinking needed to analyze a problem from first principles. 'Though many scientists talk easily and well about the particular individual hypotheses that underlie a concrete piece of current research, they are little better than laymen at characterizing the established bases of their field, its legitimate problems and methods,' Kuhn wrote.[4]

The pre-revolutionary period before the breakdown of a paradigm is marked by 'pronounced professional insecurity' caused by the persistent failure of the puzzles of normal science to come out as they should.[5] Kuhn recounts examples – the scandalous state of astronomy before Copernicus, Newton's theory of light and colour, Lavoisier's demolition of the phlogiston theory of combustion, Maxwell's electro-magnetic theories replacing theories of ether in the nineteenth century, which in turn created a paradigm crisis and Einstein's special theory of relativity in 1905.

As with a political revolution forcing a choice between two antagonistic views of politics outside the normal framework of politics, so the choice between competing paradigms in science proves to be a choice between 'incompatible modes of community life', which cannot be decided by the evaluative procedures characteristic of normal science.[6] With global warming, the link between science and politics is more than an analogy – global warming is a political idea as well as a scientific one. The science provides the feedstock for a political program; in turn, politics funds and helps propagate the science. The relationship is so deeply symbiotic that the two cannot be separated.

The risks inherent in such a situation were raised by President Eisenhower in his Farewell Address. Together with the military-industrial complex, Eisenhower identified a 'scientific-technological elite' as purveyors of miracle cures capturing government policy in warning against the recurrent temptation of believing that 'some spectacular and costly action' might offer a 'miraculous solution' to all current difficulties. Eisenhower was dismayed at the extent of government funding of scientific research. 'A government contract becomes virtually a substitute for intellectual curiosity.' The solitary inventor, 'tinkering in his shop', was overshadowed by task forces of scientists. 'The prospect of domination of the nation's scholars by Federal employment, project allocations, and the power of money is ever present and is gravely to be regarded,' Eisenhower warned.

Perhaps it wouldn't have surprised Eisenhower that the man who demonstrated that the Hockey Stick was wrong wasn't a scientist employed by government or a tenured university academic. In fact, he wasn't even a scientist, more the solitary tinkerer of Eisenhower's imagining.

In 2003, Stephen McIntyre was a Canadian businessman involved in financing speculative mineral prospecting with time on his hands. A prize-winning mathematician from Toronto University, McIntyre's interest in global warming had been aroused by the claim that 1998 was the warmest year of the millennium.

How could they know?

A claim by Mann's co-author Malcolm Hughes caught McIntyre's eye. Attempting to explain the 'weaker correlation' – in fact a negative correlation, with tree widths shrinking and temperatures rising after 1960 – Hughes said it meant past climate reconstructions were 'better than we thought they were'. Apparently, Hughes said, 'we underestimated the differences between the present century and past centuries'. The breakdown of a correlation led climate scientists to increase their confidence in its reliability? 'I could hardly believe that this sort of thing passed as science,' McIntyre recalled seven years later.

He emailed Mann asking for data sets of the one hundred and twelve proxies used in his 1998 paper. Mann replied almost immediately. He'd forgotten the exact location and would ask a colleague to follow up. It turned out the data had not been archived in a single location and the time taken to retrieve it suggested that no one had checked the data before the IPCC showcased the Hockey Stick in the Third Assessment Report.

McIntyre embarked on a painstaking forensic examination of the whole data set and compared it with the original data where he could. The data were full of errors – unjustified truncations, copying of values from one series to another, unjustified filling in of missing numbers, mis-copying of some series to a year earlier and use of obsolete data.

McIntyre teamed up with fellow Canadian Ross McKitrick, an environmental economist at the University of Guelph. Their first paper in September 2003 concluded that substantially improved quality control to Mann's dataset yielded a temperature index in which the late twentieth century is 'unexceptional compared to previous centuries'. The extent of the errors in Mann's work meant that indexes computed from it were 'unreliable and cannot be used for comparisons between current and past climates'.[7]

The political world began to notice. The US Senate was debating the McCain-Liebermann cap-and-trade bill. McIntyre and McKitrick did a briefing on Capitol Hill.

Replication of the Hockey Stick still eluded McIntyre. He continued to press Mann for details of his computer code. 'I am far too busy to be answering the same question over and over for you again,' Mann replied in November 2003, 'so this will be our final email exchange.' Thwarted, McIntyre sifted through Mann's website for clues. He found a fragment of computer code. From it, McIntyre unlocked the algorithm Mann used to standardize the various proxy data.

It became apparent to McIntyre and McKitrick that the algorithm systematically over-weighted hockey stick-shaped series, generating hockey stick results if any happened to be present. They performed a test by running statistically trendless 'red noise', simulating data from trees subject to random climate fluctuations. In ten thousand repetitions, McIntyre and McKitrick found that a conventional standardizing algorithm almost never yielded a hockey stick shape as the predominant pattern. Running the same data through Mann's algorithm produced a pre-dominant hockey stick shape over ninety-nine per cent of the time.

They ran the data from his CENSORED folder. Without the Graybill series, there was no hockey stick.

In January 2004, McIntyre and McKitrick submitted a paper to *Nature*, the journal that had published Mann's 1998 paper. The draft was sent to Mann for comments. The process was underway. They were given two weeks to revise and resubmit in the light of reviewer comments.

Then in March, the journal's editor Rosalind Cotter told them to cut the manuscript substantially. A condensed version was submitted. Months passed. In August, Cotter emailed McIntyre to say she wouldn't publish the piece. Evidently the paper was too hot for *Nature* – its mission 'to serve scientists through prompt publication of significant advances in any branch of science, and to provide a forum for the reporting and discussion of news and issues concerning science'. In the interim, it carried a *Corrigendum* to Mann's 1998 paper. 'None of these errors affect our previously published results,' it erroneously claimed.

Blocked by *Nature*, McIntyre and McKitrick put the entire record of their submission, together with referee reports, on the web. A revised version was published by *Geophysical Research Letters* in February 2005. When Berkeley physicist Richard Muller read about Mann's algorithm producing hockey sticks from red noise tests, it hit him like a bombshell. 'Suddenly the hockey stick, the poster-child of the global warming community, turns out to be an artefact of poor mathematics,' Muller wrote.[8] Bert Bolin was dismissive. 'Their analysis is hardly more reliable than the one published by Mann *et al.*'[9] Bolin had picked up the wrong end of the stick. McIntyre had not claimed to have made an alternative temperature reconstruction to Mann's, only that Mann's reconstruction depended on unreliable data and statistical methods.

Seeing that a climate journal edited by Stephen Schneider was in the process of reviewing one of Mann's papers, McIntyre contacted him. Invited to become one of Mann's referees, McIntyre asked Schneider to obtain Mann's supporting calculations and source code. Phil Jones emailed Schneider pleading with him not to. It would be setting 'a VERY dangerous precedent', Jones told Schneider.[10] 'In trying to be scrupulously fair, Steve, you've opened up a whole can of worms.'

Separately McIntyre had written to the National Science Foundation, which had financed Mann's work. Although Schneider agreed that data should be made available, Schneider and the NSF drew the line at computer code. For both, it was 'an intellectual property issue', as Schneider put it. 'The passing of time and evolving new knowledge about Earth's climate will eventually tell the full story of changing climate,' the director of the paleo-climate program at the National Science Foundation, David Verado, emailed in 2003, which begs the question why the taxpayer was funding the research in the first place.[11]

Asserting intellectual property rights was also popular with British climate scientists. In 2005, an Australian climatologist asked Phil Jones of the University of East Anglia's Climatic Research Unit for the underlying data used to compile its global temperature series. Jones refused. 'We have twenty-five or so years invested in the work. Why should I make the data available to you, when your aim is to try and find something wrong with it. There is IPR to consider.' A British government minister blocked a Freedom of Information request on similar grounds: 'Intellectual property rights must also be considered.'

Why? Are scientific data and methodologies like the recipe for Coca Cola?

Intellectual property is fundamentally different from physical property. As Alan Greenspan observed, someone's use of an idea does not make that idea unavailable to others at the same time.[12] The sole economic function in establishing intellectual property rights is to provide temporary protection from commercial exploitation by third parties, so innovation can be rewarded and incentivized.

The effect of applying the concept of intellectual property rights in pure scientific research is to attack the roots of science. 'Verification, or checking, or confirmation, is basic to every scientific enterprise, and also to every enterprise in daily life in which it is important to be sure that we are making no mistake,' Bridgman wrote in 1959. 'The meaning of truth is to be found in the operations by which truth is "verified."'[13]

Mann wasn't budging. 'These contrarians are pathetic, because there's no scientific validity to their arguments whatsoever,' he told an interviewer in the spring of 2005. The controversy spilled over onto the front page of the *Wall Street Journal*, where Mann described it as a battle of 'truth versus disinformation'. German climate scientist Hans von Storch, who had also produced findings that suggested Mann had underestimated past temperature variability, told the paper he had come under pressure from colleagues. There was a tendency in climate science, von Storch said, to 'use filters and make only comments that are politically correct'. A supporter of Mann claimed his critics were on 'some kind of witch hunt'.

It was a meme that would spread rapidly from Mann's partisans to the scientific establishment and then to politicians, turning a cover-up into a point of high principle. 'Giving them the algorithm would be giving in to the intimidation tactics that these people are engaged in,' Mann said. With this statement, Mann had thrown down the gauntlet. It was picked up by Joe Barton, the Republican chairman of the House of Representatives Committee on Energy and Commerce.

Barton decided to launch an investigation. In June, he wrote to Mann, Bradley and Hughes for details of their work, financial support and asking Mann to provide the exact computer code used to generate his results.

Quis custodiet ipsos custodes? – Who watches the watchmen?

Barton's intervention unleashed a storm. It was as if the separation of church and state had been repealed. Fellow Republican Sherwood Boehlert, chairman of the House Committee on Science, wrote to Barton, calling his investigation an illegitimate attempt to intimidate a prominent scientist. It would set a 'truly chilling' precedent. Scientific research must operate outside the political realm. 'Your enquiry seeks to erase that line between science and politics,' Boehlert wrote.

Democrat Henry Waxman, chair of the Energy and Commerce Subcommittee on Health and the Environment from 1979 until 1995, wrote to Barton complaining of his 'dubious investigation', which looked like 'a transparent effort to bully and harass climate change experts' who had produced 'highly regarded research'.

Scientists weighed in, too. Alan Leshner expressed the deep concern of the American Association for the Advancement of Science (AAAS). That the Hockey Stick formed part of the basis of the IPCC's conclusions was a reflection of it 'passing muster' in the IPCC peer review process, Leshner wrote. Twenty scientists, including James Hansen and John Holdren (a past and future White House science adviser), also expressed their deep concern. At the same time, they began downgrading the importance of the Hockey Stick. It was not an essential element but merely 'a useful *illustration* of our understanding'.

In a joint letter, the presidents of the American Geophysical Union (AGU) and the American Meteorological Society (AMS) claimed that the prospect for scientists having to defend unpopular results in the political arena could well undermine scientific progress and produce 'tainted results'. The editors of *Nature* accused Barton of cherry-picking selected information on the Hockey Stick. 'Science is, by its very nature, a process open to the questioning and overthrowing of currently accepted ideas, and the detail of Mann and his colleagues' work has been debated within the climate community,' the editorial ran, without disclosing that the journal had declined to publish criticisms of Mann's work in its own pages.

'My research has been subject to intensive peer review,' Mann wrote in response to Barton. Leshner of the AAAS agreed. The Hockey Stick had been subject to 'multiple levels of scientific peer review' to achieve publication, then 'multiple layers of the IPCC process itself'. The presidents of the AGU and AMS called on Congress to respect peer review, 'this time-tested process of scientific quality control'.

The issue not mentioned was why peer review had failed to uncover even a fraction of the problems McIntyre had found. The letters from the presidents of the AAAS, the AGU and the AMS demonstrated a stunning lack of intellectual curiosity – wasn't anyone interested in getting to the bottom of the controversy and finding out for themselves?

To the British philosopher Roger Scruton, the answer would be self-evident. The strategy of inventing experts backed up by the apparatus of scholarship, research and peer review goes back to the beginnings of the modern university in the Christian and Muslim communities of the Middle Ages. Theology was their foundational discipline and an entirely phoney one:

> The purpose of theology has been to generate experts about a topic on which there are no experts, namely God. Built into every version of theology are the foregone conclusions of a faith: conclusions that are not to be questioned but only surrounded with fictitious scholarship and secured against disproof.[14]

The dispute had escalated beyond Mann and the Hockey Stick. Ultimately it was a question of trust. Should society defer to the collective wisdom of experts or withhold scientific status to statements that cannot be independently verified, the method by which science had escaped the gravitational pull of medieval scholasticism?

In his reply to Barton's letter, Mann tried to finesse the issue. The key to replication was unfettered access to all the underlying data and the methodologies. And the code, which was the only way of verifying the actual methodology he had used? 'My computer program is a private piece of intellectual property, as the National Science Foundation and its lawyers recognized.'

In a sense, the person at the centre of the dispute was not Michael Mann, but Stephen McIntyre. It was McIntyre's insistence on seeing for himself how tree ring data were used to represent a temperature reconstruction that completely revised previous understanding that had led to this clash. 'Your statement that you have verified something is indifferent to me unless I believe that I could make the verification also,' Bridgman wrote.[15] Half a century on, Schneider described McIntyre as a 'serial abuser of legalistic attacks' for insisting on seeing Mann's computer code.[16] Something had happened to scientific standards in the intervening years.

Ralph Cicerone, president of the National Academy of Sciences (NAS), also wrote to Barton. A former research chemist at the Scripps Institution of Oceanography, now a seasoned Washington insider, Cicerone suggested that the National Academy of Sciences create an independent expert panel to conduct an assessment 'according to our standard rigorous study process' to assess Mann's area of research. 'Please let us know if we can help,' Cicerone offered.

Barton wasn't biting. 'We can't evaluate the idea without having seen it, and maybe it's a darned fine one,' committee spokesman, Larry Neal, told the *Washington Post*, 'but an offer that says "Please just go away and leave the science to us, ahem, very intelligent professionals," is likely to get the response it deserves.'

Boehlert accepted Cicerone's offer on behalf of the House Science Committee which in turn empanelled a National Research Council (NRC) chaired by Gerald North, an atmospheric physicist from Texas A&M University. North had no doubt as to the significance of the Hockey Stick. 'The planet has been cooling slowly until one hundred and twenty years ago, when, bam! It jumps up,' North told *Science* in 2000. 'We've been breaking our backs on [greenhouse] detection, but I found the one-thousand-year records more convincing than any of our detection studies.'[17]

Boehlert asked the NAS to examine three issues, specifically mentioning Mann's work, the validity of criticisms of it and whether the information needed to replicate his work had been made available. Something got lost in translation. The panel's terms of reference were re-written, dropping any reference to Mann's work.

Meanwhile Barton and Ed Whitfield, chairman of the Subcommittee on Oversight and Investigations of the Energy and Commerce Committee, announced the formation of a second panel, led by Edward Wegman, a former chairman of the NAS Committee on Applied and Theoretical Statistics and a Fellow of the

Royal Statistical Society, to perform an independent verification of McIntyre and McKitrick's critique of Mann's Hockey Stick papers.

In public, leading IPCC figures continued to promote the Hockey Stick. Giving evidence to a House of Lords enquiry in January 2005, Sir John Houghton, who had led Working Group I from 1988 to 2002, asserted that the increase in twentieth-century temperature had been 'phenomenal' compared with any variation over the whole millennium.

Privately, some of the leading figures in the IPCC were less sure. The previous October, Tom Wigley, a lead author on the Second Assessment Report and a former director of the University of East Anglia's Climatic Research Unit, emailed Phil Jones to express his doubts. Replying, Jones certainly believed that the picture of unprecedented twentieth-century warmth was true. There was 'no way' that the Medieval Warm Period was as warm as the last twenty years had been, Jones replied. Similarly there was 'no way' a whole decade during the Little Ice Age was more than 1°C on a global basis cooler than the global average from 1961 to 1990.

How could he know? These claims weren't something Jones could objectively demonstrate: 'This is all gut feeling, no science, but years of experience.'[18]

In July, Houghton addressed the Senate Energy and Natural Resources Committee. On the basis of two unpublished papers by two of Mann's colleagues, Caspar Ammann and Eugene Wahl, Houghton claimed 'the assertions by McIntyre and McKitrick have been shown to be largely false'. Concluding his prepared remarks, Houghton said he was optimistic that action to save the world from global warming would be taken. 'As a Christian, I believe God is committed to his creation and that we have a God-given task of being good stewards of creation – a task that we do not have to accomplish on our own because God is there to help us with it.'

The reality of Cicerone's 'standard rigorous study process' turned out much as Barton expected. The NRC panel 'just kind of winged it', North told a Texas A&M seminar in 2006 – 'that's what you do in that kind of expert panel'. The panel recommended strip-bark samples, i.e., Graybill's bristlecones (and foxtails) not be used as proxies in temperature reconstructions. Yet it presented four other temperature reconstructions without checking whether they also used bristlecones (they did).

At the beginning of March 2006, the panel held public hearings. One of the highlights was a presentation by paleo-climatologist Rosanne D'Arrigo. It included a mom-and-apple-pie rendition of the 'seek and ye shall find' principle, with a slide headlined 'Cherry picking'. You have to pick cherries if you want to make cherry pie, an astonished McIntyre recorded in a contemporaneous posting. D'Arrigo was as good as her word. For one of her temperature reconstructions, she had used ten out of thirty-six chronologies that had been sampled, retaining only the most 'temperature-influenced' ones and binning the rest. 'If we get a good climatic story from a chronology, we write a paper using it. That is our funded mission,' D'Arrigo's co-author, Gordon Jacoby, wrote to Schneider in his capacity as editor of *Climatic Change*.

Having smelled a rat, North's panel held its nose. Cherry-picking had become so endemic in paleo-climatology that the practitioners could no longer see – or smell – anything wrong with it. Jan Esper, another tree ring researcher, had written three years before that reducing the number of series could enhance a 'desired signal'. 'The ability to pick and choose which samples to use is an advantage unique to dendroclimatology,' Esper and four others wrote in a peer-reviewed paper. Esper's reconstruction was one of the temperature reconstructions included in the panel's final report.

D'Arrigo wasn't finished. Like Briffa's, her temperature reconstruction had a decline during the late twentieth century. Panellist Kurt Cuffey of Berkeley picked up on it. 'Oh, that's the "Divergence Problem,"' D'Arrigo responded. Cuffey probed further. How could you rely on these proxies to register past warm periods if they weren't picking up modern warmth? The matter was being studied, D'Arrigo replied.

Sitting in the audience, Richard Alley, a glaciologist at Penn State University, noticed the worried look of the panel. He urgently emailed Keith Briffa. The panel's reaction suggested that 'they now have serious doubts about tree-rings as paleo-thermometers' which Alley shared – 'at least until someone shows me why this divergence really doesn't matter'.

The NRC report, published in June, was nuanced and in places surreal. The panel concluded that it could be stated 'with a high level of confidence' that the global mean surface temperature during the last few decades of the twentieth century was higher than any comparable period during the preceding four centuries. It had less confidence in temperature reconstructions for the prior seven centuries back to AD 900.

A technical chapter reported McIntyre and McKitrick's production of hockey sticks from trendless 'red noise' using Mann's standardization procedures. Mann's results, the report said, were 'strongly dependent' on his statistical methodology. 'Such issues of robustness need to be taken into account in estimates of statistical uncertainties,' the report stated – the closest the NRC came to criticising Mann's methods.[19]

The panel entered a caveat about its lower confidence in the period before 1600. It had even less confidence in the Hockey Stick's headline claim that the 1990s was likely to have been the warmest decade and 1998 the warmest year in at least a millennium. However, it declared that surface temperature reconstructions were no longer considered 'primary evidence'. Instead climate model simulations showed that 'estimated temperature variation' in the two thousand years before the Industrial Revolution could be plausibly explained by estimated variations in solar radiation and volcanic activities.

The first two scientists to have quantified the possible effect of the Industrial Revolution on global temperatures, Sweden's Svante Arrhenius (above) and the Briton Guy Stewart Callendar (below), thought global warming would delay the return of a future ice age.
Above: *PA*; Below: *G.S. Callendar Archive, University of East Anglia*

A green movement took shape in Britain between the wars with the Kindred of the Kibbo Kift preaching a programme of pacifism, open-air education and nature-craft using symbols drawn from North American Indians, Nordic sagas and the Saxons (above left). Advocating social credit, the movement turned into the Green Shirts (above right) in the 1930s often clashing with Oswald Moseley's Black Shirts, who drew their example from the Nazis and included early proponents of organic farming (right). Above: *The Kibbo Kift Foundation*; Right: *Philip M. Coupland*

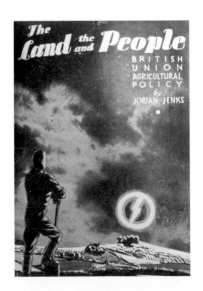

The Land and the People

BRITISH UNION AGRICULTURAL POLICY by JORIAN JENKS

In 1951, President Truman asked William S. Paley of CBS to head his Materials Policy Commission – at the centre of the picture (above), with fellow commission members. Their report, *Resources for Freedom*, argued that technology, free trade and American capitalism were the solution.
Time & Life Pictures/Getty Images

In the 1960s, two forces emerged to challenge the post-war American economic order. First, the modern environmental movement was born with the publication of Rachel Carson's *Silent Spring* in 1962.
AP/Press Association Images

Second came the emergence of the developing world. Two Argentine exiles, Che Guevara (above) and the brilliant economist Raúl Prebisch (right), were the stars of the first session of the United Nations Conference on Trade and Development (UNCTAD) in 1964, which institutionalised the economic claims of developing nations.

UNCTAD

'Whatever mark we have made in the last five years clearly bears your stamp too,' Lyndon Johnson wrote to Barbara Ward, pictured (far right) with Lady Bird Johnson at a White House reception in June 1965. It fell to Ward to reconcile the seemingly irreconcilable claims of First World environmentalism and the Third World's desire to industrialise ahead of the 1972 UN conference on the environment in Stockholm.

Francis Miller/Time & Life Pictures/ Getty Images

The other star of Stockholm was Indira Gandhi, Prime Minister of India, seen here being greeted by conference secretary-general Maurice Strong.

UN Photo/Yutaka Nagata

(Above) The Catholic convert giving a book on Buddhist economics to the born again president – E.F. 'Fritz' Schumacher presenting Jimmy Carter with a copy of *Small is Beautiful* in the White House on 22nd March 1977, two months after the new president had asked Americans to recognise America's limits.

Jimmy Carter Library

All smiles at the Venice G7 in June 1980 at which Carter (left) briefed fellow world leaders on the *Global 2000* report. It was very different a year earlier at Tokyo – 'One of the worst days of my diplomatic life,' Carter said – when German chancellor Helmut Schmidt (right) attacked Carter's energy policies and his friend, French president Giscard d'Estaing (centre), brokered a face-saving formula.

Daniel Simon/Gamma-Rapho/Getty Images

Carter's energy policies were also criticised by Ronald Reagan (above), who described himself as an environmentalist at heart. In May 1973 Governor Reagan led a campaign to conserve the John Muir Wilderness from the proposed Trans Sierra Highway in Sierra Nevada – seen here briefing the press in the mountains near Bishop, California.

Ronald Reagan Presidential Library

In 1977 Schmidt's predecessor, Willy Brandt (below right), announced he was launching a commission on international development – pictured with fellow commissioner and former British premier Ted Heath.

PA

argaret Thatcher (previous page) rejected the Brandt Report and embraced its successor, led by Norwegian ime minister Gro Harlem Brundtland – pictured together on the steps of Number 10 the day the Brundtland eport was published, 27th April 1987.

Press Association Images

June 1988 Western leaders took note of the Brundtland Report and formally adopted sustainable velopment at the Toronto G7, enthusiastically in the case of Thatcher (above left) and Canada's Brian ulroney (above right) – photographed with Reagan at a working lunch on 20th June. It put global warming gh on the international agenda.

Ronald Reagan Presidential Library

George H.W. Bush (left, seat
was elected as the environme
president, seen here signing
Clean Air Act Amendments
1990, looked on by Bill Rei
the first environmentalist
head the EPA, and ene
secretary Admiral Jan
Watkins – future protagon
in an administration sp
down the middle over glo
warming.

William K. Reilly

British soft power: to lobby Bush to go to Rio, Prince Charles dispatched the Royal Yacht Britannia to Brazil a
arranged for Reilly to meet Brazilian president Fernando Collor – pictured below with wife Rosana, togeth
with Princess Diana later in the royal couple's state visit.

AP/Press Association Images

he 1992 Rio Earth Summit was an election year political disaster for the Bush administration. A paper written
Reilly suggesting concessions to enable the US to sign the Biodiversity Convention was leaked mid summit,
hile Al Gore (seen above left with Reilly at Rio) and other leading Democrats openly attacked the administration.
William K. Reilly

onetheless, the United Nations Framework Convention on Climate Change that Bush (below) signed in Rio
3rd June 1992 remains the only global warming treaty ratified by the United States.
UN Photo

Germany's reward for undermining the US at Rio was to host the climate change convention's first Conference of the Parties (COP) in Berlin in March 1995, Chancellor Helmut Kohl (above) seen here proposing a toast to the delegates two days before the COP's successful conclusion.

AP/Press Association Images

At the beginning of the COP, Germany's environment minister, Angela Merkel (left), was elected COP president – pictured speaking at the conference's first session. The Berlin COP agreed the Berlin Mandate, which set in stone the division between North and South.

AP/Press Association Images

In October 1997, two months before Kyoto, President Clinton flew to Argentina to stand shoulder to shoulder with President Carlos Menem – seen together on the next page in Patagonia's Nahuel Huapi National Park – to seek a path around the Berlin Mandate.

William J. Clinton Presidential Library

Al Gore (above) spent only one day at the Kyoto COP in December 1997 – photographed shortly before detonating his bombshell instruction to the US delegation to show 'increased negotiating flexibility'.
UN Photo/Frank Leather

More than anyone else, the credit for obtaining agreement on the Kyoto Protocol belongs to Raúl Estrada-Oyuela (below), the Argentine diplomat who presided over the climate change negotiations and delivered the Berlin Mandate.

UN Photo/Frank Leather

2007 turned out to be the hottest year for the politics of global warming, when Al Gore together with the IPCC were awarded the Nobel Peace Prize (left to right: Dahe Qin, Sir John Houghton, Susan Solomon and Bob Watson).

Sir John Houghton

The year climaxed in December at Bali, but the strain told on the UN convention secretary Yvo de Boer (left), with UN secretary-general Ban Ki-moon and Indonesian president Susilo Bambang Yudhoyono looking on.

epa/Mast Irham

Glum faces at the December 2009 Copenhagen COP as the reality of failure dawns on Western leaders (above, left to right: EU Commission President Jose Manuel Barroso, Germany's Angela Merkel, Swedish premier Fredrik Reinfeldt, France's Nicolas Sarkozy, Barack Obama of the US and Britain's Gordon Brown), with India and China maintaining steadfast opposition.

DPA/Press Association Images

The real business at Copenhagen was negotiated by South Africa's president Jacob Zuma, Brazil's Lula, and Obama (all pictured below left), along with Chinese premier Wen Jiabao and India's Manmohan Singh (below right) – leaving the Europeans out in the cold.

AP/Press Association Images

How could the panel be so sure of the reliability of these estimates? They couldn't.

Data for natural forcings such as changes in solar radiation and volcanoes came largely from the same sources as the proxy evidence for surface temperature variations. 'These reconstructions are typically associated with as much uncertainty as reconstructions of surface temperature,' the report explained.[20]

The *New York Times* reported the Hockey Stick had been endorsed 'with a few reservations'. Mann professed himself 'very happy'. Asked about Mann and his co-authors' work, North replied, 'We roughly agree with the substance of their findings.' But Kurt Cuffey said the prominence the IPCC had given the Hockey Stick 'sent a very misleading message about how resolved this part of the scientific research was'. Statistician Peter Bloomfield commented that he'd seen 'nothing that spoke to me of any manipulation'.[21]

Cicerone penned the report's opening sentence: 'Our understanding of climate and how it has varied over time is advancing rapidly as new data are acquired and new investigative instruments and methods are employed' – a sentiment that conforms to the narrative of a linear, cumulative advance of scientific knowledge. As far as the development of theories went, Kuhn argued that this view was not supported by the history of science.

Empirical evidence should be different. Previously accepted facts about the Medieval Warm Period and the Little Ice Age gave way to new facts to fit an agenda-driven science.

Three weeks after the NRC reported, Barton released the Wegman report. Its language was as sharp as the NRC's was nuanced. Mann's work was 'somewhat obscure and incomplete'. McIntyre and McKitrick's criticisms were 'valid and compelling'.[22] There was no evidence that Mann or any of the other authors in paleo-climatology studies had 'significant interactions with mainstream statisticians'.

The isolation of the paleo-climate community was reinforced by a dense social network of authors and over-reliance on peer review 'which was not necessarily independent'. Politicization of their work meant that the paleo-climate community could hardly reassess their public positions without losing credibility. Mann's claim that the 1990s was the hottest decade of the millennium and that 1998 was the hottest year of the millennium 'cannot be supported by his analysis'.

At a congressional subcommittee hearing on July 19th, first Wegman then North were questioned on their reports.

Barton asked North whether he disputed Wegman's conclusions or methodology. 'No, we don't. We don't disagree with their criticism. In fact, pretty much the same thing is said in our report,' North replied. This didn't mean the conclusions were false – 'It happens all the time in science,' North added.

Asked for his opinion on Mann's statistical methodology, Bloomfield said that some of the choices had been 'appropriate. We had much the same misgivings about his work that was documented at much greater length by Dr Wegman.'

Holed beneath the water line, the SS Hockey Stick was listing badly.

Privately, leading climate scientists recognized it had never been seaworthy. In his exchange with Jones in October 2004, Wigley said he had read McIntyre and McKitrick's paper. 'A lot of it seems valid to me. At the very least MBH [Mann, Bradley & Hughes] is a very sloppy piece of work – an opinion I have held for some time.'[23]

The public line was denial. Asked in 2005 whether it had been unwise for the IPCC to have given so much prominence to the Hockey Stick, the IPCC chair Rajendra Pachauri replied:

> No. It is no exaggeration and it doesn't contradict the rest of the IPCC assessment. Of course you can always argue about details. But we assess all the available literature, and we found the hockey stick was consistent with that.[24]

The authors of the Fourth Assessment Report continued to tiptoe away from it. In May 2006, Mann emailed Briffa. It was 'an absolute travesty' that the Hockey Stick was being relegated. Both found it convenient to blame Susan Solomon, Working Group I co-chair. Briffa complained that 'much had to be removed'. He had been 'particularly unhappy' that he could not get the statement into the Summary for Policy Makers. 'I tried my best but we were basically railroaded by Susan,' Briffa wrote.

Some climate scientists argued that even if the Hockey Stick failed, it didn't matter. In 2005 Stefan Rahmstorf, a partisan of Mann's at the Potsdam Institute for Climate Impact Research, wrote that even if the Hockey Stick was wrong it would 'tell us nothing about anthropogenic climate change' because it post-dated the 1995 Second Assessment Report's verdict that the balance of evidence suggested a 'discernible' human influence on the climate. If so, why have any more IPCC reports?

Such revisionism needs to be set against the context of what prominent climate scientists had been saying previously. For Schneider, the one-thousand-year record did matter. Otherwise scientists could only gauge the probability that climate change in the twentieth century was natural or anthropogenic based on a single coin toss. For Gerald North, the Hockey Stick had been, bam! – the temperature jumps up.

Alternative temperatures reconstructions, telling essentially the same story, were therefore pressed into service. It was like someone was looking for sixes from rolling dice – use an algorithm that gave greater significance to series with high numbers of sixes (Mann's technique). It still left two others: pre-selecting runs that had a high number of sixes (bristlecones and foxtails were essential to most reconstructions) and through truncations and interpolations to create runs with higher numbers of sixes. According to McIntyre, 'The active ingredients in the twentieth-century anomaly remain the same old whores: bristlecones, foxtails,

Yamal' – a narrow selection of tree rings from the Siberian Yamal peninsula. 'They keep trotting them out in new costumes,' McIntyre blogged in 2007.

The one seen by more people than any other turned out to be a cross-dressed version, Mann's original dolled up in an entirely different set of clothes – starring in the most famous climate change movie of all time.

22
CLIMATEGATE

O, That this too too solid flesh would melt,
Thaw, and resolve itself into a dew.

Hamlet

Over Memorial Day weekend at the end of May 2006, four cinemas screened a movie.

'You look at that river gently flowing by … It's quiet. It's peaceful. And all of a sudden, it's a gear shift inside you. And it's like taking a deep breath and saying, "Oh yeah, I forgot about this."'

Cut to a lecture theatre.

'I used to be the next President of the United States.'

Pause.

Deadpan: 'I don't find that particularly funny.'

Smile.

Audience applause.

A week later, *An Inconvenient Truth* moved into the top-ten grossing movies playing that weekend. By mid July, it overtook *The Da Vinci Code*. At the end of the year, it had grossed nearly $50 million and won two Oscars. (Michael Moore's *Fahrenheit 9/11*, released in 2004, grossed $222 million worldwide, receiving a twenty-minute standing ovation at the Cannes Film Festival – but no Academy Awards.)

An Inconvenient Truth was the cinematic sequel to *Earth in the Balance*, fusing Gore's search for the meaning of life and saving the environment. He recalls listening to Roger Revelle as a student ('I just soaked it up'); his six-year-old son's near fatal car accident ('It just changed everything for me. How should I spend my time on this Earth?'); a tobacco farming family, losing sister Nancy to lung cancer ('It's just human nature to take time to connect the dots. I know that. But I also know that there can be a day of reckoning'); the 2000 election ('a hard blow … It brought into clear focus the mission that I had been pursuing for all these years'); and the ultimate meaning of what Revelle had told him ('It's almost as if a window was opened through which the future was very clearly visible. "See that?" he said, "see that? That's the future in which you are going to live your life"').

After the biopic segments, ice took up a large part of the movie.

The melting of the Greenland ice sheet could shut down the Gulf Stream and plunge Europe into an ice age. Viewers were left in the dark about when this might happen. In his 2007 book, Bert Bolin acknowledged that the rim of the Greenland ice sheet was melting faster than a few decades earlier, but snow was still accumulating over the ice sheet plateau. 'At some point, probably more than centuries into the future, a situation of no return may be reached and the ice sheet might disappear in a matter of a millennium or more.'[1] *An Inconvenient Truth* was 'not always adequately founded in the basic scientific knowledge that is available,' Bolin wrote. It created the impression that the ice ages were caused by variations in atmospheric carbon dioxide, 'which of course is wrong'.

Gore visited Antarctica. The ice has stories to tell us. 'Right here is where the US Congress passed the Clean Air Act.' Wrong again. Eric Steig, an isotope geologist, wrote, 'You can't see dust and aerosols at all in Antarctic cores — not with the naked eye.'[2]

Ice cores can measure the different isotopes of oxygen and figure out a 'very precise thermometer', Gore explains. 'They can count back year by year the same way a forester reads tree rings. And they constructed a thermometer of the temperature,' as a graph with a remarkable similarity to Mann's Hockey Stick flashed up on a seventy-foot digital screen. 'The so-called sceptics will sometimes say, "Oh, this whole thing is a cyclical phenomenon. There was a medieval warming period, after all,"' Gore continued. 'But compared to what's going on now, there's just no comparison.'

Gore's ice core temperature graph looked like Mann's Hockey Stick because it was Mann's Hockey Stick.

In February 2007, the IPCC started to release its Fourth Assessment Report. The attribution of rising temperatures to human activity was upgraded a further notch: 'Most of the observed increase in global average temperatures since the mid twentieth century is *very likely* due to the observed increase in anthropogenic GHG [greenhouse gas] concentrations' – 'very likely' meaning a higher than ninety per cent chance of being correct.

In October, Gore and the IPCC were jointly awarded the 2007 Nobel Peace Prize for their role in creating 'an ever-broader informed consensus about the connection between human activities and global warming'. The timing was propitious. In December, negotiators were to meet in Bali and adopt the Bali Road Map with a deadline of agreeing a follow-on deal to the Kyoto Protocol in Copenhagen in December 2009.

A month before Copenhagen, over a thousand emails to and from scientists at the UEA's Climatic Research Unit, together with three thousand other documents, found their way onto the internet. It was soon dubbed Climategate.

The origins of Climategate lay in the IPCC's attempt to quarantine the Hockey Stick without repudiating it. Doing so meant subverting IPCC principles. All parts of the IPCC assessment process, Sir John Houghton wrote in 2002

> need to be completely open and transparent. IPCC documents including early drafts and review comments have been freely and widely available – adding much to the credibility of the process and its conclusions.[3]

The aim of IPCC reports is to summarize the latest science. The IPCC was therefore duty bound to report McIntyre and McKitrick's critique of the Hockey Stick

– unless they could find a more recent study that sidelined their conclusions. So that's what they set out to do.

They consulted two unpublished papers by Caspar Ammann and Eugene Wahl. After a couple of attempts, the first paper was turned down by *Geophysical Research Letters*. Stephen Schneider facilitated provisional acceptance of the second in *Climatic Change* to meet the IPCC deadline. The published (and reviewed) version had verification statistics that corroborate McIntyre and McKitrick's critique, but fell outside the IPCC's deadline. So the final IPCC text claimed McIntyre and McKitrick were unable to reproduce Mann's results (they could, using Mann's algorithm) but that Wahl and Ammann could when using Mann's original methods (without mentioning that Wahl and Ammann's published paper reported that the reconstruction failed a verification test).

As an IPCC reviewer, McIntyre prodded the lead authors to show the decline on the Briffa temperature reconstruction. Don't stop in 1960 – 'then comment and deal with the "divergence problem" if you need to.'

When published, the comment was there, but the decline wasn't.

After publication, McIntyre wanted to see how his comments had been handled. Rejected, still considered inappropriate to show the recent section of Briffa's reconstruction, was the editorial response. The trail led back to the UK and to the Chapter Six review editor, John Mitchell of the Met Office.

On June 21st 2007, David Holland, a semi-retired engineer, submitted a Freedom of Information Act request to Department for Environment, Food and Rural Affairs (Defra), the UK government department responsible for oversight of the IPCC review records. He followed in January 2008 with requests to the Met Office.

The story Holland was told kept changing. First; Mitchell's records had been deleted; then, Mitchell's involvement in the IPCC had been in a personal capacity; finally, release of Mitchell's emails would prejudice relations between the UK and an international organization and that the IPCC – contrary to its policy of openness and transparency – refused to waive confidentiality.

In May, Holland sent Freedom of Information requests to Reading and Oxford universities and to the University of East Anglia about the change in the deadline for the Wahl and Ammann paper. Within two days, Phil Jones emailed Mann:

> Mike, Can you delete any emails you may have had with Keith re AR4 [the Fourth Assessment Report]? Keith will do likewise ... Can you also email Gene [Wahl] and get him to do the same? I don't have his new email address ... We will be getting Caspar to do likewise.[4]

Mann forwarded the email to Wahl, who told a Department of Commerce investigation that he 'believes he deleted the referenced emails at that time'.[5]

On the morning of November 17th 2009 someone tried to upload 160MB of data from the CRU server onto the RealClimate website. They must have had a sense of humor – RealClimate was established with a mission to fight McIntyre in the blogosphere Hockey Stick wars. Later that day, an anonymous poster, FOIA, appeared on the climate sceptic Air Vent website.

> We feel that climate science is, in the current situation, too important to be kept under wraps. We hereby release a random selection of correspondence, code, and documents.

The blogosphere lit up. The mainstream media was slower, apparently queasy at publishing private emails and unsure whether they were genuine. Three days later, the University of East Anglia confirmed they were.

It is hard to conceive of a less favorable run-up to the Copenhagen climate change negotiations. The CRU's initial reaction of going to ground only made it worse. The Met Office's Vicky Pope was left branding the release of the emails 'a shallow and transparent attempt to discredit the robust science undertaken by some of the world's most respected scientists'. George Monbiot in the *Guardian* thought otherwise. 'Confronted with crisis, most of the environmentalists I know have gone into denial,' Monbiot wrote. 'There is a word for the apparent repeated attempts to prevent disclosure revealed in these emails: unscientific.'[6]

Two prominent climate scientists agreed. 'The whole concept of "we're the experts, trust us" has clearly gone by the wayside with these emails,' Judith Curry of Georgia's Institute of Technology, told Andrew Revkin of the *New York Times*. Mike Hulme of the University of East Anglia told Revkin that 'the IPCC itself, through its structural tendency to politicise climate change science, has perhaps helped foster a more authoritarian and exclusive form of knowledge production'.

The following weekend, the *Sunday Times* carried a story that the University of East Anglia had thrown away records from which the temperature reconstructions had been derived. The original numbers had been dumped to save space. Roger Pielke Jr of Colorado University discovered the loss when he asked for the original records. 'The CRU is basically saying, "Trust us,"' Pielke said.

The disclosure fed speculation that the CRU had been fiddling its surface temperature reconstruction, one of the three principal ones used in climate studies derived mainly from the Global Historical Climatology Network.* There was another explanation. Proper record keeping was hardly one of the CRU's

---

* The Global Historical Climatology Network (GHCN) contains eight thousand series whereas the CRU only uses two thousand. Despite the importance of global warming and therefore the monitoring of temperature trends, by 2005, the GHCN sample size has fallen by over seventy-five per cent from its peak in 1975 to less than any time since 1919. Ross McKitrick, 'A Critical Review of Global Surface Temperature Data Products' (unpublished, 2010) http://papers.ssrn.com/sol3/papers.cfm?abstract_id=1653928

strengths. In 2002, historian James Fleming, writing a short biography of Guy Stewart Callendar, was shocked to find that Callendar's papers at the CRU hadn't been stored properly. As a result, the CRU agreed to loan the papers to Colby College, Maine, where they were organised into archive boxes.[7]

In December, the Met Office wrote to seventy scientists at British universities asking them to collect signatures for a statement to 'defend our profession against this unprecedented attack'. Over one thousand, seven hundred scientists signed.

> The evidence and the science are deep and extensive. They come from decades of painstaking and meticulous research, by many thousands of scientists across the world who adhere to the highest levels of professional integrity. That research has been subject to peer review and publication, providing traceability of the evidence and support for the scientific method.

It all looked somewhat desperate. 'The response has been absolutely spontaneous,' Julia Slingo, the Met's chief scientist improbably claimed. An anonymous scientist spoke of being put under pressure to sign. 'The Met Office is a major employer of scientists and has long had a policy of only appointing and working with those who subscribe to their views on man-made global warming.'[8]

At the beginning of December, the University of East Anglia announced it was establishing an independent review into the CRU emails headed by Sir Muir Russell, a retired civil servant. Meanwhile the House of Commons Science and Technology Committee started an inquiry of its own. It became clear that the Met Office's December statement on behalf of the UK science community was not universal. The Royal Society of Chemistry argued that the nature of science dictated that research be transparent and robust enough to survive scrutiny. 'A lack of willingness to disseminate scientific information may infer that the scientific results or methods are not robust enough to face scrutiny, even if this conjecture is not well founded.'[9]

The Royal Statistical Society argued for publication of data. 'Science progresses as an ongoing debate and not by a series of authoritative and oracular pronouncements.' Commercial exploitation was not a valid reason to withhold data. If a company is granted a patent, the details of the invention must be revealed – 'it cannot justifiably seek reimbursement for that knowledge and not make it available'. All parties needed to have access to the facts. 'It is well understood, for example, that peer review cannot guarantee that what is published is "correct."'

The Institute of Physics was concerned about the integrity of scientific research. The emails constituted '*prima facie* evidence of determined and coordinated refusals to comply with honourable scientific traditions and freedom of information law'. A wider inquiry into the integrity of the scientific process in the CRU's field of research was needed, the Institute argued.

Of all the congressional and parliamentary hearings on global warming, the Science and Technology Committee's hearing on March 1st 2010 was the most important since Hansen's appearance in 1988. The inquiry had been prompted by Labour's Graham Stringer. By the time it began, it was on a compressed time-table ahead of the May 2010 general election, with a single extended hearing. The hearing didn't resolve the scientific issues. Its importance was in raising a question mark over apparently settled science.

It was also Phil Jones' first public appearance since the story broke. Fleet Street's parliamentary sketch-writers reported from the packed committee room. For the *Guardian*'s Simon Hoggart, Jones brought back memories of the weapons scientist David Kelly shortly before his suicide in the run-up to the Iraq war. 'The resemblance was disturbing and painful,' Hoggart wrote. Jones looked 'taut, nerv-ous, often miserable. At times his hands shook'. His low spot, Hoggart thought, came when Stringer, with a Ph.D. in chemistry, asked Jones about not making data available because all they'd do is find something wrong with it. 'Yes, I have obviously written some very awful emails.'

Jones' appearance hadn't been made easier by the committee's first witness. 'With the aggression of someone who used to go eyeball to eyeball with Margaret Thatcher,' Hoggart wrote that former chancellor Nigel Lawson

> spoke with dripping contempt. 'Proper scientists, scientists with integrity, wish to reveal their data and all their methods. They do not require freedom of information requests!'[10]

According to the *Telegraph*'s Andrew Gimson: 'One hammer blow followed an-other, culminating in the crushing accusation from Lord Lawson that for the period before 1421, the scientists "relied on one single pine tree" to establish how hot the world was, "which was more than it [the pine tree] could bear."'[11]*

Professor Edward Acton, the university's vice chancellor, provided Quentin Letts with comic relief, a younger version of Professor Calculus from the Tintin books, Letts wrote in the *Daily Mail*, nodding and beaming at everything Jones said,

> his eyeballs bulged with admiration for the climate change supremo. His eyes were pulled so wide in wonderment they must have nearly split down the seams like banana skins. Others, watching the tremulous Professor Jones, will have been less impressed. He may be right about man-made climate change. But you do rather hope that politicians sought second, third, even

---

> * It was actually worse than Lawson implied. Mann interpolated values for the first five years (1400–1404) of the Gaspé series, the only series going back to the fifteenth century because there were zero trees for this period. Stephen McIntyre & Ross McKitrick, 'Corrections to the Mann et al. (1998) Proxy Data base and Northern Hemispheric Average Temperature Series' in *Energy & Environment* Vol. 14, No. 6 (2003), Table 5.

twentieth opinions before swallowing his theories and trying to change the world's industrial output.[11]

More than a year later, what surprised Stringer most about the hearing was Jones' 'astonishing' admission that his temperature series wasn't reproducible. Although the data-sets remain available, the computer programs had changed. Scientists – and Jones himself – couldn't reproduce his graph. 'That is just a fact of life in climate sciences,' Jones had told the committee.[12]

23
STATE OF DENIAL

Q. So you are not prepared to comment on whether the events at CRU undermine the IPCC or not?
A. If we take Chapter Six ... This is the most robust peer review process that you will see in any area of science.

Professor Julia Slingo, Chief Scientist, the Met Office
(March 1ˢᵗ 2010)

After the sketchwriters left, the MPs cross-questioned a trio of government scientists – Sir John Beddington, the government's chief scientific adviser, the Met Office's Julia Slingo and Bob Watson, who had moved from the White House, via the World Bank and chairing the IPCC, to Whitehall as Defra's chief scientific adviser.

Before the hearing, Beddington had sent the committee a memorandum. 'The integrity of British science is of the highest order,' he wrote. Certainly the British government had a big exposure to climate change science. Of the nineteen countries that sent and paid the fifty-one lead authors of the four key chapters of the Fourth Assessment Report, the US and the UK provided eighteen, an analysis by David Holland shows. Together they provided half the review editors and one third of the authors of the summaries for policymakers. '[The] IPCC, which has a tiny budget, is dominated by the hugely funded climate research organizations of the USA and the UK,' Holland concluded.

Climategate heaved a large rock into this pond. Had it damaged the image of UK science, one MP asked?

'No, I do not think UK science has been damaged,' Beddington replied.

Did the emails give him any cause for concern?

Yes. 'I think some of the wording is unfortunate.'

Had the events at the CRU undermined the IPCC? Slingo told MPs they should remember that the IPCC peer review process was 'much greater than any other science ever receives'.

Didn't the Hockey Stick and the fact McIntyre had been in a tiny minority give her some cause for worry about peer review, Stringer asked?

'Not at all, no,' Slingo replied. 'The controversy around the original methods of Mann *et al* has been fully addressed in the peer review literature and I think those issues are now largely resolved.'[1]

In some ways, Slingo's answer was the most extraordinary of the day. True, the Wegman report had endorsed McIntyre's critique of Mann's handling of statistics and orally Gerald North had agreed with Wegman's findings. So in that sense, Slingo was correct to have said the issue had been resolved. But the version furnished by the IPCC in the Fourth Assessment Report – the most robust peer review process in any area of science, Slingo claimed – was a falsification, being silent on Wegman, relying instead on a paper that hadn't been published, requiring a retroactive change in IPCC rules, the published and peer-reviewed version of which contained verification statistics that corroborated McIntyre.

Scientists' power in debates on global warming derives from their monopoly in the presentation of scientific matters. Underpinning the idea of global warming is that scientists should be trusted to represent the science in a balanced and objective way. At issue in this exchange was not scientific knowledge itself, but the process of documenting and agreeing it. Here, Britain's most powerful climate scientist was presenting her interpretation of a process which scientist and non-scientist alike could see was so detached from reality as to be a fairy tale.

Watson downplayed the importance of the Hockey Stick. The Third Assessment Report's hit single, '1998 – the hottest year of a thousand years', had been superseded by the not-as-catchy 'Model that heat'. 'A lot was made of the Hockey Stick, but a much more important issue is what has happened in the last half-century,' Watson told the committee. The evidence for attributing cause and effect was derived from 'theoretical modelling' by trying to explain observed changes in the climate.

The biggest question mark raised by Climategate was the evidence of intent to delete emails subject to a Freedom of Information request. What information did top climate scientists want to suppress, to the extent of breaking the law to do so? A cover-up had brought down President Nixon. Jones was luckier. By the time it had come to light, prosecution was time-barred.

Earlier in the session, Stringer charged that the university was attempting to dodge allegations of malpractice, in particular by deleting emails. 'It is hard to imagine a more clear-cut or cogent *prima facie* piece of evidence, is it not, and yet you have taken the opposite view?' he challenged Acton. 'To my mind there is *prima facie* evidence,' Acton replied. 'Why else did I set up the Muir Russell independent review?'

Russell, the former civil servant, took a different view. 'The Prime Minister doesn't want the truth, he wants something he can tell Parliament,' Yes Minister's Sir Humphrey Appleby says. 'I wasn't going to put the review into the position of making the sort of quasi-judicial prosecutorial investigative judgment,' Russell told the committee in October 2010. To have done so would be 'alleging that there might have been an offense, and that really wasn't the thing that my inquiry was set up to do … If we ducked or avoided, I plead guilty to that', Russell added.

Russell recognized that the tactic of the Fourth Assessment Report's Chapter Six to distance the IPCC from the Hockey Stick without disowning it involved a degree of subterfuge. The second order draft of the Chapter cited Wahl and Ammann's work to refute McIntyre and McKitrick. This provoked the US government reviewer: the reference to Wahl and Ammann should be deleted; it did not comply with Working Group I deadlines and substantial changes subsequently made to the paper, including the insertion of tables of statistical verification tests, showed Mann's reconstruction had failed. The comment elicited a tart response: 'Rejected – the citation is allowed under current rules.' Russell took the view that the ends justified the means. 'We are trying to get to the big issue end point,' he told MPs. 'We are really just saying, be all this as it may, that the end point seems to us to be a sensible place.'

Meanwhile the UEA had commissioned a second panel under Lord Oxburgh, an eminent geologist and prominent advocate of alternative energy. At the March hearing, Acton told the committee its purpose was to 'reassess the science and make sure there is nothing wrong'. In October, five months after he had submitted his report, Oxburgh told the committee that what Acton had told them was 'inaccurate'. Instead Oxburgh decided to focus on the honesty of the scientists and the integrity of the researchers, in doing so, taking care to tiptoe around the elephant in the room – the practice of deleting emails in response to Freedom of Information requests. Honesty was one thing, what about competence? Liberal Democrat Roger Williams pointed out that a scientist could be honest but incompetent and honest but misled. 'That is a complicated question,' Oxburgh replied.

The five-page Oxburgh report, written in around three weeks, found no evidence of deliberate scientific malpractice. Nor did it find anything to alter the overall results of the CRU's instrumental temperature reconstruction. But Stringer found the report 'quietly devastating'. It remarked on the 'inappropriate statistical tools' used by some other paleo-climatologist groups and their potential for producing misleading results, 'presumably by accident rather than design'. The potential for misleading results arising from selection bias was 'very great'. It reiterated criticisms made in the 2006 Wegman report:

> It is regrettable that so few professional statisticians have been involved in this work because it is fundamentally statistical. Under such circumstances there must be an obligation on researchers to document the judgmental decisions they have made so that the work can in principle be replicated by others.[2]

The Oxburgh report described as 'regrettable' the practice of the IPCC to neglect to highlight the divergence issue (it had been the decision of CRU's Briffa to underplay the divergence). Muir Russell agreed, describing the truncation as 'misleading'.

By a three to one vote (Stringer being the dissenter), the House of Commons committee found no reason to challenge the scientific consensus on global warming.[3] However it criticized the withholding of data and codes, a practice it called problematic 'because climate science is a matter of global importance and of public interest, and therefore the quality and transparency of the science should be irreproachable'.[4]

At a press conference, committee chair Phil Willis went further, calling it reprehensible. 'That practice needs to change and needs to change quickly,' he said.

As for Jones, there was no reason why he should not resume his post as CRU director: 'He was certainly not co-operative with those seeking data, but that was true of all the climate scientists.'[5]

Some climate scientists spoke out against the committee's emphasis on disclosure. Oxford University's Myles Allen said climate scientists had used their professional judgment to distinguish between professional scientists and activists and members of the public. 'The big implication in all this for science is that the [FOI Act] is taking away our liberty to use our own judgment to decide who we spend time responding to.'

The committee argued that Jones had brought the FOI requests on himself by his failure to respond helpfully to data requests, which was bound to be viewed with suspicion. Had all the available raw data been available online from an early stage, these kinds of unfortunate email exchanges would not have occurred, the committee concluded. One of those was an email from Mann to a colleague at the CRU in 2003 – the subject: 'Reconstruction errors':

> I'm providing these for your own personal use, since you're a trusted colleague. So please don't pass this along to others without checking w/ me first. This is the sort of 'dirty laundry' one doesn't want to fall into the hands of those who might potentially try to distort things...[6]

Jones had told the committee that the published emails represented one tenth of one per cent of his output, implying one million emails. 'Further suspicion could have been allayed by releasing all the emails,' the committee thought.

After the general election, it fell to David Cameron's Coalition government to respond to the committee's report. The Climategate emails did not provide evidence to discredit evidence of global warming; the rigor and honesty of the scientists were not in doubt; the peer review process had not been subverted or the IPCC process misused. Openness and transparency should be the presumption, but there were good reasons for not making data available immediately or, indeed, at all. Scientists might not be legally allowed to give out data; commercial rights should be respected; there might be 'security considerations'.

The Coalition's arguments highlighted the contradictory nature of 'state science'. The advance of scientific knowledge is based on destructive challenge. Because it is always possible to achieve agreement between a theory and observational evidence, Popper argued that the supreme rule for deciding on the subsidiary rules of empirical science was specifying that they do not protect any statement in science against falsification. The general attitude of promoting a search for falsifying instances was, according to the American philosopher of science John Losee, the most important in the history of science, typified by the eighteenth-century astronomer William Herschel's view that the scientist must assume the role of antagonist against his own theories, a theory's worth being proved only by its ability to withstand such attacks.

Barriers preventing such attacks are barriers to science. By carving out exemptions for the publication of data and code, the government was inviting scientists to find ways of subverting Popper's supreme rule. If the government's interest had been in genuine scientific knowledge, it would have stipulated that it would only fund research based on data and methods which are freely available. That such a course would have been unthinkable for a government fully signed up to global warming illustrates the incompatibility between the presentational constraints of state science and the epistemological requirements of the real thing.

In January 2010, HRH the Prince of Wales visited the CRU. Its patron for nearly twenty years, the prince was briefed by Jones and discussed 'the appalling treatment they had endured'.

The CRU's refusal to disclose data was raised during Acton's second appearance before the Science and Technology Committee. He spoke of his keenness to become an exemplar on disclosure and freedom of information. 'I want my university to be whiter than white on it,' Acton averred. Even as he was speaking, Acton's university was engaged in a fourteen-month battle before the Information Commissioner to prevent Professor Jonathan Jones, a physics lecturer at Oxford University, being copied data the CRU had given Georgia Tech.

The Oxford Jones had first requested the data in July 2009. The university refused. In October an internal inquiry upheld the refusal. In December, Jones complained to the Information Commissioner which started its analysis of the case in March 2010. The UEA used every conceivable excuse – including mutually incompatible ones, such as the information was already in the public domain but disclosure might adversely affect Britain's international relations.

On June 23rd 2011 the Information Commissioner ruled that UEA had been in breach of the Freedom of Information Act. Subject to appeal, the university was given thirty-five days to comply or face the prospect of being held in contempt of court.

Meanwhile, the IPCC's standing was dealt a further blow in January 2010 with media uproar about its claim, recycled from a WWF report that the Himalayan glaciers could disappear by 2035 – centuries earlier than the most pessimistic forecasts. A second claim, that fifty-five per cent of the Netherlands lay below sea level (the correct figure being twenty-six per cent), led to a storm in the Dutch media. The Dutch legislature stated that the previously accepted reliability of the IPCC was now at issue and instructed the government to carry out a review.

The Netherlands Environmental Assessment Agency published a report in July. It focused on the Working Group II report on impacts. Seven of the thirty-two conclusions in the summaries could not be traced to the main report. It also found that the Working Group II summary focused only on potential negative impacts of climate change, rather than presenting policymakers with a complete picture.

Such was the damage to the IPCC's credibility that in March 2010 the UN secretary-general and chair of the IPCC asked the InterAcademy Council (IAC) to review IPCC procedures and management. The panel, chaired by an economist, Princeton's Harold Shapiro, delivered a carefully worded report. Scientific debates always involve controversy, Shapiro wrote, describing climate science as 'a collective learning process'. In Shapiro's judgment, whether the IPCC remained 'a very valuable resource' was conditional on it highlighting 'both what we believe we know and what we believe is still unknown' – a fine-grained description of the epistemological issues at the heart of the debate on the science of global warming.[7]

The IAC also highlighted the problematic relationship between science and politics. Although scientists determined the summaries for policymakers, the final wording was negotiated with government representatives 'for clarity of message and relevance to policy'. At the same time, the panel warned that 'straying into advocacy' could only harm the IPCC's credibility. When chapter lead authors are sitting next to their government representatives, it could put the author in the position of either supporting a government position at odds with the working group report or opposing their government. 'This may be most awkward when authors are also government employees,' the panel suggested.

The effect of dependence on government patronage for policy-driven research can be seen in Penn State's investigation into allegations of misconduct by Michael Mann following the Climategate emails. The report exonerated Mann and showered him with praise. Since 1998, Mann had received research funding from two government agencies. In a description of Mann's work and conduct on the Hockey Stick that the *Atlantic*'s Clive Crook said defied parody, both agencies had an

> exceedingly rigorous review process that represents an almost insurmountable barrier to anyone who proposes research that does not meet the highest prevailing standards, both in terms of scientific/technical quality and ethical considerations.[8]

Success in getting funding 'clearly places Dr Mann among the most respected scientists in his field'. The committee found only one ground for criticism of Mann's conduct. Sharing someone else's manuscripts without their permission had been 'careless and inappropriate'.

Nearly five decades after President Eisenhower had spoken of the baleful prospect of the domination of America's scholars by federal employment, the investigatory criterion used by Penn State provides evidence of his prescience. There was another consequence of taxpayer-funded science. Since James Hansen's 1988 testimony, the single individual who has had the most impact on the course of the debate on the science of global warming was not drawn from the legion of government scientists, but the solitary Stephen McIntyre.

Dissenters such as Richard Lindzen, who disagree with the consensus about the physical processes and likely effect on atmospheric temperatures of rising levels of carbon dioxide, were sidelined. Unlike them, McIntyre's disagreement was not about the physical mechanisms of global warming. His work focused on the methodological and procedural mistakes underpinning the findings adopted by the consensus. His demonstration that Mann's one-thousand-year temperature record was flawed forced the IPCC to change direction. Thus the 2007 Fourth Assessment Report retreated from the 2001 Third Assessment Report to a position closer to the 1995 Second Assessment Report – a shift from empirical evidence derived (by flawed means) from nature to justification based on theoretical computer modelling.

The impact of McIntyre's work in undermining public confidence in the scientific consensus was magnified by the reaction of the IPCC and the national scientific elites, particularly those in Britain and America. Rather than acknowledge there had been a problem, the IPCC embarked on a strategy of denial. So the Fourth Assessment Report was stitched up to avoid undermining the credibility of the IPCC's previous pronouncements, an exercise in which the British side took the lead.

It turned out to be a disastrous misjudgment. When the cover-up was blown open with the release of the Climategate emails, more than climate scientists' credibility was called into question. Their integrity was, too.

24
TIME'S WINGÈD CHARIOT

And yonder all before us lie
Deserts of vast eternity.

Andrew Marvell

Anyone who believes in indefinite growth of anything physical on a physically
finite planet is either a madman or an economist.

Kenneth Boulding

Time. The measurable aspects of science and civilized life of the more fundamental passage of nature, A.N. Whitehead defined it.

Time Present could play tricks. 'Global warming is happening' as a statement of physical reality in the one hundredth of a second that constitutes the psychological present is not verifiable. The most that can be said is global warming *has* happened between two dates in the past.*

Time Past tripped up climate scientists, intent on re-writing climate history.

Time Future poses an insuperable problem. We cannot know it until it has happened. The expected impacts of global warming occur at a glacial speed compared to the quick march of human cohorts across the plain of terrestrial existence. The differential tempo led advocates of action on global warming to speed it up to accommodate it to humanity's timescales, the IPCC notoriously bringing forward a melting of the Himalayan glaciers by over three centuries.

Ancient civilizations moved so slowly as to be barely perceptible to the modern eye. The Egypt of the Pharaohs lasted around twenty-six centuries. In one hundred and twenty paces, the British Museum's Egypt gallery traverses more than two thousand years of artefacts bearing the stamp of a way of life that barely changed over centuries and millennia.

Adjacent galleries portray the explosive growth of a younger civilization. Ancient Greece developed from its archaic to its classical period in a couple of centuries. Small, unstable city states, dialog with other cultures and a culture that prized innovation gave Greek civilization a dynamism that made it impossible to know where it was going next. The accuracy of long-range forecasts – or prophecy as it was once called – depends on the assumption that societies don't change; a valid assumption in the case of the Egypt of the Pharaohs but invalid for ancient Greece, or our own.

---

* It is not possible to verify whether the global temperature anomaly now is higher or lower than it was one hundredth of a second before. By contrast, the velocity of a moving object can be determined now, a concept that defeated the ancient Greeks but was solved by Galileo.

It fell to economists to systematically bring together the geophysical and human timescales. In doing so, they risked the mockery of nature and of future generations because it depends on assumptions of determinism and causality for the global climate system and the development of human civilization.

'Time is the ultimate constraint on mortals,' British economist Charles Good-hart has written. 'If time were unlimited and costless, wealth could always be augmented by more work.'

Time is money. It plays a central role in the Austrian school of economics' theory of capital. The time from production of intermediate goods to the final sale to a consumer is a factor that accumulates in the cost schedule of consumer goods. It's a small step from this nineteenth-century insight to valuing capital investments as a discounted stream of expected returns underpinning modern financial theory as a way of recognizing the cost of time.

The practical consequences of not including the value of time can be seen in the economies of the pre-1990 Soviet bloc. As disciples of Marx's labour theory of value, capital had no value other than the labor embodied in it. Time was excluded. As a result, communist economies squandered capital on a colossal scale, destroying their economies in the process.

The middle years of the first decade of the twenty-first century saw concern about global warming reach a fever pitch. Participants at the January 2007 Davos World Economic Forum voted climate change the issue with the greatest global impact in the coming years.

Twenty months later, Lehman Brothers filed for bankruptcy. The world was condemned to a different future. 'A fall in the pit, a gain in your wit,' China's premier Wen Jiabao told the 2009 Davos Forum, 'Shaping the Post-Crisis World'.

Why had none of you predicted it, the Queen asked an economist in 2009? Like natural scientists, economists are – or were – believed to be gifted with special powers to pierce through the inscrutable character of the future. Unlike them, economists have a reputation for arguing with each other. Whereas the history of the natural sciences is characterized by successions of dominant paradigms, economics is typified by clashing schools – classical economists against Keynesians, neo-Keynesian against neo-classical schools, Austrians and Marxists and neo-Marxists.

Schumpeter explained why he thought economists were so disputatious compared to scientists. Arguing was endemic in economics. 'Nature harbours secrets into which it is exciting to probe; economic life is the sum total of the most common and drab experiences,' he wrote. 'Social problems interest the scholarly mind primarily from a philosophical and political standpoint.'[1] An economist needed to have a vision designed to give expression to certain facts of the world in which we live and different visions give rise to different interpretations of economic processes. Keynes was the pre-eminent example, a vision derived from the special characteristics of England's 'ageing capitalism' as seen by an English intellectual, Schumpeter thought.

British-born economist Kenneth Boulding had vision in abundance. Economics was too narrow to understand what it was trying to describe, Boulding, the founder of evolutionary economics, argued. What was needed was an approach that unified natural and social sciences: systems science or systems analysis. Boulding's brilliance was recognised by Keynes, who published a paper of his in 1931, when Boulding graduated from Oxford. Boulding then settled in the US. In 1949, he became the second winner of the American Economic Association Bates Medal awarded to the most distinguished young economist, the first being Paul Samuelson and Milton Friedman the third.

With his white shoulder-length hair and Liverpudlian accent, Boulding became something of a visionary and guru in later life. 'Does man have any responsibility for the preservation of a decent balance in nature, for the preservation of rare species, or even for the indefinite continuance of his race?' he asked four years before *Silent Spring* popularized ecological issues. A committed pacifist, in 1965 Boulding helped organize the first anti-Vietnam War teach-in. The following year, Boulding was the star at a Resources For Freedom forum in Washington, DC, when he picked up Adalai Stevenson's Spaceship Earth and spoke of the open 'cowboy economy', and the closed economy of the spaceman. 'The shadow of the future spaceship, indeed, is already falling over our spendthrift merriment,' Boulding warned. 'Oddly enough, it seems to be in pollution rather than exhaustion that the problem is first becoming salient.'[2]

This raised a further question. Why conserve, Boulding asked? What has posterity ever done for me? Unless the individual identified with some inter-generational community, 'conservation is obviously "irrational"'. Even if it were conceded that posterity is relevant to present problems, 'we still face the question of time-discounting' – that is, the price put on the value of time, 'and the closely related question of uncertainty discounting'. It could be argued that the ethical thing to do is not to discount the future at all – this would be the approach taken by the British government's Stern Review forty years later – that putting a cost on time was 'mainly the result of myopia' and was 'an illusion which the moral man should not tolerate'.

Such reasoning did not satisfy Boulding. Time-discounting might be 'a very popular illusion' but it had to be taken into consideration in the formulation of policies. So conservationist policies almost always had to be sold under some other pretext which seemed more pressing, Boulding told the forum.

Of recent economists, Yale's William Nordhaus – 'about the most reasonable person I know', according to fellow economist Jeffrey Sachs – has the longest professional interest in the economics of global warming. As the first environmental wave was cresting in the early 1970s, Nordhaus wrote an eviscerating critique of Jay W. Forrester's 'World Dynamics' model used in the Club of Rome's *The Limits*

to Growth. 'Can we treat seriously Forrester's (or anybody's) predictions in economics and social science for the next one hundred and thirty years?' Nordhaus
asked in his review, 'Measurement without Data'.[3] Forrester's lack of humility in
predicting the future was not warranted when placed beside economists of great
stature who had also got the future badly wrong, Nordhaus wrote. Marx had predicted the immiserization of the working class under capitalism; Keynes guessed
that capital would have no net productivity in 1973; Galbraith, that scarcity was
obsolete. 'And now, without the scarcest reference to economic or empirical data,
Forrester predicts that the world's standard of living will peak in 1990 and then
decline. *Sic transit Gloria.*'

In 1979, Nordhaus was one of the first professional economists to raise the
issue of global warming. There was 'widespread evidence' that combustion of
fossil fuels, in causing the build-up of atmospheric carbon dioxide, 'will be
the first man-made environmental problem of global significance'.[4] But the
conclusions he drew were tentative and circumspect. Climatologists thought
that a 0.6°C rise over the previous hundred years had led to major, but not
catastrophic, results. Such temperature changes were 'rather trivial', according
to Nordhaus. 'Mean temperature changes of this size are not economically significant.'[5] He qualified his recommendation about the desirability of policies to
slow down the consumption of fossil fuels as

> *deeply unsatisfactory, from both an empirical and a theoretical point of view.* I
> am not certain that I have even judged the direction in the desired movement
> in carbon dioxide correctly, to say nothing of the absolute levels.[6]

Nordhaus also grappled with the problem of future time. In a 2007 paper, he
challenged economists' assumption that generations living many years from now
would have the same tastes and preferences as people today, when they would
be consuming goods and services largely unimagined in a vastly different world.
Future generations might come to love the altered landscape of a warmer world.
Perhaps, Nordhaus suggested, economists should incorporate uncertainty about
future preferences, an approach that was 'largely uncharted territory' in economic
growth theory.

From the late 1980s, more economists became interested in global warming.
In 1991, *The Economic Journal* carried a special issue on the subject. The American
economist William Cline, a specialist in trade and capital flows, endorsed the
science. 'Overall, greenhouse science holds up well to scrutiny,' Cline wrote. 'Its
logic and physics are compelling. Although warming to date is less than predicted,
the shortfall is within the range of natural variability.'[7] The principal shortcoming
in the scientific and policy debate, Cline thought, had been the failure to extend
the time horizon of the analysis to two hundred and fifty to three hundred years.
'For purposes of planetary management, a horizon of thirty-five years is woefully
inadequate.'

In his contribution, Nordhaus wrote that climate change was likely to produce a combination of gains and losses – 'with no strong presumption of substantial net economic losses'.[8] This wasn't an argument in favor of climate change or a laissez-faire attitude, but rather for a careful weighing of costs and damages 'if we are to preserve our precious time and resources for the most important threats to our health and happiness'.

The vast majority of economists expressing a view followed Cline in taking the consensus on the science as given and not to be questioned. As Nordhaus put it in a 2007 seminar, 'We social scientists are downstream: we collect the debris from science as it comes by us, the good models, the bad models, the good studies and the bad studies.'[9] This attitude of uncritical acceptance came under sustained challenge in a series of papers and articles by David Henderson, the former OECD chief economist. Henderson argued that it was 'unnecessary and imprudent' for economists to arrive at such confident and sweeping conclusions. There was 'pervasive uncertainty' and 'sheer lack of knowledge' in relation to the climate system. It was 'misleading' to speak in general terms of 'the science' in a way that suggested there were 'no significant doubts, queries or gaps'.[10] A study of the contribution of Working Group I to the Fourth Assessment Report found that the terms 'uncertain' and 'uncertainties' appeared more than one thousand, three hundred times.

As the rhetoric of alarmism ratcheted up in the first decade of the new century, Henderson criticized received opinion as 'seriously over-presumptive' characterized by a 'lack of awareness of today's prevailing over-statement, over-confidence, and ingrained bias'. The IPCC assessment reports, Henderson cautioned, were far from being models of 'rigor, inclusiveness, and impartiality'. Economists were inadvertent in ignoring issues of professional conduct that had come to light over the Hockey Stick. The strong commitment to the official consensus led its upholders to react to any form of criticism or dissent as undermining established science, non-subscribers being portrayed as members of a 'denial lobby' and treated as Thought Criminals.

Should economists, as Nordhaus suggested, simply take whatever natural scientists decided to float down the river or follow Henderson and, as he put it, exhibit the lack of credulity economists would deploy in analyzing any other policy issue? Economists should be in a better position than others to make their own assessment of the science because much of it is about statistics and modelling.

To Australian economist Ross Garnaut, the answer was a no-brainer. 'The outsider to climate science has no rational choice but to accept that, on a balance of probabilities, the mainstream science is right in pointing to high risks from unmitigated climate change,' Garnaut wrote in his 2008 government review for the states and Commonwealth of Australia.

Canadian economist Ross McKitrick is critical of the credulity of professional economists in accepting the claims of climate scientists when they are generally better trained and equipped in the handling of statistics. 'The typical economist has way more training in data analysis than a typical climatologist,' McKitrick told the author. 'Once they start reading climate papers they start spotting errors all over the place.'

In a 1939 review 'Professor Tinbergen's Method,' Keynes wrote a scathing attack on the shortcomings of multiple correlation analysis to quantify the effect of a single factor. Suppose a model takes account of three factors. It is not enough, Keynes argued, that these should be causal factors: 'There must be no other significant factor.' If there were 'then the method is not able to discover the relative quantitative importance of the first three'. For carbon dioxide to be the only possible explanation for heightened global temperatures, as a matter of logic, scientists must first be able to quantify every single feature of the changing climate. Yet large-scale changes in the climate, such as the causes and timing of ice ages, are still not well understood.

Economics should be central to deciding what, if anything, to do about global warming. Yet only the US had conducted an economic appraisal ahead of the Rio Earth Summit. No official economic analysis had preceded the Kyoto Protocol – a conceptual disaster, according to Nordhaus, lacking political, economic, or environmental coherence. Instead, the policy debate was framed as a binary question: *Is global warming happening? Scientists tell us it is, so we must do something.* That 'something' – attempting to return the developed world to the emissions level at the beginning of the 1990s – was about symbolism, not economic rationality.

Not until 2005 did any national body outside the US begin to consider the economic dimension. Early that year, the House of Lords Economic Affairs Select Committee held hearings on the economics of climate change. The committee included two former Conservative chancellors (Nigel Lawson and Norman Lamont), a former governor of the Bank of England (Robin Leigh-Pemberton) and economist and Labour peer Richard Layard.

When they cross-examined a senior Treasury official, it emerged that the Treasury had not conducted any appraisal of the economic costs and benefits of global warming to the British economy. Earlier, American economist Robert Mendelsohn had told the committee that global warming would be beneficial for regions such as the UK that were 'too cold'.

Neither had the Treasury conducted an economic appraisal of the costs of meeting the UK's target for a sixty per cent cut in carbon dioxide emissions by 2050. 'The target was quite consciously introduced over a very long period because we understand the costs of making these changes are much smaller if they are planned over substantial periods of time,' Paul Johnson, the Treasury's chief micro-economist, said. It had left the economic analysis to the IPCC, with some input from Defra. Lawson expressed astonishment. 'In my time at the Treasury as Chancellor it would have been unthinkable for the Treasury not to spend quite a lot of time on a serious economic analysis of an issue as important as this.'

The committee's chairman Lord Wakeham, a former Conservative energy secretary, was determined to get a united report and restricted discussion on the science to avoid splitting the cross-party committee. Wakeham was dismayed by what he saw of the IPCC, a 'magic circle' creating certainties out of massive uncertainties. 'Wickedly bad' was his verdict on the IPCC's process, an institution he thought intent on its self-perpetuation.

It was evident to the committee that the IPCC was prone to deep-seated politicization. In his evidence, environmental economist Richard Tol of the University of Hamburg said that, although he had been involved in the Third Assessment Report, he had not been nominated by the German government to work on the Fourth. German government policy meant only those with close connections to the Greens would be nominated to Working Groups II (on the impacts of climate change) and III (on policy responses). 'Things have become more and more politicized,' Tol observed.

In its report published at the beginning of July 2005, the committee criticized the Blair government's lack of candour. The costs of its decarbonization policies were 'unhelpfully vague'. Its claim that costs prior to 2020 were 'negligible' was 'wildly optimistic'. It called on the Treasury to be 'more active', concluding that unless it was 'we do not see how the Government can argue that it has adequately appraised its long-term climate targets in terms of likely costs and benefits'.

25
STERN REVIEW

If we don't act, the overall costs and risks of climate change will be equivalent to losing at least five per cent of global GDP each year, now and forever.

Nicholas Stern

Forecasts tell you little about the future but a lot about the forecaster.

Warren Buffett

Be careful what you wish for.

Two weeks after the Lords committee reported, Chancellor Gordon Brown announced that Sir Nicholas Stern would lead a major review on the economics of climate change.

Stern had been appointed to the Treasury to head the Government Economic Service in 2003. He had come to Treasury with very fixed ideas about taxation, but Brown wasn't interested. He had enjoyed relatively short spells at his previous employers, the World Bank and the European Bank for Reconstruction and Development. He didn't seem to have much else to offer and was shunted off to work on Tony Blair's Africa Initiative.

Global warming was Stern's lucky number. Other economists had thought longer and harder about the economics of global warming; none would rival Stern's public impact or attain his status as a global warming guru. Stern's academic background was in development economics. His advocacy of more aid flows and central planning brought him into conflict with the free-market economist Peter Bauer. 'Preoccupation with the analysis of market failure and the theory of corrective intervention is a notable feature of development economics,' Bauer wrote in a 1984 riposte to Stern.[1]

To these was added certainty about the science of global warming. Asked in 2009 about the possibility that 'the science' might be wrong – ten years into a period of no statistical rise in observed global temperatures – Stern answered, 'It's very, very remote.' Less than one in a hundred? 'Oh, much, much less.'[2]

Nonetheless Stern's grasp of the science was poor. In the same interview, Stern stated that the greenhouse effect could be observed experimentally in greenhouses – 'and most people have observed the greenhouse effect themselves in greenhouses. Yes?'[3] No. In a greenhouse, rising warm air is prevented by glass from escaping – a convection effect. The physics of the atmospheric (so-called) greenhouse effect are entirely different: the absorption by water and carbon dioxide molecules of long-wave radiation from the Earth's surface – a radiation effect.*

In arguing that humans could not adapt to a warmer world, Stern asserted that the risk was of a rise in temperature greater than any since the end of the last ice age 'ten or twelve million years ago'.[4] In fact, the last glaciation ended, and the current inter-glacial began, around eleven thousand, five hundred years ago. (The oldest Palaeolithic cave paintings in central and southern France are thought to date back twenty-three thousand years ago.)[†]

Science – or 'The Science' – was the occasion of Stern's first skirmish with fellow economists. Henderson assembled a team that included two members of the Lords Economic Affairs Committee (Nigel Lawson and Robert Skidelsky), McKitrick and five other economists. Stern had made a 'premature and injudicious choice' in basing the review on 'an unbalanced and technically defective account of 'the science', they wrote. His use of the Hockey Stick created a misleading impression of a dramatic departure from a stable thousand-year norm. 'By taking as given hypotheses that remain uncertain, assertions that are debatable or mistaken, and processes of inquiry that are at fault, the Review has put itself on a path that can lead to no useful outcome,' Henderson and his colleagues argued.[5]

In a swift rebuttal, Stern gave no quarter. 'This is not a theory that is fraying at the edges,' he wrote. Falling into Professor Tinbergen's trap, Stern argued that unless the role of anthropogenic greenhouse gas emissions was recognized the temperature increase of the previous forty or fifty years 'cannot convincingly be explained' by climate models. 'In fact, the latest science suggests that the risks could be substantially greater than previously seen,' a claim that was not based on any real world evidence.[6]

The Stern Review was published in October 2006. For the first time outside the US, public debate about global warming shifted to its economic consequences – in apocalyptic terms. Failure to act could create economic and social disruption on a scale associated with the First and Second World Wars and the Great Depression of the 1930s. The costs and risks of climate change would be equivalent to losing at least five per cent of global GDP every year 'now and forever'. If a wider range of risks and impacts was taken into account, 'The estimate of damage could rise to twenty per cent of GDP or more.'[7] These were eye-popping numbers.

How had he done it?

As with other cost-benefit analyses of global warming, the Stern Review used a model to integrate physical impacts and a range of economic scenarios and variables to derive a social cost of carbon dioxide. The PAGE integrated assessment model used by Stern and his team had been developed by Chris Hope, a

---

*   Stern was in good company in mis-ascribing the effect of convection to radiation. In a primer on global warming, Sir John Houghton, the first chair of the IPCC's Working Group I, wrote of glass in a greenhouse as having 'somewhat similar' properties to the atmosphere in absorbing radiation which is re-emitted back into the greenhouse, the glass acting as a 'radiation blanket'. *Global Warming: The Complete Briefing* (1994), p. 21.

†   Of the many interglacials of the Quaternary Period (there have been at least nine), the current one is but the most recent. Its maximum period of warmth is now passed – in North West Europe, some five to seven thousand years ago.

Cambridge academic. Although the details differed, it shared the economic and utilitarian principles of the other two widely cited models, Nordhaus's DICE and Tol's FUND models.

PAGE used a time horizon stretching out to 2200 – a compromise between natural scientists who wanted to go out to 2400 or 2500, and economists who, as Hope put it, were unhappy to go beyond 2050, knowing that economic models were lucky to hold up much beyond five to ten years. But if the modelling stopped before 2050, it would be too early to show benefits from capping greenhouse gas emissions.

The review team was especially attracted to PAGE's ability to handle uncertainty, using a variant of Monte Carlo probability analysis, with ten thousand to one hundred thousand model runs to derive a single result. The model incorporated the possibility of catastrophic climatic events, such as the melting of the West Antarctic ice sheet occurring *before* 2200 based on some work by Hans Joachim Schellnhuber and Stefan Rahmstorf of the Potsdam Institute. In Hope's words, these were 'just slightly more than illustrative' – the evidence base for them being 'very thin'.

The single largest impact on estimating a value for the social cost of carbon is the climate sensitivity of carbon dioxide. The Review took a triangular distribution ranging from 1.5°C to 4.5°C, with 2.5°C as the most likely value. The Review also ran a 'high climate' scenario to demonstrate a sizeable probability that the climate sensitivity of carbon dioxide was higher than previously thought.

The second most important input was the pure time preference rate; in layman's terms, the cost of time. Stern rejected the approach used by economists such as Nordhaus to derive the cost of time from market data, such as from bond yields, returns on investment and the amount people saved from their incomes. For Stern, it wasn't a question of what the pure time preference rate *is*, but what it *ought* to be. 'We take a simple approach in this Review: if a future generation will be present, we suppose that it has the same claim on our ethical attention as the current one.'[8]

The distinction has profound implications for what the Review was about. Was it about economics or was it, above all, about ethics? 'If you care little about future generations you will care little about climate change,' Stern argued, a position which did not have 'much foundation in ethics and which many would find unacceptable.' By the same token, it could be said; if you care little about the present, you will care a lot about climate change. There is a way of balancing the two. The discount rate, embodying the cost of time, is the mathematical bridge linking the present and the future.

At the heart of the Stern Review is an ethical argument, one that Stern articulated in almost identical terms as Boulding had suggested – but rejected – forty years earlier. Stern quoted Frank Ramsey, the brilliant young Cambridge mathematician and philosopher who, in the 1920s, described pure time discounting as 'ethically indefensible' and Oxford economist Roy Harrod, who called it a 'human infirmity' and 'a polite expression for rapacity and the conquest of reason by passion'.

These views neglect mortality as the most certain fact of an individual's existence. Stern did recognize the relationship between an individual's mortality and the cost of time. 'The allocation an individual makes in her own lifetime may well reflect the possibility of her death and the probability that she will survive a hundred years may indeed be very small,' Stern conceded.[9] Individuals' preferences had 'only limited relevance for the long-run ethical question associated with climate change', Stern argued. The Review was thus couched firmly within a collectivist perspective – what matters is the welfare of the anthill, not the individual ant. In his modelling, Stern used a pure rate of time preference of 0.1 per cent a year to reflect the possibility of humanity being wiped out, implying an almost ten per cent chance of the extinction of *homo sapiens* before 2100.

There was a downside to Stern's ethics – his treatment of the world's poor and the transfer of wealth from them to the richer generations of tomorrow. Sir Partha Dasgupta, Frank Ramsey Professor of Economics at Cambridge, criticized Stern for taking a very inegalitarian attitude to the distribution of wellbeing when futurity was not at issue. The ethical parameter Stern had adopted to reflect inequality and risk in human wellbeing was, Dasgupta wrote, 'deeply unsatisfactory'. Overall, Stern's assumptions would require the current generation to save 97.5 cents of every dollar it produced – 'so patently absurd that we must reject it out of hand'. Economists had taken the threat of climate change seriously, Dasgupta concluded, 'but the cause is not served when parameter values are so chosen that they yield desired answers'.[10]

For Stern, it was all about the future. It was difficult to assess the impacts over a very long time period because future generations 'are not fully represented in current discussions'.

Who should then represent them and who knows what they will think?

The implication is Stern and like-minded successors, in a dynasty of economic Pharaohs, presiding over the global economy for decades and centuries ahead. In a 2007 critique, Nordhaus wrote of Stern, stoking the dying embers of the British Empire, to apply colonial-style Government House utilitarianism from 'the lofty vantage point of world social planner'.[11]

Nordhaus also solved the puzzle of how Stern had derived his numbers. By using his DICE model, he could convert a one per cent reduction in output over the next century to a 14.4 per cent reduction 'now and forever'. On Stern's methodology, more than half the estimated damages 'now and forever' occur after 2800. 'The large damages from global warming reflect large and speculative damages in the far-distant future magnified into a large current value by a near-zero time discount rate,' Nordhaus wrote.[12]

Having answered the 'How?' what of the 'Why?'

Just as with the British government's assessment of weapons of mass destruction, Nordhaus suggested the Stern Review should be read primarily as 'a document that is political in nature and has advocacy as its purpose'.[13] Was this overly harsh? An answer can be found in the Review. 'Much of public policy is actually about changing attitudes,' Stern wrote, in an authentic expression of the Blair-era style of governance as PR.[14]

Thus there is nothing in the Review that might put a question mark over whether Kyoto was a good deal for Britain or for the world. Despite mounting evidence to the contrary, Stern argued that countries would meet their Kyoto obligations. 'Governments make and respect international obligations because they are in line with perceptions of responsible and collaborative behaviour,' Stern wrote, 'and because domestic public opinion supports both the objectives and mechanisms for achieving them' – a finding contradicted by Scott Barrett's authoritative analysis.[15]

Neither did Stern examine what the costs and benefits of global warming might be for Britain. If a warmer climate benefited Britain, then cutting emissions is a lose-lose proposition.

Stern was dismissive of adaptation. 'An inherent difficulty for long-term adaptation decisions is uncertainty, due to limitations in our scientific knowledge of a highly complex climate system,' Stern wrote.[16] While uncertainty was an impediment to adaptation, the same lack of knowledge wasn't a barrier to governments. 'There speaks the true bureaucrat' was Nigel Lawson's stinging comment on Stern's presumption of the superiority of government wisdom.[17]

By presenting an ethical argument – that future generations should have as much weight as the present – in the garb of an economic cost-benefit analysis, the Stern Review was mis-sold by its author and by its sponsoring government. 'Tackling climate change is the pro-growth strategy,' the review claimed.[18] 'In broad brush terms,' the review stated, 'spending somewhere in the region of one per cent of gross world product forever could prevent the world losing the equivalent of five to twenty per cent of gross world product forever.'[19]

There's something morally suspect about a proposition couched in terms of making you better off financially and morally. Ethics come into play when following financial self-interest leads to a morally bad outcome. On the other hand, ethics are cheapened when used in support of someone's supposed material self-interest. If Stern had had the courage of his (moral) convictions, he should have made the case that taking action against global warming is costly and would hurt financially, but that the current generation had a moral duty to later generations.

How long would the pain last? Deep in the Review are two charts showing the costs of stabilizing emissions at Stern's target of 550 ppm of carbon dioxide equivalent. On the assumption that this would cost one per cent of Gross World Product (GWP), the chart showed a break-even point around 2080; if the cost was four per cent of GWP, the benefits would not exceed the costs until around 2120.[20]

The midpoint of the two implies there being no net benefit during the twenty-first century. A young person's great-grandchildren might conceivably benefit, but not their children or grandchildren.

For countries committed to cap their emissions, the break-even point – assuming there is one – would be pushed into the twenty-second century and probably beyond. Stern reckoned that non-Annex I parties were likely to be responsible for over three quarters of the rise in energy related carbon dioxide emissions. Agriculture and changes in land use alone accounted for forty-one per cent of emissions, with Indonesia and Brazil accounting for half these.

Since the 1995 Berlin Mandate, the world community had been locked on a course that excluded non-Annex I parties from emissions caps. While endorsing Kyoto, Stern did not analyze the emission cuts required by Annex I countries to meet his 550 ppm stabilization target or the impact on their economies. For citizens of Annex I countries, the prospect of action on climate change ever producing a net benefit was very remote, even if Stern's assumptions on the science turned out to be right.

26
SELLING SALVATION

'Solutions' are at hand, given wise collective decisions and actions. It is the combination of alarmist visions with confidently radical collectivist prescriptions for the world as a whole which characterizes global Salvationism.

David Henderson, 2004

On October 30th 2006, flanked by Tony Blair and Gordon Brown, Stern launched his Review. Blair spoke first. Within seconds, the hyperbole was in full flow. If the science was right, the consequences for the planet were literally disastrous – not in some science fiction future many years ahead, 'but in our own lifetime'.

The Review was the most important report on the future produced by the government, Blair said. 'We know it is happening. We know the consequences for the planet.' Stern showed the economic benefits of strong early action easily outweighed the costs. 'For every £1 we invest now, we can save at least £5 and possibly much more by acting now.' Investment to prevent it 'will pay us back many times over'.

Blair spoke for less than ten minutes.

Brown then took the stage – for an interminable sixteen. Nothing was too small to escape mention. A rainforest initiative with Brazil, Papua New Guinea and Costa Rica. A plug for biofuels. 'I'm determined that we use biofuels from palm and rape oil, to soya and sugar and eventually cellulosic biofuels,' he said, announcing that Al Gore was joining Alan Greenspan as his personal adviser. 'Environmental policy is economic policy,' Brown insisted.

It was also about politics. As Bagehot in the *Economist* noted, thanks to the Conservatives' new leader David Cameron putting green issues at the heart of his attempt to re-brand the Conservative party, 'Stern has acquired a significance for British politics that Mr Brown would almost certainly not have predicted.'

Blair's spin won.

'Blair: World needs to act on climate change now,' read the headline in the next day's *Daily Mail*. Stern's calculation of spending one per cent of GDP to save potential costs of five to twenty per cent of GDP were 'certainly compelling', *The Times* said in its leader. Given the risk of 'something really catastrophic', like the melting of the Greenland ice sheet, 'the costs are not huge. The dangers are', the *Economist* concluded.

To the *Daily Telegraph*, Stern's pinning a price tag on global warming was 'invaluable'. According to the *Financial Times*, the economic benefits of action would eventually far outweigh the costs by as much as $2.5 trillion a year, although the paper didn't say when that might be. How convincingly had the review made its case, asked the *FT*'s lead economic commentator Martin Wolf? 'Sufficiently so,' he wrote, putting his finger on the issue of the discount rate. 'The review argues, sensibly, that there is no reason why the welfare of our generations should be intrinsically more important than those of our grandchildren,' Wolf thought.[1]

There was some isolated dissent. To economist and energy expert Dieter Helm, the one per cent cost of action seemed a 'terribly low' estimate and hard to square with the 'substantial lifestyle changes that many environmentalists argue are necessary and the scale of the taxes that appear necessary'.[2]

Most scathing was the *Daily Mail*. Even if Stern was right, on its own, Britain could do precious little about it. 'But never fear – Superman is here, in the shape of our posturing prime minister,' the paper said. Calling him deluded, 'anyone listening to him yesterday would think he could save the world single-handedly'.[3] To Nigel Lawson, the near universal and credulous acceptance of the Stern Review was the result of two things coming together: 'The ever-present media appetite for alarmism and the uncritical respect accorded by the innumerate majority to anything with numbers in it, however dubious.'

The real target was overseas. The Review would be used as 'a vehicle to take on the doubters internationally' and enable Tony Blair to persuade a sceptical George Bush, a Whitehall source told the *Financial Times*.

There was a great deal more skepticism and suspicion on the other side of the Atlantic. The *Washington Post*'s economic commentator Robert Samuelson dismissed the review as 'a masterpiece of misleading public relations'.[4] Stern's headlined conclusions were 'intellectual fictions' fabricated to justify an aggressive anti-global warming agenda, Samuelson argued. 'Anyone serious about global warming must focus on technological progress – and not just assume it,' as Stern had done. 'Otherwise, our practical choices are all bad: costly mandates and controls that harm the economy, or costly mandates and controls that barely affect greenhouse gases. Or, possibly, both,' Samuelson concluded.[5]

The widespread acceptance of the Stern Review revealed a paradox. The politics of the moment demanded a call for action which collectively the governments of the world wouldn't sign up to and politicians were not capable of delivering. On the other hand, most mainstream economic analysis pointed to an optimal path of a gradually tightening of greenhouse gas emissions, a more achievable goal, but one that did not satisfy the apocalyptic temper of the times. William Nordhaus's DICE model suggested a social cost of carbon of $35 a ton in 2015 rising to $206 a ton at the end of the century. By contrast, the Stern Review estimate of the current social cost of carbon at $350 a ton was ten times higher.

The snow lay thick on the ground when Stern met his economist peers at Yale in February 2007. The Yale symposium on the Stern Review was a 'really serious occasion', Chris Hope recalled. The US was engaging with Stern intellectually rather than politically and had rolled out its big guns. With several hundred people in the hall, it was the first time in fifteen years Hope felt the need to rehearse a presentation word for word.

Chaired by Ernesto Zedillo, the former president of Mexico, symposium participants included Jeffrey Sachs, Scott Barrett, Robert Mendelsohn, William Cline and Nordhaus. Each offered penetrating comments. Of them, Sachs, Cline and

Barrett were most sympathetic to Stern's conclusions. Sachs went furthest. By 2010, the world would have agreed a post-Kyoto target to stabilize atmospheric carbon dioxide at 500 ppm. 'It will take two years to ratify, and go into effect January 1st 2013,' Sachs predicted.

Although Cline thought the Review 'very much on the right track', he laid bare the questionable mechanics of how Stern had arrived at his conclusions. Stern had to modify his zero cost of time assumption by factoring in a small allowance for mankind's self-implosion, otherwise the Review's use of an infinite time horizon 'simply explodes'. Putting Stern's welfare equation into a spreadsheet would show that ninety-three per cent of all future welfare occurs after 2200. In Cline's view, Stern probably should not have extrapolated the damage rate into the infinite future. 'The combination of near-zero pure time preference with an infinite horizon probably balloons the value of damage avoided unreasonably,' Cline told the symposium.[6]

Welcoming the decisive role of ethics in Stern's analysis, Barrett disagreed with Stern's conclusion that the case for 'strong, early action' was clear cut. 'I am not saying his conclusions are wrong; I am saying that other conclusions can be supported.'

Others were more critical. Mendelsohn said the review was not an economic analysis. It had merely asserted that 550 ppm was the least cost option without comparing it against alternatives other than doing nothing at all. It was like saying: 'I have a great policy for educating children and look, it is better than closing all schools.'

Stern had ignored the environmental damage implied by his proposals; one plan involved a combination of two million windmills, ten million hectares (24.7 million acres) of solar cells and five hundred million hectares of biofuel (1.2 billion acres) – more than ten times the surface area of California and more than forty times California's 2007 farm acreage. Taking away that amount of cropland for renewable energy would very likely have a much larger impact on agriculture than climate change, Mendelsohn suggested. 'One can look at the Stern Review as a fairly complete argument why aggressive near term abatement does not make sense.'[7]

Nordhaus was particularly critical of Stern's central assertion that the cost of climate change was the equivalent of a twenty per cent cut in per capita consumption 'now and forever'. Stern's formulation inflicted 'cruel and unusual punishment' on the English language because the Review's 'now' didn't mean 'today'. A 'distant rumble' of a three per cent consumption loss in 2100 on Stern's extreme-extreme-extreme case was exaggerated by a factor of one thousand per cent, not because of the estimates of the damages, but primarily through discounting.[8]

When it came to his response, Stern made a partial concession. Although the 'now and forever' language was accurate, 'Perhaps it wasn't particularly felicitous, and on reflection, we might have used some other wording.'[9] Grudging, perhaps.

Belated, certainly. But collectively, economists acquitted themselves far better than had the natural scientists with their cover-up of the Hockey Stick. And, unlike the author of the Hockey Stick, Stern on this occasion was prepared to reason and debate with his critics.

In a subsequently published postscript to the Review, Stern gave further ground. 'Ethical positions cannot be dictated by policy analysts,' he wrote.[10] For the first time, Stern provided numbers that demonstrated the huge sensitivity of his conclusions to the discount rates assumed – an increase in the discount rate from 1.4 per cent to 3.5 per cent would reduce the estimates of the economic cost of greenhouse gas emissions from five per cent to 1.4 per cent of world GDP.[11] At this higher, but still modest discount rate, Stern's clarion call for immediate and drastic action simply evaporated.

Stern's methods had not withstood scrutiny, but his call to action based on catastrophist assessments remained fashionable. In a July 2008 paper, Harvard economist Martin Weitzman argued that Stern was right, but for the wrong reasons. Stern's cost benefit analysis was completely flawed. Moreover, the use of cost benefit analysis was inappropriate for an issue such as climate change.

Based on a survey of twenty-two studies on the climate sensitivity of carbon dioxide, Weitzman focused on the fifteen per cent of cases in which temperature increases substantially greater than 4.5°C 'cannot be excluded'. Weitzman superimposed 'the non-zero probability' of greenhouse gas self-amplification, such as methane-out-gassing precipitating a cataclysmic runaway-positive feedback warming. If temperatures rose by around 10–20°C, there could exist 'truly terrifying' consequences such as the disintegration of the Greenland and at least part of the Western Antarctic ice sheets – 'horrifying examples of climate-change mega-disasters [that] are incontrovertibly possible on a time scale of centuries', Weitzman wrote.[12]

From small probabilities of huge climate impacts occurring at some indefinite point in the remote future, Weitzman conjured up 'fat tails' – low probability, high impact events – conjoining with fat tails begetting yet fatter tails. Unlike traditional 'thin-tailed' cost benefit analysis, Weitzman acknowledged that it was 'much more frustrating and much more subjective – and it looks much less conclusive – because it requires some form of speculation (masquerading as an "assessment")'.

Weitzman's analysis gained extra currency following the bankruptcy of Lehman Brothers and the near collapse of the West's banking system. All of a sudden, risks which financial models suggested were extremely remote became fat-tailed risk events. Even before the banking crisis struck, Martin Wolf in the *Financial Times* was strongly influenced by the fat-tail analysis: 'Above all,' Wolf wrote, 'I find persuasive the argument of Professor Martin Weitzman of Harvard University that it is worth paying a great deal to eliminate the risk of catastrophe.'[13]

The trouble was no amount of money could completely eliminate the risk of climate catastrophe. If climatic 'tipping points' exist, no one can be sure at what point they tip. Even if emissions fell to zero, past emissions would mean that the world is 'committed' to many decades of further warming.

Weitzman himself drew a different conclusion. The existence of fat-tailed potential catastrophes meant paying more attention to how fat the bad tail might be. If thought to be fat, it meant being much more open to drastic action, including 'serious mitigation' (emissions cuts) and possibly using geo-engineering technology to slim down the fat tail fast.

Lawson for one believes Weitzman's position to be absurd.

> In a world of inevitably finite resources, we cannot possibly spend large sums on guarding against any and every possible eventuality … The fact that a theoretical future danger might be devastating is not enough to justify substantial expenditure of resources here and now, particularly since there are many other such dangers wholly unconnected with global warming.[14]

Weitzman's fat-tails mark an important milestone in the development – or more accurately, the disintegration – of the idea of global warming and its call to cut carbon dioxide emissions. The assumption that there is some kind of stable, quantifiable relationship relating the amount of carbon dioxide in the atmosphere to global temperatures and thence the impacts on societies and the environment provides a quantified rationale for preventative action.

Under the fat-tail analysis, there is no level of emissions reduction that can completely remove the risk of catastrophe over the indefinite future. Even in its own terms, a policy of doing nothing becomes a rational response, as it is no longer possible to quantify the avoided costs and risks that emission cuts are meant to be buying.

Catastrophic events occur. As readers of Asterix and Obelix know, the only thing the Gauls feared was the sky falling on their heads. And it seems that it did – a massive comet struck Southern Bavaria sometime in the third and fourth centuries BC, scarring the culture of the Celts. To live life under the shadow of the possibility of another such event suggests a somewhat morbid sensibility. *You only live once*, as they say.

The optimal economic response points in the direction of abandoning policies that retard economic growth, specifically those purportedly designed to slow down or arrest global warming. What constitutes a natural catastrophe to one society might be a minor inconvenience to a richer one. Bert Bolin wrote in 2007 that the IPCC was not yet able to tell from climate models which countries would be most affected. 'The issue could thus not then be resolved adequately beyond the obvious conclusions that poor countries were more vulnerable than rich ones and less able to protect themselves.'[15] Bolin had stumbled on the insight that Frédéric Bastiat had derived when he asked why mid nineteenth-century Europe no longer suffered from famines. The answer? Rising prosperity.

No one knows what the climate of the future will be. Still less can we have any idea how societies of the future will respond to these unknown changes. But history shows that wealth insulates societies from nature and that wealthier societies are better equipped to overcome natural disasters than poor ones.

Whether the cause is natural, man-made or some combination of the two – the answer is the same. Economic growth.

27
Cucumbers into Sunbeams

America is addicted to oil … By applying the talent and technology of America, this country can dramatically improve our environment, move beyond a petroleum-based economy and make our dependence on Middle Eastern oil a thing of the past.

George W. Bush, January 31st 2006

We will harness … the soil to fuel our cars.

Barack Obama, January 20th 2008

Economists assumed governments would adopt optimal policies to reduce greenhouse gas emissions at the lowest possible cost. What actually transpired bore greater resemblance to what Gulliver witnessed on his travels.

At the grand academy of Lagado, Gulliver saw a professor working to extract sunbeams from cucumbers. In another eight years, he should be able to warm the governor's garden. Would Gulliver give him something to encourage him in his ingenuity? Gulliver's master had given him money for this very purpose 'because he knew their practice of begging from all who go to see them'. (In the age of global warming, tendering financial inducements would be obligatory to encourage renewable energy.)

A second professor was engaged in a project to return human excrement to its original state as food. (Recycling human effluent was a particular fixation during the first environmental wave.) A third was contriving a method of building houses starting at the roof and working downwards. (New houses in England have small windows to reduce their carbon footprint.)

But it was the political school where the professors appeared wholly out of their senses – 'a scene which never fails to make me melancholy', Gulliver says. Economists urged politicians to set a mechanism to price carbon dioxide emissions and let the market find the most efficient means. But politicians couldn't stop themselves from prescribing what people and businesses should do – irrespective of the cost per ton of carbon dioxide saved or whether it actually reduced emissions. European politicians were especially partial to wind farms, producing some of the world's most expensive electricity (solar power, which German politicians favoured most of all, was even more costly).

Neither could politicians evade the global logic of global warming. Without a global agreement, unilateral action was pointless. Governments committed to massively costly renewable energy programs thus had incentives to inflate the prospects for successive rounds of international climate change negotiations. Living under its Kyoto bubble, politicians in the European Union would point to the faintest possibility of a son-of-Kyoto at the end of the rainbow. In 2006,

California's politicians passed a state-wide cap on greenhouse gas emissions, inviting competing states to help themselves to some of California's GDP.

The global warming policy spectrum ranged from petty infringements of personal freedom, such as the ban on incandescent light bulbs in the EU, Australia and Canada, to policies harming the world's poorest and damaging the environment. When the EU leaders discussed the light bulb ban in 2007, Angela Merkel complained that the green replacements weren't as good as the old ones. 'Most of the light bulbs in my flat are energy-saving bulbs,' Merkel said. 'They're not yet quite bright enough. When I'm looking for something I've dropped on the carpet, I have a bit of a problem.' Merkel wasn't alone. In the UK, the prospect of withdrawal of high-wattage incandescent bulbs from supermarkets sparked a stampede.

Further along the spectrum were attempts to create markets for permission to emit carbon dioxide. Government-planned markets are inherently problematic. As Martin Wolf has written, 'Wherever there is a gap between the market value of something and an official price or the price government is prepared to allow, there is an incentive to cheat and bribe.'[1] Because it does not cost anything to emit carbon dioxide, when politicians legislate a value on it, they created a universe of opportunities for fraud and organized crime.

The EU's Emissions Trading Scheme (ETS) is the most ambitious and complex cap-and-trade scheme – a Common Agricultural Policy for the atmosphere, a sort of Gosplan meets Milton Friedman. The largest ETS scam is carousel trading. Fraudsters buy carbon credits in one country without paying VAT, then sell them in another with VAT, pocketing the difference, costing European taxpayers an estimated €5 billion. According to Europol, up to ninety per cent of all 2009 trades could have been carousel trading.

Analysts at investment bank UBS reckoned that in 2012 the ETS will cost consumers €18 billion, but will have only reduced power sector emissions by between one and three per cent since its inception. By 2025, the cumulative windfall to electricity producers could total €210 billion. The ETS also pumps billions of euros to countries not subject to emissions caps via the Kyoto Protocol's Clean Development Mechanism, which failed for the reasons Raùl Estrada anticipated in his 1998 critique.

South Korea turned out to be the biggest winner of the CDM jackpot, scooping eighteen per cent of all the credits issued worldwide. The French owner of a factory in Onsan was expected to reap more money than the whole of Africa. 'We didn't make the rules of the game,' an executive of Rhodia SA told the *Wall Street Journal*. Rhodia could expect to earn more than $1 billion over seven years from destroying laughing gas, but the real joke was on European consumers and taxpayers. 'It's good business, because the Western world is basically desperate,' a consultant explained, as it needed to buy emission reductions credits.

Biofuels are at the most harmful end of the policy spectrum. When more than three quarters of the three-year increase in world corn production is diverted to nonfood use, economics predicts that the price of corn will move in one direction: up. Between January 2002 and June 2008, food commodity prices more than doubled, with nearly half that increase occurring in the eighteen months from January 2007.[2]

An internal World Bank paper estimated that most of the seventy to seventy-five per cent increase in food prices not accounted for by higher energy costs and the fall in the dollar was 'due to biofuels and the related consequences of low grain stocks, large land use shifts, speculative activity and export bans'. In March 2008, the director of the UN World Food Programme told the European Parliament that the shift to biofuels production has diverted lands out of the food chain. 'In the short term, the world's poorest are hit hard.'

Spiralling food prices led to rioting across the globe. In Haiti, the government fell. Rioting spread across vulnerable Muslim countries, including Egypt, Yemen, Mauritania and Bangladesh. There were street protests in Italy at the price of pasta, although the government in Rome didn't fall.

Global warming was predicted to raise food prices and create instability in vulnerable parts of the world. 'Warming of a few °C is projected to increase food prices globally, and may increase the risk of hunger in vulnerable populations,' the IPCC warned in 2001. Temperatures overall hadn't risen in the first decade of the new century, but food prices and hunger did. The cause was global warming policies – in a brutal form of the Global Warming Policy Paradox (GWPP).

Global warming policies damage the environment. Rising global temperatures were also predicted to harm ecosystems and reduce biodiversity. Cutting down tropical rain forests to turn into palm oil plantations certainly does destroy wildlife habitat and reduce biodiversity. The GWPP is not an accident or down to bad luck. It is an inescapable outcome of the economic and political dimensions of the global warming idea.

Some of the most vociferous proponents of the global warming orthodoxy, including leading scientists and a large number of NGOs such as Greenpeace, Friends of the Earth, Oxfam and the WWF, reject mainstream economics and dismiss concerns about costs. The catastrophist conclusions of the Stern Review reduce the near-term costs of cutting carbon dioxide emissions relative to the potential gains (in terms of losses avoided) in a far distant future. To the extent such costs aren't accorded their due weight, harmful policies will be adopted.

Hyping global warming as a planetary catastrophe and driven by a deep-seated hostility to fossil fuels, NGOs shared responsibility for the ensuing policy insanities. Having campaigned tirelessly for bio-energy, in 2010 Greenpeace was caught out by the policies it had advocated. After an outcry at the destruction of Indonesian rainforest, Greenpeace UK's John Sauven denounced subsidized palm oil power stations as 'orang-utan incinerators'. The plantations weren't 'sustainable'. According to Liz Bossley of energy consultants Consilience Advisory, 'While there is a food shortage in the world, there will always be a question mark over the use of land to grow plants for power generation.'

The Royal Society for the Protection of Birds (RSPB) is a vociferous supporter of wind farms, although they kill eagles and other raptors. 'Switching to renewable energy now, rather than in ten or twenty years, is essential if we are to stabilize greenhouse gases ... Wind power is the most advanced renewable technology,' the RSPB claims. In similar vein, America's Audubon society 'strongly supports' wind farms, while acknowledging that its stance 'will not be without some impact' on birdlife.

More concerned with fundamentally changing the economic basis of modern urban civilisation than preserving the countryside and wildernesses, environmental NGOs ended up supporting the despoliation of rural areas and uplands with arrays of alien wind farms. It represented the negation of what their spiritual forebears stood for, transcendentalists such as Emerson and Thoreau and the trio of John Muir, Gifford Pinchot and Teddy Roosevelt. The attitude of environmental NGOs to the environment became that of the American major in the Vietnam War: it became necessary to destroy the village to save it.

'It all sounded so, well, intelligent,' *New York Times* columnist Thomas Friedman wrote in December 2009 on Denmark's renewable electricity. While the Danes claim that thirty per cent of their electricity comes from renewables, on average, wind power – the largest one – provided less than ten per cent of the country's needs.* According to Paul-Frederik Bach, who integrated wind power into the Danish grid, 'maintaining the myth of the successful integration of wind power may be good public relations, but refusing to face realities is self-deception'.[3]

Although Denmark installed around three thousand megawatts of wind power capacity and its Vestas wind turbine company became a world leader, it had not cracked the fundamental problem of generating electricity from wind. Wind farm electricity is unpredictable. Its output profile is different from the profile of electricity demand.

While onshore wind farms generate around one fifth of their theoretical capacity, during calm periods, output plunges. During one twenty-four-hour period, output from all Denmark's wind farms fell to less than one quarter of one per cent of capacity. Too much wind at the wrong time – and surplus electricity has to be given away to neighboring countries at spot prices close to or actually zero. When spot electricity prices are negative, Denmark *pays* other countries to take their surplus wind power.

Germany made the largest commitment to renewables. In 2008, a combined wind and solar capacity of twenty-nine thousand, two hundred and eleven megawatts put Germany just ahead of the US, with twenty-five thousand, nine hundred megawatts. A 2009 study by the University of Bochum described Germany's

---

*   In reality, Denmark's electricity market is so tightly integrated into Germany's that, in practical terms, it is part of the German market. Wind power accounts for only seven per cent of the combined market. Paul-Frederik Bach, *The Variability of Wind Power* (2010), p. 26.

renewables experience as 'a cautionary tale of massively expensive environmental and energy policy that is devoid of economic and environmental benefits'.[4]

Its renewable regime was established in 1991, with generous feed-in tariffs, which in 2000 were extended for up to a further twenty years. Under it, the least efficient technologies receive the highest subsidy.* For producing a fraction of one per cent of Germany's electricity, solar power producers can expect to receive €53 billion in subsidies – two and a half times the €20.5 billion for wind power. German feed-in subsidies run in parallel with the EU's Emissions Trading Scheme. Thus marginal reductions in German emissions increase the quantity of tradable emissions certificates to electricity producers in other European countries. The principal beneficiaries turn out to be the coal-fired power stations of Italy's ENEL and Spain's Endesa. Far from creating a German win-win, the Bochum study concluded that Germany's renewables program 'imposes high costs without any of the alleged positive impacts on emissions reductions, employment, energy security, or technological innovation'.

Spain, second after Germany in its commitment to solar power, took a different path. Rather than charge electricity users the full cost of renewables, electricity generators received a 'tariff deficit' credit from the government. By the end of 2012, they are projected to have accumulated €21 billion of IOUs from the government. Aside from the dire fiscal consequences as Spain struggles to shrink its budget deficit, a 2009 study by Gabriel Calzada Alvarez of Madrid's Universidad Rey Juan Carlos put a price tag on green jobs: €570,000 per job since 2000. Each green job cost Spain 2.2 real jobs, Calzada estimated.[5]

America was not spared similar follies. California has its own cap-and-trade regime and, in 2005, seven states – Connecticut, Delaware, Maine, New Hampshire, New Jersey, New York and Vermont – formed the Regional Greenhouse Gas Initiative (Massachusetts under Governor Mitt Romney dropped out at the last moment, but re-joined under his successor).

Like H.J. Heinz, federal renewable and energy efficiency programs come in fifty-seven varieties. Some, like a $10 million tribal energy program, date from the Carter-era 1975 Energy Policy and Conservation Act. Others came about as a result of Energy Policy Act of 2005 and the Energy Independence and Security Act passed two years later. With their help, American politicians developed a serious bioethanol habit, even though seventy-four to ninety-five per cent of its energy content comes from coal and other conventional energy sources. It promised them a double hit – a national security high and a global warming one. Bioethanol wasn't a fossil fuel, so it was green, and it didn't come from the Middle East, but pushed up food prices, stoking instability in the Middle East.

---

* In 2005, solar received €54.53 per kWh – over six times the €8.53 per kWh for wind. Manuel Frondel, Nolan Ritter, Christoph M Schmidt, Colin Vance, *Economic Impacts from the Promotion of Renewable Technologies: The German Experience* (2009), Table 1.

Bioethanol producers were enriched with tax credits, grants, loan guarantees and import controls. Most of all they benefited from the Energy Policy Act's renewable fuel standard which requires bioethanol – costing a multiple of conventional hydrocarbons – to be blended into transportation fuel; 13.95 billion gallons in 2010 rising to 36 billion gallons in 2022.

Other than Australia, the US was the only developed nation not to ratify the Kyoto Protocol – and Australia did after John Howard lost the 2007 election. Some that ratified Kyoto treated it as little more than a statement of intent. Canada's finance minister John Manley suggested it would be all right to ratify but not comply. 'The bailiff isn't going to arrive to seize our property,' he remarked in November 2002.[6] And when Canada found Kyoto too burdensome, the Harper government decided to withdraw from it. For the US, Kyoto would have been a legal commitment that could have been litigated through the courts. The case of *Massachusetts v. EPA*, which the Bush administration lost in the Supreme Court, illustrates how judicial rulings can have unforeseen and potentially far-reaching effects.

The political debate in the US was also fundamentally different. In parliamentary systems, legislators are subject to strong party discipline. Even in Canada, with its federal system and hydrocarbon-rich provinces such as Alberta, the Prime Minister Jean Chrétien obtained every vote of his Liberal party caucus to ratify Kyoto. None of Canada's major parties challenged the scientific rationale for action.

In the US, there was a debate.* In July 2003, Senator James Inhofe of Oklahoma, chair of the Senate Committee on Environment and Public Works, directly attacked the scientific consensus. 'If the relationship between public policy and science is distorted for political ends,' Inhofe argued, 'the result is flawed policy that hurts the environment, the economy and the people we serve.'[7] He ridiculed science writer David Appell for claiming global warming would create greater chaos than the two world wars, and cited political scientist Aaron Wildasky's description of global warming as the mother of all environmental scares. 'Much of the debate over global warming is predicated on fear, rather than science,' Inhofe said. 'After studying the issue over the last several years, I believe that the balance of the evidence offers strong proof that natural variability is the overwhelming factor influencing climate.' Concluding, Inhofe asked: 'Could it be that man-made global warming is the greatest hoax ever perpetrated on the American people? It sure sounds like it.'

---

*   In Australia, political argument between the parties broke out in 2009 after opposition leader Malcolm Turnbull, who supported the Rudd government's climate change policies, was deposed and replaced by sceptic Tony Abbott.

At the beginning of the nineties, corporate America responded to the rising wave of environmentalism in much the same way as big business elsewhere: it rode it. Coca Cola's Don Keough provided Maurice Strong with some top marketing and public relations executives for the 1992 Rio Earth Summit. Strong was thrilled. 'Since no organization has a more sophisticated capacity for marketing, this was a gift indeed,' Strong wrote. 'Thus were our key messages promulgated throughout the world.'[8]

Strong asked Swiss industrialist Stephan Schmidheiny to mobilize business support. Schmidheiny set up the Business Council for Sustainable Development. It produced a business manifesto, *Changing Course*, co-signed by forty-seven chairmen and CEOs, including two from oil companies (Chevron and Shell) and three auto manufacturers (Volkswagen, Nissan and Mitsubishi). The largest number was from US corporations.*

Changing Course could have been vintage 1972 Barbara Ward in arguing that environmental degradation and rich countries' over-consumption prevented the development of poorer ones. 'These two sets of alarming trends – environment and development – cannot be separated.' The world was moving on. 'The national income of Japan's one hundred and twenty million people is about to overtake the combined incomes of the 3.8 billion people in the developing world,' the authors of *Changing Course* wrote, just as Japan was entering two lost decades at the end of which, by one measure, it had been over-taken by China as the world's second largest economy.[9]

Global warming was acknowledged as politically the most difficult issue. 'We cannot return to the lower energy scenarios of the past nor change our energy systems drastically,' i.e., little change of course in *Changing Course*. 'We foresee a gradual shift to a more sustainable mix of energy sources,' – hardly a ringing endorsement of the prospects for a post-fossil fuel economy.

After Rio, a split began to develop within the US business community and with the rest of the world. Leading the challenge was the Global Climate Coalition, comprising oil companies, automakers, utilities and trade associations such as the US Chamber of Commerce.

Formed in 1989 in response to James Hansen's testimony, initially it had a watching brief. That changed after the 1992 presidential election. To win support for Al Gore's BTU tax, the Clinton administration did a deal with Senator Robert Byrd – a cut in taxes on coal in exchange for higher taxes on oil. According to William O'Keefe, who led the Global Climate Coalition, it fostered the perception that the new administration was deeply hostile to the use of liquid petroleum products.

---

* They were Kenneth Derr of Chevron, Maurice Greenberg of AIG, Charles Harper of Conagra, Allen Johnson, retired chairman of 3M, Samuel Johnson of SC Johnson, Frank Popoff of Dow Chemical, former EPA Administrator, William Ruckelshaus, of Browning-Ferris, Paul O'Neill of ALCOA and treasury secretary under George W. Bush, and Edgar Woolard of DuPont.

In the run-up to Kyoto, there was a twin-track campaign highlighting the economic consequences of capping carbon dioxide emissions. The first was an advertising campaign through the summer of 1997: 'The UN Climate Treaty isn't Global … and it won't work.' The second focused on the Congress and the Senate in particular. Byrd asked the coalition to help find co-sponsors for what became the Byrd-Hagel resolution, ending up with sixty-four. 'We communicated a lot with members of the Congress to support the resolution,' O'Keefe recalled.

The ninety-five to zero vote in favor of Byrd-Hagel was the coalition's high point. Environmental NGOs battered its members with shareholder proxy campaigns and membership haemorrhaged. Before Kyoto, BP broke ranks and quit. In 1999, Ford Motor Company left. Ford chairman, life-long environmentalist William Ford, told a Greenpeace business conference the following year, 'I expect to preside over the demise of the internal combustion engine.' Shell, Texaco and Daimler Chrysler also left and it was disbanded in 2002.

The coalition was advised by Don Pearlman, a Washington attorney with a wide range of clients, including OPEC members. Pearlman had a reputation for his ferocious intelligence and unrivalled knowledge of the labyrinthine negotiating texts spawned by the Rio Convention. 'I respected how well he did his job,' Bob Watson, IPCC chairman at the time, told the author. Helping bring about Watson's ouster turned out to be the coalition's parting shot. There was considerable opposition to Watson's reappointment within the US business community. The incoming Bush administration was persuaded that, in O'Keefe's words, Watson was 'too much of an advocate'.

Watson won some votes, but lost those of Asia, Africa and Eastern Europe. The US had already announced its support for India's Rajendra Pachauri. The choice would have unexpected consequences. By turns aggressive and dismissive, Pachauri's reaction to widely publicized mistakes in the Fourth Assessment Report did little to enhance the IPCC's battered credibility.

In the mythology of global warming, fossil fuel firms prevented the US from joining the EU and other developed nations in leading the world to a virtuous low carbon future. There was a simpler explanation. There were never enough votes in the Senate for a simple majority, let alone the sixty votes needed to avoid a filibuster, to pass cap-and-trade. In 2003, the McCain Lieberman Climate Stewardship Act was defeated fifty-five to forty-three. By voice vote two years later, the Senate passed a resolution calling on Congress to enact a comprehensive and mandatory market-based scheme to stop or slow the growth of greenhouse gas emissions without inflicting 'significant harm' on the economy. It was a hollow statement. An amendment to the Energy Bill, calling for voluntary action sponsored by Chuck Hagel, was carried by sixty votes to twenty-nine. A competing one sponsored by McCain and Lieberman that called for mandatory caps went down by thirty-eight to sixty votes.

Two years later, the Senate debated another cap-and-trade proposal, this time targeting a sixty-three per cent reduction by 2050 (similar to the UK's Climate Change Bill). The legislation never made it out of committee.

By then, an alternative path had opened up. In a five to four judgment, the Supreme Court required the EPA to regulate greenhouse gas emissions under the 1972 Clean Air Act. 'Perhaps the most important decision ever handed down in the annals of environmental law,' Lisa Jackson, EPA administrator in the Obama administration has called *Massachusetts v. EPA*. It was also perhaps the most stupid. The Court might as well have instructed King Canute to reverse the incoming tide.

While there is a clear rationale to regulating what comes out of automobile tail pipes to reduce local air pollution, global warming is a global phenomenon. In the Court's opinion, because US motor vehicles accounted for six per cent of global emissions of carbon dioxide (four per cent of global greenhouse gas emissions, as the Chief Justice wrote in his dissent), they made a 'meaningful contribution' to greenhouse gas concentrations. 'While it may be true that regulating motor vehicle emissions will not by itself *reverse* global warming, it by no means follows that we lack jurisdiction to decide whether EPA has a duty to take steps to *slow* or *reduce* it.'[10]

Having the jurisdiction is one thing. Having the means is something else. Even if EPA regulations reduced fuel burn by fifteen per cent from 2016, and assuming US vehicle emissions remain at the same percentage of global emissions for a hundred years, it would delay the full effect of a century's worth of global warming by ten months, from January 1st 2116 to October 25th 2116.*† This assumes people don't change their behavior. If, in response to finding that a tank of gas taking them farther, Americans decided to drive more miles, the smaller the effect will be on emissions. If they don't, lower demand for oil will mean lower oil prices, making it cheaper for drivers elsewhere in the world to use their cars more and emit more carbon dioxide.

Massachusetts v. EPA revolved around whether the EPA's refusal to regulate gas emissions presented a risk of harm to Massachusetts that was both 'actual' and 'imminent'. In the opinion of the Chief Justice, arriving at such a conclusion required such an attenuated chain of causation that it went to the very outer limit of the law.

To demonstrate injury to Massachusetts, the Court cited a report by the National Research Council that identified a number of environmental changes that, in the Court's opinion, had 'already inflicted significant harms'. These included the retreat of mountain glaciers, reduction in snow cover extent,

---

* EPA regulations for new vehicles project average light vehicle emissions of 295g of CO_2 for the 2012 model year falling to 250g per mile for the 2016 model year. *EPA and NHTSA Finalize Historic National Program to Reduce Greenhouse Gases and Improve Fuel Economy for Cars and Trucks* (April 2010), p. 4.

† On the (more correct) basis of the Chief Justice Robert's dissenting opinion that US motor vehicles account for four per cent of global greenhouse gas emissions, the full effect would be postponed by less than seven months.

earlier spring melting (harmful to the Commonwealth of Massachusetts?) and accelerating sea level rise.* A 10–20 cm rise in sea levels during the twentieth as a result of global warming, had already begun to 'swallow' Massachusetts' coast land, the Court asserted.

In his dissent, Chief Justice Roberts described this as 'pure conjecture'. Boston's rising sea level had been caused by land subsidence. The anticipated sea level rise over the course of the twenty-first century could be less than the margin of error of the model used to estimate the elevation of coastal land. 'It is difficult to put much stock in the predicted loss of land.'

Should carbon dioxide – without which there would be no plant life and only single celled life-forms and funguses – be defined as a pollutant, as defined by the Clean Air Act? For the Court, the matter was straightforward: carbon dioxide is a physical and chemical substance emitted into the ambient air. By the same definition, so is the most important greenhouse gas in the Earth's atmosphere – water vapor.

On December 7th 2009 – the opening day of the Copenhagen climate conference – the Obama administration announced its endangerment finding on carbon dioxide and other greenhouse gases, as required by the Supreme Court. 'This administration will not ignore science or the law any longer,' Lisa Jackson said. The EPA was now 'authorized and obligated' to reduce greenhouse gas emissions under the Clean Air Act.

As president-elect, Barack Obama videotaped some remarks on climate change. A federal cap-and-trade system would cut emissions by eighty per cent by 2050. Not only would this save the planet, 'it will also help us transform our industries and steer our country out of this economic crisis'. With its upside down economics, the speaker might have been one of the professors of the political school of the grand academy of Lagado.

---

*   Nowhere in the Court's opinion is there any discussion of the potential benefits of longer growing seasons and stronger plant growth that might arise from having more carbon dioxide in the atmosphere.

28
HUGGING HUSKIES

There is money to be made and there are jobs to be created. To do that we must meet the challenge of Kyoto.

Tony Blair, December 14th 1997

I want us to be the greenest government ever.

David Cameron, May 14th 2010

Concluding his visit to the academy, Gulliver relates that he had seen nothing to invite a longer continuance. He began thinking of returning to England. For a traveller wearying of the policy follies of global warming, England could offer little respite.

From 2002, a Renewables Obligation was imposed on energy companies, costing energy consumers £7.3 billion to March 2011. It proved a spectacularly inefficient way of reducing carbon dioxide emissions, costing up to £481 per tonne of carbon dioxide displaced from gas-fired power stations – nearly twenty times more than the £25 per tonne estimate used by the government.

Harming the many, the obligation was highly lucrative for the few – wind farm developers and some of Britain's largest landowners. The development of off-shore wind farms would result in a windfall to one of the largest – £38 million a year to the Royal Family according to one newspaper report, as the seabed within Britain's territorial waters is owned by the Crown Estate. Nonetheless, in 2011 Prince Philip told a wind farm developer that wind farms were absolutely useless, completely reliant on subsidy and an absolute disgrace. 'You don't believe in fairy tales, do you?' the Duke asked Mr Wilmar of Infinergy, who expressed surprise at the Duke's 'very frank' views.

Earlier that year, a senior executive of a German-owned utility confirmed to the author that the Duke was right: the only economic function of wind farms was to collect tax revenues. Seen that way, wind farms are a throwback to a distant era. Tax farms were used before the English Civil War. In 1641, Parliament passed a bill confiscating the estates of some of the biggest tax farmers, later commuted to a heavy fine. They were no more popular in pre-revolutionary France; Lavoisier was guillotined for the same crime.

In June 2008, the House of Commons debated the Climate Change Bill to write into law the government's target for a sixty per cent cut in carbon dioxide emissions by 2050, something no other country was doing. In his third term, Tony Blair had come under pressure from 'The Big Ask Campaign', spearheaded by Friends of the Earth, for legislation to mandate year-on-year cuts in emissions.

The decisive pressure came from the new approach of the Conservative opposition under David Cameron. Green issues were central to Cameron's re-branding of the Conservative Party. The shift began after the 2005 election during Oliver Letwin's tenure as shadow Defra secretary. In October, Letwin and his Liberal Democrat counterpart, Norman Baker, issued a joint letter urging 'independent verification' of 'year on year carbon reduction requirements'.

One of the progenitors of the poll tax in the 1980s, Letwin was also David Cameron's chief policy adviser in his campaign for the Conservative Party leadership. On a visit to the Eden Project in Cornwall at the end of that month, Cameron declared: 'With global poverty and terrorism, climate change is one of the three greatest challenges facing mankind today.' The Conservatives should 'show real leadership on this vital issue' and committed the party to annual targets underpinned by a statutory framework if he won.

Once elected, Cameron consistently outbid Labour on commitments to rapid decarbonization. In March 2006, he criticized Gordon Brown: 'In a carbon-conscious world, we have got a fossil fuel Chancellor.' A month later, he flew to Norway to be filmed hugging a huskie and inspecting a retreating glacier, a trip that helped define his leadership.

In August, a letter from a cross-party group of MPs called for a bill setting annual targets to be introduced in the next parliamentary session. In a coordinated move, Cameron urged cuts in emissions of at least sixty per cent by 2050, without any apparent attempt to estimate the cost. Certainly no cost estimates were made public.

Labour felt obliged to respond. It committed to a Climate Change Bill in the Queen's Speech of November 2006. Again, Cameron upped the stakes. 'I hope that it will be a proper bill, and not a watered-down bill,' he said. 'Government have got to give a lead by setting a proper framework; that must mean an independent body with annual targets and an annual report from Government on its progress.' This was much more interventionist than Blair, who in February 2007 warned of the impracticality of annual targets. The bidding war was now in full swing.

The bill was introduced in the Lords in November 2007. There was little real debate. It passed its Third Reading without a division on March 31st 2008 and reached the Commons with a strong cross-party consensus behind it, where there were only a handful of dissenters. The most long-standing of them was Andrew Tyrie, Conservative MP for Chichester. He had been skeptical about the economics of rapid decarbonization since early 1990 when, as John Major's advisor in the Treasury, he had been involved in an inter-departmental working group, chaired by the deputy chief economic advisor, John Odling Smee. In 2005 and 2006, Tyrie tried to dissuade the front bench from making expensive commitments to legislate on carbon reduction, to little avail.

Tyrie was determined to register his dissent and pressed ahead with opposition at the bill's Second Reading by ensuring there was a division. But he was able to rally the bare minimum needed to force a vote. No other country would be foolish enough to consider such a measure, he said in the debate: 'Although UK emissions will fall, they will reappear, probably at even higher levels, as the industries that we close down with our higher costs base reopen in China and elsewhere.' The bill would give China every incentive to delay an international agreement. 'Why should they rush to agree anything when they can acquire our industrial base and those of other countries silly enough to go it alone?' Tyrie enquired.

The all-party consensus resumed as soon as Tyrie sat down. Labour's Desmond Turner said Tyrie was a member of the Flat Earth Society. 'The issue is not counting beans but the survival of the species,' Turner declared.

Shortly before the debate, Peter Lilley obtained a copy of the bill's regulatory impact statement. He was stunned. At first, he thought he had been reading the figures back to front. At best, the benefits might exceed the costs by £52 billion; at worst, the costs would be £95 billion greater than the benefits – and those excluded transitional costs of up to one per cent of GDP until 2020 and the leakage of carbon-intensive business abroad. It exposed the black hole at the heart of the bill. It was probably the most expensive legislation ever put before Parliament.

Lilley intervened when the junior environment minister, Phil Woolas, who subsequently became the first MP to be disqualified by an election court for lying about his opponent, claimed the benefits would exceed the costs. 'He makes an interesting point and we will be able to debate it,' the minister replied. 'It is not a fundamental principle.'[1] The cost estimates were not debated then and the bill went through the whole of its Commons committee stage without any discussion of its supposed costs and benefits.

The whips on both sides packed the debate and told any doubters to abstain rather than force a division. Pressure was brought to bear on the remaining dissenters. At both Second and Third Reading, just five MPs voted against the bill, including the two tellers for the noes. In addition to Tyrie and Lilley, Christopher Chope, Philip Davies, and Ann Widdecombe constituted the entire complement of dissenters. At the Third Reading, the Whips packed the debate, so that Tyrie, the leading dissenter, was not called to speak.

During passage of the bill, the 2050 target was raised from sixty per cent to eighty per cent. Given that the target has a base year of 1990, and using reasonable growth assumptions, an eighty per cent reduction is equivalent to about ninety per cent on levels that would have been achieved in 2050; in other words, almost total decarbonization.

The eighty per cent target required a new regulatory impact assessment. The government produced a new estimate in March 2009. The benefits of cutting emissions had risen nearly tenfold from those estimated when the bill was presented to Parliament – to £1,020 billion. Lilley wrote to Ed Miliband, the new climate change secretary, a Cabinet post created six months earlier, asking him to confirm that the cost of the Climate Change Act amounted to £16,000 to £20,000 for every UK household.

Miliband replied a month later. Climate change was a potential public health catastrophe. The new estimates were predicated on a global agreement to cap emissions. 'Showing leadership through the Climate Change Act, the UK will help to drive a global deal,' Miliband wrote, a claim that would be put to the test less than eight months later at Copenhagen.

The impact statement had been frank about the consequences of not reaching agreement. 'The UK continuing to act while the rest of the world does not, would result in a large net cost for the UK,' as the benefits of UK action would be distributed around the world, but the UK would bear all the costs.[2] In his letter, Miliband rejected Lilley's claim of the cost of the legislation to UK households and distorted what the impact statement said. 'The impact statement shows that the benefits to UK society of successful action on climate change will be far higher than the costs,' Miliband claimed.

He was wrong. The impact assessment's £1,000 billion estimate was of the benefits to the world of UK action within the context of an international agreement to cap emissions, not the benefits to the UK. The study had not examined the possible local effects of global warming on Britain, which it is quite plausible to believe might be beneficial to the UK even if it was harmful to the rest of the world.

Miliband should have known that what he had written was untrue. As the responsible minister, six weeks before writing to Lilley, he had certified that he had read the one-hundred-and-twenty-six-page document and that it represented a 'reasonable view' of the costs and benefits of the legislation.

Subsequently Lilley raised the matter with Ed Davey, Miliband's successor but one as climate secretary. 'We are not aware of evidence that would have allowed us to have made a reliable estimate of the distribution of those benefits, including the UK,' Davey replied in April 2012.

Just as the British government became a climate change outrider, Britain's largest oil company adopted the opposite strategy to its American peers. In May 1997, three weeks after Tony Blair's election, BP chief executive John Browne signaled the new strategy in a speech at Stanford. While all science is provisional, Browne declared, 'It falls to us to begin to take precautionary action now.' 'We came out of denial,' he said of the speech on its tenth anniversary. Other companies had followed, 'But the old church is now a pretty small place.'

Five months later, Browne addressed the second annual Greenpeace business conference. BP's close relationship with Greenpeace was evidence of the deepening relationship between the environmental movement and business. 'We support that effort, which can only be beneficial,' Browne said. BP hired Tom Burke, formerly of Friends of the Earth, as its environmental adviser. Burke is at the nexus of green power in Britain, with its overlapping circles of NGOs (Burke had also been director of the Green Alliance for nine years), government (special adviser to three Conservative environment secretaries), academia (visiting professor at

Imperial and University Colleges, London), and business (later moving from BP to be Rio Tinto's environmental policy adviser).

At the 1998 Davos World Economic Forum a month after Kyoto had been agreed, Browne made an impassioned plea for other companies to support it. When he'd finished, the Global Climate Coalition's O'Keefe got up. The US petroleum industry would not support Kyoto and the US would not ratify it, O'Keefe said. Browne turned his back on him. A London colleague called to say that he had never seen Browne so angry.

The following year, Browne was at the UN headquarters in New York to receive an Earth Day award. 'People want large and successful corporations to use their skills and their know-how to address the environmental agenda,' he told the ceremony.

After acquiring three oil companies (Amoco in 1998; ARCO in 1999; and Burmah Castrol in 2000), BP adopted the tagline 'Beyond Petroleum 2001.' Apparently BP was no longer an oil company. It was now an energy company. Extensive internal polling after the Amoco merger showed that sixty per cent of BP staff saw addressing the environmental issue as the single most important issue in defining the quality of the company they worked for.[3] Safety did not top the list. Perhaps it was a warning of what was to come.

Not so Exxon Mobil. It was an oil company and continued to behave like one. After the demise of the Global Climate Coalition, Exxon Mobil continued to support groups that challenged the science and economics of global warming. In its 2005 Corporate Citizenship Report, Exxon Mobil described the IPCC's conclusions on the attribution of recent warming to increases in greenhouse gases as relying on expert judgment rather than objective, reproducible statistical methods.

This provoked an irate letter from the Royal Society's Bob Ward accusing Exxon Mobil of being 'very misleading'. The IPCC's 'expert judgment' was, Ward wrote, based on objective and quantitative analyses. Despite Ward's protestations, at issue was not whether the IPCC's conclusions were based on subjective judgment, which was clearly the case, but the scientific standing of such judgments. Even so, there is quite a lot of make-believe in the Royal Society's characterization of the IPCC's conclusions, especially as it relates to the IPCC's Summary for Policy Makers. Britain's former top civil servant Andrew Turnbull said they would be more accurately described as a summary *by* policymakers, not for policymakers:

> The scientists prepare a draft but this is redrafted in a conclave of representatives from the member Governments, mostly officials from environment departments fighting to get their ministers' views reflected.[4]*

---

* Regarding the summaries for policymakers in the Third Assessment Report, the Royal Society had stated in March 2005: 'Each sentence of which was agreed sentence by sentence at meetings of the governments from member countries of the IPCC.' The Royal Society, *A guide to facts and fictions about climate change* (March 2005).

Ward went on to express concern about Exxon Mobil's support for skeptic organizations that were 'misinforming' the public on the science. According to the *Washington Post*, the 'testiest moment' at Exxon Mobil's 2006 annual shareholder meeting came when chief executive Rex Tillerson was questioned about the company's skeptical stance in the face of a growing scientific consensus. 'Scientific consensus' was an 'oxymoron', Tillerson replied.

By then, Exxon Mobil was preparing to change its posture. It wasn't in business to be on the losing side in a battle of ideas. In 2005, it decided to cease funding groups such as the Competitive Enterprise Institute, one of the most effective skeptic organizations. 'The fact that we were supporters of some of those groups had become a real distraction to the issue at hand, which is how do we produce the energy the world needs without more greenhouse gas emissions,' Kenneth Cohen, the company's VP for public affairs, told journalists in February 2007. According to the Natural Resources Defense Council's Dan Lashof, 'They found that it was untenable to be in a position of casting doubt on whether global warming is happening and whether pollution is responsible for that.' The episode is evidence refuting the widely held belief that fossil fuel companies are to blame for America not signing up to binding emissions controls.

BP and Exxon Mobil disagreed on virtually every other aspect of global warming. BP's Browne supported Kyoto; Exxon Mobil's Tillerson argued that every nation would need to participate ('developed nations cannot go it alone'). Browne supported emissions trading ('one of the most promising of all the options'). Tillerson thought it would result in volatile prices for emissions allowances, create economic inefficiencies and invite market manipulation. Instead Tillerson supported a revenue-neutral carbon tax.

They differed on investment strategy. BP made a big bet on solar. In 1998, Al Gore opened BP's first solar manufacturing facility in the US. The next year, BP bought Solarex to become the world's largest solar energy company. BP's solar capacity increased from just over twenty megawatts in 1997 to two hundred in 2006; in 2007, Browne said that he anticipated this would rise to over seven hundred megawatts. 'We plan to invest at $1 billion per annum in alternative energy sources such as these,' Browne pledged.

Four years later, BP announced the complete closure of the business. 'We have realized that we simply can't make any money from solar,' the company told the *Financial Times* in December 2011.

Exxon Mobil stuck closer to its hydrocarbon DNA. In July 2009 it signed a deal with Synthetic Genomics of La Jolla, California, to research and develop biofuels derived from algae by using sunlight to convert carbon dioxide into oils and long-chain hydrocarbons. Later in 2009, it deepened its commitment to extracting hydrocarbons from shale rock with the acquisition of XTO Energy.

What explains the different responses of the two? Leadership doubtless played a part. But perhaps the most important is the different political environment the two operated in. No American oil company has as intimate relationship with the

US federal government as BP's with the British government. Ever since 1913, when Winston Churchill took a controlling stake in BP's original predecessor, the interests of BP and the British state were seen in Whitehall as virtually indistinguishable.

In the 2000s, BP was Blair Petroleum. One of Blair's top aides, Anji Hunter, left Downing Street to work for Browne. It would have been unthinkable for BP to have opposed Kyoto and the policy of the British government. For a time, being ahead of the curve on climate change appeared commercially smart.

Investors thought so, too.

Aggressive cost cutting – necessary to compensate for BP's green investments that yielded no return – helped propel BP's share price upwards. From the date of Browne's Stanford speech in May 1997 to April 20th 2010, BP's share price rose from 359.5p ($5.84) to 655.4p ($10.00) – a seventy-one per cent increase in dollar terms, compared to a thirty per cent increase in Exxon Mobil's stock price. After the markets closed on April 20th, there was an explosion on the Deepwater Horizon drilling rig in the Gulf of Mexico.

It was BP's third – and most serious – accident caused by safety shortfalls, the first being an explosion at BP's Texas City refinery (2005) and then a ruptured pipeline spewed oil across Alaska's North Slope (2006).

Six months after the Deepwater Horizon explosion, BP's share price closed at 377.2p ($5.95), giving up ninety-seven per cent of the dollar gain it had made since Browne's Stanford speech.

29
DANGEROUS CLIMATE CHANGE

I'm more worried about global warming than I am of any major military conflict.

Chief UN weapons inspector Hans Blix, March 13ᵗʰ 2003

The world has already reached the level of dangerous concentrations of carbon dioxide in the atmosphere and immediate and very deep cuts in the pollution are needed if humanity is to survive.

Rajendra Pachauri, chairman of the IPCC, January 2005

The prodigious expenditure of effort and resources corralled by governments trying to cut emissions of carbon dioxide yielded puny results. Unable to make good on their promises to cut carbon dioxide emissions, and as global temperatures remained more or less flat, the rhetoric of alarm was ratcheted up.

In 2000, Annex I carbon dioxide emissions were 538 million tonnes lower than in 1990 – a fall of 3.6 per cent. Emissions from the ex-Communist countries of the Soviet bloc more than accounted for this – down 1,752 million tonnes – or forty-four per cent (a figure which excludes the former East Germany).[1] The collapse of communism turned out to be the most effective decarbonisation policy of all time.

In 2007, Annex I carbon dioxide emissions were 593 million tonnes higher than 2000 – a rise of four per cent. The onset of the global recession cut Annex I emissions by 1,3650 million tonnes (a nine per cent fall) – the second most effective global warming policy. Virtually all this was offset by increased non-Annex I emissions. The net effect of the worst global recession since the 1930s was a reduction of less than twelve million tonnes of carbon dioxide.

Kyoto's bifurcation between 'rich' North and 'poor' South actually helped push up emissions. According to Dieter Helm, part of Britain's apparent reduction in emissions came about through de-industrializing. 'Driving up our energy prices drives energy-intensive production overseas,' Helm wrote in February 2012.[2] This led to higher emissions. In the three years to 2009, non-Annex I countries emitted fifty-seven per cent more carbon dioxide (642g) to generate one kWh of electricity than Annex I nations (408g). And the trends were diverging. While Annex I countries were reducing the carbon-intensity of electricity production – emissions per kWh fell by eight per cent between 2000 and 2009 – non-Annex I countries increased their emissions per kWh by 3.5 per cent.

Doing nothing would have been more rational than what governments actually did. Rationality was not the yardstick. During the middle years of the first decade of the new century, climate change became *dangerous* climate change.

To UN weapons inspector Hans Blix, climate change presented a bigger risk than a major war. In his 2005 valedictory address as president of the Royal Society, Lord May, a zoologist, compared climate change to weapons of mass destruction. To Sir Richard Mottram, a former permanent secretary at Britain's Ministry of Defence and chairman of the Joint Intelligence Committee, 'Climate change is a vastly greater threat to civilization than terrorism.'* In January 2004, Sir David King, Tony Blair's chief scientific adviser, described climate change as a more serious threat than terrorism. Eighteen months later, suicide bombers killed fifty-two people and injured seven hundred on the London Underground and on a double-decker bus.

These are all examples of climate change derangement syndrome: otherwise perfectly sane people making statements that in any other context would be regarded as absurd. According to John Ashton, the UK's first climate change envoy, 'There is every reason to believe that as the twenty-first century unfolds, the security story will be bound together with climate change.' The Foreign Office was given the strategic goal to create the right political conditions to reach international agreement on tackling global warming. It was going to do this by persuading world political leaders that 'a stable climate is essential for their national security and prosperity' – similar language to Tony Blair's justification of the Iraq War. It was downgraded after Labour lost the May 2010 election and William Hague became foreign secretary.

James Lee of American University's Inventory of Conflict and Environment Project predicted that people would respond to climate change 'by building bomb shelters and buying guns'. The prospect of war between the US and Canada over rights to the Northwest Passage was not farfetched, Lee wrote in the *Washington Post*. 'Anyone convinced that the United States and Canada could never come to blows has forgotten the war of 1812.' When the article appeared in January 2009, Americans were too busy buying record numbers of snow blowers to cope with that winter's heavy snowfalls to be worrying about invasion from the north.

A 2003 study for the Pentagon hyped up the threat to 'dramatize the impact' of global warming.[3] There was too much 'Imagining the Unthinkable' – the study's subtitle – and a dearth of 'Recognising the Obvious'. Abrupt climate change 'could potentially de-stabilise the geo-political environment, leading to skirmishes, battles and even war due to resource constraints'. Drawing on archaeological and ethnographical data from pre-industrial eras, it offered a doom-laden, neo-Malthusian prognosis of resource shortages leading to war. 'Humans fight when

---

* For a semi-anonymous civil servant, Mottram shot to national fame for telling a fellow civil servant: 'We're all fucked. I'm fucked, you're fucked, the whole department's fucked. It's been the biggest cock-up and we are all completely fucked.' He wasn't talking about climate change but the resignations of his department's senior media handlers. David Graves and George Jones, 'Sixsmith stands by story as infighting continues' in the *Daily Telegraph*, February 25th 2002.

they outstrip the carrying capacity of their natural environment. Every time there is a choice between starving and raiding, humans raid.'

The Pentagon's 2010 Quadrennial Defense Review stated that 'climate change and energy are two key issues that will play a significant role in shaping the future security environment'. Intelligence assessments indicated that climate change could have 'significant geopolitical impacts' around the world by acting as an 'accelerant' of instability and conflict.

So to flood and drought, famine and pestilence was added the spectre of resource wars. Just as predictions of food shortages become self-fulfilling, threats to the international system and world peace are more likely to come about because of global warming policies than global warming itself. In potentially the most destructive example of the Global Warming Policy Paradox, global warming policies risk undermining the international trading system and the economic structure of mutual advantage that would render a resource war a self-defeating strategy.

Digging around ancient burial sites for potential causes of war in the twenty-first century, the lessons from the single most important event of the twentieth century were lost. In an open global trading system, wars of conquest are economically insane. The Second World War was a struggle for control of territory and resources, Gerhard Weinberg, one of the conflict's foremost historians, has written.

Germany and Japan didn't need to control a resource to benefit from it. The post-war performance of the German and Japanese economies disproved the efficacy of militarist/economic resource war theories of conquest and national survival. Neither country needed to conquer to obtain resources to prosper. Both performed incommensurately better as part of the international trading system than they had in attempting to overturn the world order.

Thus the first lesson is economic. Trade is based on voluntary exchange, in which both sides have an incentive to maintain the exchange. The gains from trade are greater than those from plunder. The plunderer, in the act of expropriation, destroys the incentive of the people who are plundered to keep producing. After the Nazi-Soviet Pact of 1939, the two totalitarian regimes agreed to deepen economic ties. For around a year, the Soviet Union accounted for the bulk of German overseas trade, supplying critical war materials that the Nazi war machine then used against the Soviet Union. Indeed, Germany obtained more from the Soviet Union through trade than it did from pillage.

The second lesson is geo-strategic. Power is not based on control of resources, but the ability of an economy to add value to those resources. The Allies paid for their war supplies. Where they could, the Axis plundered.

The third is political. A country can't invade and then trade. An economic policy based on conquest will result in the exclusion of that country from the international trading system. Conquering countries to get hold of primary resources – food, minerals or energy – with the idea of re-exporting them in manufactured goods is self-evidently absurd.

These incentives and penalties pre-suppose an international trading system. Countries which are part of it enjoy the benefits of membership. Those which are not, do not. Its existence therefore constitutes a powerful incentive for good behavior and for states to resolve their differences peaceably.

The post-war trading system was the product of one of the greatest acts of statesmanship in history. It was conceived by the US during the Second World War. Cordell Hull, President Roosevelt's secretary of state, was determined that the disasters of protectionism, which he saw had imperilled peace and prosperity, would not be repeated. American leadership had to overcome stiff opposition from Britain, which wanted to keep imperial preference, and protectionist interests at home, to incorporate America's former enemies into an expanding international trading system.

If Roosevelt and Hull were right about the link between peace, prosperity and an open trading system, global warming policies that fragment global markets undermine world peace. At an EU summit in March 2008, the EU threatened the US and China with trade sanctions if they didn't commit to ambitious cuts in greenhouse gases. Should international negotiations fail, 'appropriate measures' could be taken to protect European industry, European leaders declared. France's President Sarkozy went further. A mechanism was needed to 'allow us to strike against the imports of countries that don't play by the rules of the game on environmental protection'.

The EU's 2008 Climate Action and Renewable Energy Package provides one. In the absence of an effective international agreement, the European Commission is to bring forward proposals for green tariffs equivalent to the price of emissions credits under the Emissions Trading Scheme, even though the EU allocated credits worth tens of billions of euros for free. In 2012, EU plans to extend its Emissions Trading Scheme to non-EU airlines drew threats of retaliation from China. The continent which started the Second World War hadn't learnt its lessons.

On November 18th 2004, UN secretary-general Kofi Annan received the Russian Federation's instrument of ratification. Ninety days later, on February 16th 2005, the Kyoto Protocol came into force. The EU had traded Russia's membership of the World Trade Organisation, ironically swallowing the continuation of energy subsidies for Russian businesses, for its desire for Kyoto to take effect.

Russia's ratification unleashed a wave of climate change alarmism. Finding a successor to Kyoto became the focus of international climate change negotiations. The thirty-four months to the Bali conference in December 2007 marked the most febrile phase of the fever. The UN Environment Programme's Klaus Töpfer spoke of a terrifying vision of a 'planet spinning out of control'. Global temperatures could rise by 5.8°C by the end of the century, Töpfer said, a recently published report in *Nature* claiming the rise might be even higher as early as the middle of the century.

Kyoto going live isolated two non-ratifiers – George W. Bush and Australia's John Howard. 'Until such time as the major polluters of the world, including the United States and China, are made part of the Kyoto regime, it is next to useless and indeed, harmful, for a country such as Australia to sign the Kyoto Protocol,' a defiant Howard told the Australian Parliament. 'The more I studied it, the more I became convinced that Kyoto was very Eurocentric,' Howard told the author in 2011, down to the choice of the 1990 base year. The collapse of the command economies meant that the progressive Left had to find other causes to fight, climate change fitting the bill, becoming a substitute religion.

Unlike Australia, US participation in a son-of-Kyoto was essential. The UK held the presidency of the G8 in 2005 and between July and December also the EU presidency. The opportunity for Tony Blair's brand of messianic Salvationism was irresistible.

Before Russian ratification, Blair focused his efforts on Putin. In July 2004, the Russian Academy hosted a seminar in Moscow on climate change and Kyoto with a British delegation led by Sir David King. It was an ill-tempered affair. At a press conference, Andrei Illarionov, Putin's principal economics adviser, denounced the British and singled out King. 'At least four times during the course of the seminar ugly scenes were staged that prevented the seminar from proceeding normally.'

By this stage, Illarionov was on the losing side inside the Kremlin. Illarionov criticized Putin's decision to ratify Kyoto as 'motivated purely by politics – not by science or economics'.

Having squared Putin, Blair switched his focus. 'To describe George as a sceptic on climate change would be an understatement,' Blair wrote in his memoirs. 'As time progressed he shifted his thinking, but did so too slowly – a quality of conservatives I don't admire.'[4] As Paula Dobriansky, undersecretary of state for democracy and global affairs, put it, Blair and Bush had a 'vibrant and active exchange' during the British G8 presidency.

Despite Blair's characterization, Bush showed a stronger grasp of the problem than Blair. His administration's policy was a determined, cogent and well-developed effort to press the reset button on the two components of the international climate change negotiations that were visibly failing – its heavy reliance on emissions reductions and the Annex I bifurcation. What Blair and other world leaders achieved was to push Bush – against his better instincts – into accepting the principle of Kyoto-style emissions caps.

Bush's attempt to bridge the Annex I bifurcation was taken forward by his successor. It was defeated at Copenhagen. But the policy Bush inherited had come to a dead end. He had to try something different. To have carried on where Clinton had left off would have meant falling into Einstein's definition of insanity – doing the same thing over and over again expecting a different result.

The Bush reset began with a speech in June 2001, just before his first trip as president to Europe. 'America's unwillingness to embrace a flawed treaty should not be read by our friends and allies as any abdication of responsibility,' Bush said. 'To the contrary, my administration is committed to a leadership role on the issue of climate change.'[5] He pledged to work within the UN framework – and elsewhere – to develop an 'effective and science-based' response to global warming. He also spoke of the uncertainties:

> We do not know how much effect natural fluctuations in climate may have had on warming. We do not know how much our climate could, or will change in the future. We do not know how fast change will occur, or even how some of our actions could impact it.[6]

From May 2001, Paula Dobriansky was responsible for leading the US team in the climate change negotiations, filling the post previously occupied by Tim Wirth and Frank Loy. On taking up her position, Dobriansky talked to both, the latter being particularly engaged (the two had previously worked together and had the same special assistant, Nigel Purvis, at the State Department). There had been questioning of the science at Cabinet-level meetings at the White House. So the National Academy of Sciences was brought in. Working closely with Dobriansky in the State Department, it assessed all types of issues relevant to climate change and ended up restating the consensus view.

The initial phase came to an end on 9/11, when the focus switched to counter-terrorism. Subsequently a two-pronged strategy was developed: prioritizing technological solutions and developing complementary diplomatic avenues to those established under the UN climate change convention. By 2007, the Bush administration had spent $37 billion on climate science, technology and incentive programs. In parallel, it developed a series of international initiatives, including the Asia-Pacific Partnership on Clean Development and Climate, and collaboration on projects such as carbon sequestration and the development of hydrogen fuel-cell technology. It culminated in May 2007 with the inception of the Major Economies forum to bring together the world's largest emitters.

American negotiators were frustrated at the incapacity of the UN process to produce results – 'very challenging', as Dobriansky called it. Indeed, the Major Economies initiative was similar to that set out in a June 2007 article by Todd Stern, a former Clinton administration official and later Obama's climate envoy, and William Antholis of the Brookings Institution. They called for an ecological E8 of world leaders, half from the developed world, and half from developing nations.

In September, Boyden Gray, America's ambassador to the EU, wrote in the *Financial Times* of the 'sclerotic UN process' hobbled by the participation of nearly two hundred countries. It had been US leadership of a small group of major countries that had driven through the Montreal Protocol and delivered ten times the greenhouse gas reduction of Kyoto, Gray reminded suspicious Europeans.

However, the differences between the Montreal and Kyoto protocols were substantive as well as procedural. When the Montreal Protocol was concluded, there were ready substitutes for CFCs and the main emitters of ozone-depleting substances were the industrialized nations. Attempting to repeat the success of the Montreal process pre-supposed that the principal emitters of carbon dioxide were willing and able to cut their emissions. With China adding new coal-fired power stations by the week and rapidly overhauling the US as the number one producer of carbon dioxide, the big question was: supposing China was not willing?

For Tony Blair, the obstacle wasn't China and the other large emerging economies. It was President Bush. 2005 was the year to heighten the pressure ahead of July's Gleneagles G8.

At the beginning of February, the British hosted a three-day conference, 'Avoiding Dangerous Climate Change', at the Met Office's headquarters in Exeter. There were two ghosts at the party: Izrael and Illarionov. 'Anyone who is frightened about the prospect of global warming is welcome to come and live in Siberia,' Illarionov told a journalist.

Other participants stuck to the script. Chris Rapley, director of the British Antarctic Survey, was alarmed by the melting of the West Antarctic ice sheet.* Antarctica was no longer the 'slumbering giant' of the Third Assessment Report: 'I would say that this is now an awakened giant.' Bill Hare of the Potsdam Institute told delegates that the 3°C increase expected by 2100 would kill off all the frogs and spiders in South Africa's Kruger National Park and leave more than 3.3 billion people living in countries suffering large crop losses. Lord Oxburgh, chairman of Shell's UK holding company, warned that unless governments took urgent action there 'will be a disaster'.

Was the amount of carbon dioxide in the atmosphere reaching danger level? The previous month, Rajendra Pachauri was unequivocal. Carbon dioxide concentrations had already reached dangerous levels and called for immediate and 'very deep' cuts in the pollution if humanity is to 'survive', he told a UN conference in Mauritius.

Collectively the scientists and assorted experts gathered in Exeter to talk about avoiding dangerous climate change weren't so obliging. 'That's a value judgment to be made by policymakers,' said Bert Metz of the Netherlands Environmental Assessment Agency. The conference concluded that the risks of global warming were 'more serious' than previously thought. 'Avoiding more serious climate change' didn't have quite the same ring. In 2006, Nicholas Stern would be more reliably on-message with forecasts of catastrophes, generating many times the PR impact of the Exeter conference.

Writing in the *Times of India* later that month, Swaminathan 'Swami' Aiyar sounded a note of caution. The best scientific assessment says global warming

---

*   It was hearing from the British Antarctic Survey about the ozone layer that had persuaded Margaret Thatcher to throw her support behind international efforts to cut emissions of CFCs and other ozone-depleting substances.

is happening, 'yet never in history have scientists accurately predicted what will happen one hundred years later'.[7] He had nearly been convinced by photos of the rapid retreat of an Andean glacier publicized by Greenpeace. When Swami visited it, he found others had shown little movement and one glacier had advanced. Greenpeace and other ecological groups had well-intentioned people with high ideals. But as crusaders, 'they want to win by any means, honest or not. I do not like being taken for a ride, by idealists or anyone else'.

A month before the Gleneagles G8, the national science academies of the G8 nations plus those of Brazil, China and India issued a joint statement:

> The scientific understanding of climate change was sufficiently clear to justify taking prompt action to reduce net global greenhouse gas emissions.[8]

The academies told the G8 leaders that they should 'acknowledge that the threat of climate change is clear and increasing'. At the beginning of July, the Royal Society published a sixty-page report saying that increased levels of carbon dioxide would cause the oceans to acidify.

Bush was being set up in a pincer movement between the science and the threat of G8 isolation. Although he acknowledged the threat of global warming, he qualified it by reference to the scientific uncertainties, an escape hatch the science academies were trying to close off. 'Tony Blair is contemplating an unprecedented rift with the US over climate change at the G8 summit next week, which will lead to a final communiqué agreed by seven countries with President George Bush left out on a limb,' the *Guardian* reported. France and Germany preferred an unprecedented split communiqué to a weak one.

According to the report, the US objected to drafting that described climate change as a serious and long-term challenge – wording that got into the final communiqué – and that which said there was strong evidence that 'significant' global warming was occurring with human activity contributing to it – wording that did not. The US made a significant concession in accepting that 'we know enough to act now' to justify action to stop and then reverse the growth of greenhouse gases. Inch by inch, the Bush administration was being cornered into accepting the principle of emissions caps.

If 1988 was global warming's *annus mirabilis*, 2007 was the *ne plus ultra*, with a concatenation of events culminating in Bali at the year's end. Australia's voters obliged, voting out the other Kyoto hold-out. Nature was unbiddable. Observed global temperatures stubbornly showed no discernible upward trend since the turn of the century.

At the beginning of February 2007, the IPCC released a twenty-page summary of its Fourth Assessment Report. It declared global warming 'unequivocal' and human activity its main driver. Compared to the 2001 Third Assessment Report, the IPCC raised its confidence level in its projections from 'likely' (meaning sixty-six to ninety per cent) to 'very likely' (better than ninety per cent), although there exists no empirical means of verifying either the forecasts or the confidence levels surrounding them.

'February 2nd will be remembered as the date when uncertainty was removed as to whether humans had anything to do with climate change on this planet,' Achim Steiner, Töpfer's successor at UNEP, claimed. 'The evidence is on the table.' In reality, the 'evidence' was the product of three days and nights of wrangling between teams of government officials from more than a hundred countries and the report's lead authors, the *New York Times* reported.

A month later, the EU agreed a 2020 package to cut emissions to eighty per cent of 1990 levels by 2020 and to derive twenty per cent of its energy from renewable sources by the same year. 'We can avoid what could well be a human calamity,' said Angela Merkel, chair of the two-day summit.

Groundbreaking, bold and ambitious, Blair described the deal. That was certainly true for Britain and came as a surprise to the rest of the British government. They thought Blair was committing Britain to deriving fifteen per cent of its *electricity* production from renewables, the maximum amount thought possible. Instead Blair committed Britain to fifteen per cent of its total *energy* production, including home heating and transport, from renewables. Sometimes important details slipped Blair when he was after the big picture. He had not known whether the forty-five-minute readiness-for-use claim in the intelligence dossier on Iraqi weapons of mass destruction applied to battlefield or strategic weapons. Had Blair made a similar slip at his final EU summit as PM? In 2011, Tony Blair's office told BBC's *Panorama* that the decision hadn't been a gaffe but a decision to protect the environment and help energy security.

On September 24th, the UN convened a high-level meeting in New York. Bush gave it a miss. Arnold Schwarzenegger took centre stage. California was pushing the US beyond debate and doubt to action. The responsibility of all nations was 'action – action, action, action'. The pressure was piling up on Bush, as Britain's environment minister Hilary Benn made clear. The US had to end its opposition to mandatory caps on emissions. 'It is inconceivable that dangerous climate change can be avoided without this happening,' Benn said.

Among the eighty heads of state and governments, there was a lone dissenting voice – Václav Klaus, president of the Czech Republic. 'The risk is too small, the costs of eliminating it are too high and the application of a fundamentalistically [sic] interpreted "precautionary principle" is a wrong strategy,' Klaus told the conference.[9] Mingling with other world leaders, several congratulated Klaus for speaking out and said how much they agreed with him.

In the debates on global warming and environmentalism, Klaus is the anti-Gore. The two are a study in contrasts: Gore, the Southern Baptist preacher invoking the terrors of the Earth if mankind did not repent from its breach with nature, in Bali beseeching delegates to find grace and feel joy; the other, speaking in gentler cadences and the precision of a former econometrician, the central European who had learnt the value of freedom and classical liberalism from its absence in post-war Czechoslovakia.

No other world leader challenged Gore. They first crossed swords in a televised debate in February 1992 during the run-up to the Rio Earth Summit. 'I disagreed with almost everything he was saying at that time,' Klaus wrote later. Subsequently Klaus described a lecture by Gore as 'utterly absurd' and 'scaremongering'.

More than his disagreement with the scientific consensus and the economics of the proposed solution, Klaus's opposition is philosophical. At stake was human freedom. 'If we take the reasoning of the environmentalists seriously, we find that theirs is an anti-human ideology,' Klaus wrote in 2008. 'It sees the fundamental cause of the world's problems in the very existence of *homo sapiens*.' Socialism was no longer the greatest threat to freedom, democracy and the market economy, Klaus argued. It was 'the ambitious, arrogant, unscrupulous ideology of environmentalism'.

Three days after the UN conference, the US hosted a meeting of the sixteen country Major Economies grouping, including seven non-Annex I nations. 'We've come together today because we agree that climate change is a real problem and that human beings are contributing to it,' secretary of state Condoleezza Rice told the gathering. The next morning, President Bush came to the State Department to address the delegates. He looked exhausted, stumbling over the names of the key people in the forthcoming climate conference in Bali. A new approach was needed, Bush said. It should involve the world's largest greenhouse gas emitters, developed and developing nations alike:

> We will set a long-term goal for reducing global greenhouse gas emissions. By setting this goal, we acknowledge there is a problem. And by setting this goal, we commit ourselves to doing something about it.[10]

His pledge was a substantial concession. It didn't earn him any applause from the delegates.

What did was Bush's assurance that the US would advance negotiations under the UN climate change convention. Delegates applauded for opposite reasons. The Europeans were wedded to belief in the efficacy of multilateral institutions. At the UN conference, Angela Merkel had spoken of the centrality of the UN process and called for global emissions to be halved by 2050 – seemingly oblivious that the UN process was incapable of delivering such commitments. Non-Annex I delegates applauded because the UN process provided them with the surest guarantee of not having to control their emissions while minimising any political fall-out.

On October 12th, the Norwegian Nobel Committee announced it was awarding the 2007 Nobel Peace Prize jointly to the IPCC and Al Gore. The committee wanted to contribute to a sharper focus on the processes and decisions that appear to be necessary to protect the world's future climate, and thereby to reduce the threat to the security of mankind.

A month later in Valencia, Spain, UN secretary-general Ban Ki-moon launched the IPCC's Fourth Assessment Report. 'Already, it has set the stage for a real breakthrough,' Ban said. At the UN meeting in September, political leaders had been clear: 'We cannot afford to leave Bali without such a breakthrough.'

On November 24th, Australians went to the polls and delivered their Bali breakthrough. Climate change was a 'perfect storm', according to Howard. It hadn't been much of a political issue until 2006. There was a long drought and an early start to the bush fire season. The Stern Review generated a lot of publicity. So did Al Gore's *An Inconvenient Truth*, which Howard thought 'spiced with attacks' on the Bush administration.

Howard tried to adjust to the new political climate in a January 2007 speech. Describing himself as a 'climate-change realist', Howard said he accepted the broad theory about global warming. 'I am skeptical about a lot of the more gloomy projections,' he told the National Press Club in Canberra.

Howard was not going to surf the climate change wave like Labor's Kevin Rudd. He rejected Rudd's claim that climate change was the overwhelming moral challenge facing Australians. 'It de-legitimizes other challenges over which we do have significant and immediate control.' Neither should Australia set a target based on the needs of other countries. 'I will not sub-contract our climate change policy to the European Union,' Howard said, rejecting Labor's call for a sixty per cent cut in Australia's emissions by 2050. Looking back four years on, Howard though the politics of global warming contributed to his defeat; it was a case of 'John Howard didn't seem that interested' when the issue had come to dominate Australian politics.

Ten days after taking office, Kevin Rudd flew to Bali and handed Ban Ki-moon Australia's instrument ratifying Kyoto. Moments later, he addressed the conference and 'all people of goodwill committed to the future of our planet'. Climate change was the defining challenge of the age, Rudd said. It was imperative for the conference to agree to work together on a global emissions goal, one that recognized the core reality: 'We must avoid dangerous climate change.'

30
BALI

This is a historic moment, long in the making ... Now, finally, we are gathered together in Bali to address the defining challenge of our age.

UN secretary-general Ban Ki-moon

There is no doubt that the fate of our civilization hangs in the balance.

The Prince of Wales

As Berlin was to Kyoto, so Bali was to Copenhagen.

Twelve years on, the climate change negotiations had accreted multiple layers of complexity. The Berlin Mandate had been relatively straightforward, covering three pages. Bali was the thirteenth meeting under the UNFCCC (COP13) *and* the third conference of the parties serving as the meeting of the parties to the Kyoto Protocol (COP/MOP3). The world's largest economy was a party to the convention but not the Protocol, so the COP/MOP had to decide how to manage this fissure.* In addition, there were meetings of subsidiary bodies, dozens of contact groups (a way to get around UN rules that permit no more than two meetings at the same time) and informal consultations. The resulting Bali Road Map was not defined in a single document; rather it set up a series of processes with the aim of agreeing a comprehensive regime in December 2009 at Copenhagen.

The degree of complexity was in inverse relation to the probability of reaching an effective agreement. It all pointed to the essential unreality of attempting to create a global regime to regulate the quantity of ubiquitous, naturally occurring gases.

Carbon dioxide is released into the atmosphere not only from burning fossil fuel (which along with cement production accounts for seventy-five per cent of the increase in atmospheric carbon dioxide). Carbon dioxide is also released from burning wood and animal dung (providing a source of heat for millions in the developing world) and through bacteria breaking down organic matter. It is absorbed by growing vegetation and by the oceans. The IPCC estimated that the remaining twenty-five per cent came from deforestation, turning grassland into cropland and changing agricultural practices.

Methane, the second most important 'man-made' greenhouse gas, is released from fossil fuel production, but also by farm animals and rice paddy fields. In the Fourth Assessment Report, the IPCC thought it 'very likely' that observed

---

* It agreed on a twin-track approach, establishing a new Ad Hoc Working Group on Long-Term Cooperative Action under the convention in addition to an existing Ad Hoc Working Group on Further Commitments for Annex I Parties, under the Protocol.

increases in nitrous oxide had been driven by increased fertilizer use and more intensive agricultural practices, as well as fossil fuel combustion.

Agreement to slow down global warming would require regulating not only energy production but also agriculture and land use. The regime would need to last decades and even centuries. For developing countries experiencing rapid industrialization, still heavily reliant on agriculture, with food accounting for a high proportion of household budgets, that included countries which were custodians of the vast majority of the world's tropical forests, the logic of such an agreement was not an enticing prospect.

Yet for the true believers, Bali carried a huge burden of expectation. It was going to change the course of history. 'It is our chance to usher a new age of green economics and truly sustainable development,' Ban Ki-moon told the conference. One hundred and fifty corporate CEOs put their names to the Bali Communiqué organized by the Prince of Wales. It called for global emissions to be more than halved by 2050, a proposition supported by brands such as GE, DuPont, Shell UK, Coca Cola, Nike, Nestlé, British Airways, NewsCorp, Nokia, Volkswagen and Tesco. The shift to a low carbon economy would create significant business opportunities worth billions of dollars.

To Al Gore, these were the foothills. The greatest opportunity of solving the 'climate crisis' was not new technology and sustainable development. It was in finding 'the moral authority' to solve all the other crises and unleashing 'the moral imagination' of humankind.* 'We are one people on one planet. We have one future. One destiny,' Gore told the packed hall. They should feel privileged 'to be alive at a moment when a relatively small group of people could control the destiny of all generations to come'.

The proximate obstacle to the realization of this vision was not a Lockean argument in favour of freedom and popular sovereignty. Neither was it doubts over the science or the machinations of shadowy vested interests undermining the consensus and somehow preventing governments from acting.

If capping carbon dioxide emissions really was 'the pro-growth strategy', as the one hundred and fifty corporate leaders asserted, why had the G77 plus China been so hostile to acquiring anything that appeared like Annex I-style obligations right from the start of the climate change negotiations? The blanket exemption of non-Annex I countries had been the principal cause of America's non-ratification of the Kyoto Protocol. Without extending emissions caps to the major emerging economies, Annex I countries would face deeper emissions cuts and higher car-

---

*   Gore's speech at the Bali conference on December 13th 2007 is one of the most significant he made on this or any other subject. At the time of writing there is no transcript on Gore's website at http://blog.algore.com/2007/12/ but there are a number of websites with video of the speech and a fairly complete transcript, from which the quotes used in this chapter have been checked, can be found at http://www.irregulartimes.com/gorebalispeech.html

bon prices. As long as the price of carbon in the rest of the world was zero, even more economic activity would be diverted from them to non-Annex I countries.

With the US intent on getting developing world commitments onto the table, the outcome was a real world test of the concept of sustainable development. Did sustainable development have any genuine content or was it a masterstroke of branding to buy Third World acquiescence for First World environmentalism, as long as it was lubricated with copious aid flows and did not constrain their economic aspirations?

The dynamics between the three key players – the US, the EU and the G77 plus China – were almost unchanged from Kyoto. Bali was the last chance for the Bush administration to overcome the original sin of the climate change negotiations so that a future agreement should contain bankable commitments from key members of the G77 plus China. Conversely, the objective for the G77 plus China was, as far as possible, to avoid assuming such commitments and to deflect attention from this by playing up the alleged inadequacy of the Annex I nations meeting their obligations.

In this, the G77 plus China was aided and abetted by the EU, together with the usual supporting cast of assorted NGOs and scientists proclaiming the end of the world if the US did not commit to drastic emissions cuts. Then there was Al Gore, who reprised the role he had played in Kyoto in making a dramatic appearance to under-cut US negotiators.

Unlike Kyoto, Bali was to be the beginning of a process leading to an agreement on emissions caps two years later. However the EU decided it wanted to start at the end, by negotiating the overall quantum of emissions cuts. The EU wanted to corner the US into a putting a number on its emissions cut right away. The Bali conference would be meaningless if it did not set clear targets, in the words of Sigmar Gabriel, the German environment minister.

Based, it was said, on the Fourth Assessment Report, the EU demanded agreement that global emissions be cut by twenty-five to forty per cent below 1990 levels by 2020. The IPCC was not meant to give explicit policy advice. At the February 2007 launch of the summary, Susan Solomon, the American Working Group I co-chair, refused to be drawn on its policy implications. 'It would be a much better service for me to keep my personal opinions separate than what I can actually offer the world as a scientist,' she had told a press conference. 'People are going to have to make their own judgment.'[1]

Politicians sought the cover provided by what scientists said should be done. The latter obliged with the Bali Declaration. Signed by more than two hundred scientists, the declaration said that a new climate treaty should limit temperature increases to no more than 2°C above pre-industrial temperatures, a number already adopted by the European Union. A 'fair and effective' agreement, in the opinion of the scientists, would require greenhouse gas emissions in 2050 to be no more than half their 1990 level.

Unwisely, the UN's Yvo de Boer aligned the secretariat with the EU by circulating a four-page text containing the twenty-five to forty per cent figure. Its preamble cited the 'unequivocal scientific evidence' that required Annex I nations to cut their emissions by this magnitude, although there was no evidence as such, only computer simulations based on a series of unverifiable assumptions. Defining the emissions caps upfront was dismissed by Harlan Watson of the US negotiating team. 'In our view that pre-judges the outcome of the negotiations over the next two years.' In this, the US had the support of Canada, Japan and Russia. De Boer disagreed. It would be a 'critical issue' for the negotiations.

By trying to turn Bali into a showdown with the US over the quantum of emissions reductions, the EU was repeating the mistake of Kyoto. The critical issue was not the quantum of emissions reductions, but the extent to which non-Annex I nations would be subject to them. A treaty that didn't include commitments for China and the other large non-Annex I economies would be dead on arrival in the Senate. Were the EU and the secretariat trying to disprove Einstein's definition of insanity?

In part, the EU's negotiating obtuseness reflected its immense institutional inertia in having obtained agreement among its twenty-seven member states. In part, it was because lead responsibility lay with member states' environment ministries and the European environment commissioner who saw their principal constituency as environmental NGOs. Then there was the superficially attractive narrative that framed President Bush, who would be leaving the White House in thirteen months, as the principal obstacle to reaching agreement.

In a lightning visit, Bush's opponent in the 2004 election endorsed the EU narrative by attacking the Bush administration for undermining attempts to agree stringent emissions caps. Global momentum would make emissions caps a reality whatever the opposition of Bush or from Congress. 'This is going to happen,' Kerry told reporters. 'It's going to happen, because it has to.'[2]

By contrast with the EU's environment ministry-led approach, the US negotiating position was the product of an inter-agency process convened by the National Security Council and involved the Departments of Agriculture, Commerce, Defense, Energy, Justice, State, and Treasury, the White House Council on Environmental Quality, the National Economic Council and the EPA. The State Department led the negotiations, institutionally vastly more experienced in international diplomacy than Europe's environment ministries.

Before heading to Bali, Dobriansky and her team in the State Department met negotiators who had served in the Bush I and Clinton administrations. Several from the Clinton administration emphasized the importance of obtaining high-level commitments from non-Annex I parties. This, they argued, needed to be set forth in Bali. One former senior State Department official asked for a private meeting with Dobriansky to give her some insights before Bali. 'Very helpful, very useful,' Dobriansky recalled.

Thus there was a consensus among current and former executive branch policymakers of both parties. Furthermore, the US had reached agreement on a text with key parties prior to Bali, according to Dobriansky. Yet what transpired in the Bali conference hall suggested something rather different and made the denouement at Bali the most dramatic of all the COPs before Copenhagen.

Snow lay thick on the ground in Kyoto ten years before, and snow would fall on the delegates in Copenhagen two years later. Lying eight degrees south of the equator, Bali was better located for a December meeting on global warming. It drew nearly eleven thousand participants. They included more than three thousand, five hundred government officials, outnumbered by five thousand, eight hundred representatives of NGOs, UN bodies and agencies and nearly one thousand, five hundred accredited members of the media.

The Indonesian government brought some welcome colour. COP President Rachmat Witoelar and others on the conference platform wore tropical shirts. Indonesian President Susilo Bambang Yudhoyono wrote a song for the occasion – its lyrics a marked improvement on the Dutch poet laureate's efforts at The Hague in 2000:

> *Mother Earth is getting warmer*
> *Climate change is tragedy for all*
> *Together we must find answers*
> *Don't let it destroy our life*

Ill-at-ease government leaders and ministers were invited to sing the chorus as a video of smiling children, burning forests and trees uprooted by storms was beamed into the hall:

> *We all gather in Bali*
> *We all gather in Bali*
> *We want to save our planet*
> *We want to save our planet*
> *We are all united here in Bali*
> *For a better life, a better world*
> *For you and me*

The Indonesians also broadened the talks beyond environment ministers to bring in trade and finance ministers. These included talks aimed at removing trade barriers and tariffs on environmental goods and services. There was little doubt about the geopolitical orientation of the conference chair as a leading member of the G77. In 1955, Sukharno, Indonesia's first president, hosted the Bandung Conference, the first major Asian/African conference, and in 1961 founded the Non-Aligned Movement together with Tito (Yugoslavia), Nasser (Egypt), Nehru (India) and Nkrumah (Ghana).

'We are all united here in Bali,' the chorus sang.

Up to a point. The delegates were united in their determination to present the outcome of the COP as a success, a sentiment which the G77 plus China used to isolate the United States in the COP's final minutes.

The underlying motivations of the key players can be assessed by their responses to attempts at Bali to find breaches in the Berlin Wall separating the Annex I parties from the rest. There was a renewed effort on voluntary commitment, which China and the G77 had chased off the agenda at Buenos Aires in 1998. At Bali, it was Belarus' proposal to 'legitimize' its participation in Kyoto's first commitment period, supported by Russia and the Ukraine. It sank without trace.

On the conference's third day, there was discussion of a long-standing Russian proposal to enable developing countries to take on voluntary emission limits. It attracted the support of the EU and other Annex I parties. India and Saudi Arabia voiced their opposition.

There had been ongoing discussion about how to carry out a review of the Protocol. The fifty-four-member African Group and China warned against 'undermining the Protocol', even though it specifically required the review. India went further, wanting to rule out new commitments for developing countries.

At the beginning of the conference's second week, there was another meeting on the review. Annex I parties wanted it to focus on the effectiveness of meeting the Protocol's objective. Growth of non-Annex I emissions would inevitably call into question the Annex I bifurcation. Russia, Canada and Australia wanted to establish a working group.* They were opposed by South Africa, China, India and Saudi Arabia. Joined by the EU, the three Annex I parties then proposed requesting proposals on how to amend the Annexes to the Protocol, i.e., to provide some form of graduation mechanism. They were opposed by a solid phalanx of China, India and Saudi Arabia.

The consistent pattern of opposition by the G77 plus China provides context for their public statements in front of the cameras. At the high-level segment a couple of days later, the US restated President Bush's position that a future agreement should include a long-term global emissions goal and national plans with measurable mid-term goals. In front of the TV cameras, South Africa declared that it would take serious mitigation actions (i.e., limits on greenhouse gas emissions) that were measurable, reportable and verifiable. If made in good faith, why had South Africa – with China and India – acted to block discussion that might result in a developing nation becoming an Annex I party or acquiring similar obligations? It was a question that would hang over the climactic events during the COP's grand finale.

Before that came the conference highlight. The hall was packed and security tight. Many delegates were forced to watch the proceedings on TV. For a time, Rajendra Pachauri was locked out of the hall.

---

*   The US was not present, as it was not a party to the Protocol.

'Fresh from receiving the Nobel Peace Prize in Oslo alongside the IPCC, Academy-award winner, best-selling author, former Vice President, Senator and Congressman from the United States of America and climate change's single most effective messenger to the world, I present to you Al Gore,' Cathy Zoi – an employee of Gore's Alliance for Climate Protection – told the cheering hall as Gore strode across the platform.

'We, the human species, face a planetary emergency,' Gore intoned. 'That phrase still sounds shrill to some ears but it is deadly accurate as a description of the situation that we now confront.' He spoke of his shock as scientists had repeatedly brought forward estimates for the date when the entire north polar ice cap would disappear. Years ago, they had thought it might be gone towards the end of the twenty-first century. Only three years ago, they thought it could happen by 2050. 'Now, this week, they tell us it could completely disappear in as little as five to seven years.'

He compared people who believed in the threat of climate change, but did nothing about it, to victims of Nazi death squads.

> 'First they came for the Jews, and I was not a Jew, so I said nothing. Then, they came for the Gypsies, and I was not a Gypsy, so I said nothing,' and he listed several other groups, and with each one he said nothing. Then, he said, they came for me.

Those who thought that the climate crisis would only affect their grandchildren – and, as the crisis got closer to them, their children – were wrong. It would get them too. 'It is affecting us in the present generation, and it is up to us in this generation to solve this crisis.' Quoting Churchill, most world leaders were like the appeasers of the 1930s and 'decided only to be undecided, resolved to be irresolute, adamant for drift, solid for fluidity'.

Speaking ten years and four days after his appearance at the Kyoto conference where he had publicly instructed American negotiators to make concessions, Gore dropped a second COP bombshell. 'I am not an official of the United States and I am not bound by the diplomatic niceties. So I am going to speak an inconvenient truth.' His voice tightened as he wiped sweat from his face. 'My own country, the United States, is principally responsible for obstructing progress here in Bali.' The hall went wild with applause and cheering. There were others, Gore said, who could also help move the process forward, but they weren't named – and would the audience have cared? They had just heard what it was convenient to believe.

Those who had just applauded his 'diplomatic truth' had two choices. They could direct their anger and frustration at the United States. Or they could decide to move forward without the US; do all of the difficult work and save a large, blank space in the document and footnote it: 'This document is incomplete, but we are going to move forward anyway.'

The negotiations would culminate in Copenhagen in two years' time. 'Over the next two years the United States is going to be somewhere it is not now,' Gore told the delegates. 'You must anticipate that.' He could not guarantee that the next president would have the position he assumed – 'but I can tell you that I believe it is quite likely'.

Gore spelt out the issues that needed to be decided in Bali. Targets and time-tables, of course, together with that blank space for the next president of the United States to ink in; a plan for a fully-funded, ambitious adaptation fund (not a tough sell in a hall packed with delegates from countries who expected to be its beneficiaries); and a deforestation plan ('it is difficult to forge such an agreement here') – and not a word on the single biggest lesson from the Kyoto Protocol: its failure even to contemplate the prospect of major developing economies eventually being subject to a global agreement to limit their emissions.

A global agreement was not going to work unless it included the world's largest economies. Rather than use his standing to highlight the hole in the heart of the climate change treaties, Gore chose to isolate the US and the Bush administration as it attempted to fill the hole that the Clinton-Gore administration had bequeathed it. 'We have everything we need,' Gore bellowed, 'save political will. But political will is a renewable resource.'

It was scarcely Gore's finest hour.

New Scientist environment blogger Catherine Brahic described what happened next:

> The audience rises to its feet, cheers, whoops some more, Gore makes his way down the aisle, drenched in sweat, shakes the hands that are reaching towards him. The last one he shakes is that of a grinning government representative sitting just behind me – he's from China. I ask him if political will really is a renewable resource. 'We will see,' he smiles.[3]

China's answer came two days later. Ministers and officials had wrangled for much of the day over the precise wording that might – or might not – provide a basis for some form of construction to bridge the gaping divide in the climate treaties. A smaller group met until the early hours of Saturday morning, reaching apparent agreement on the most contentious issues.

Shortly after 8am, Rachmat Witoelar gavelled the resumed session to order. Even minor changes to the text would compromise the meeting's ability to come to an agreement, Witoelar warned.[4] He invited the COP to adopt the draft text. Portugal, on behalf of the EU, supported the text and called for all parties to do the same. Witoelar scanned the hall. 'India, please come forward.'

At a small meeting mandated by Witoelar, the Indian delegate explained, the G77 plus China had agreed a modification of the text in respect of the scope of possible obligations placed on developing country parties: nationally appropriate mitigation actions by developing country parties in the context of sustainable development, supported by technology, financing and capacity building, in a measurable, reportable and verifiable manner. The language hid a shift in emphasis. Measurable, reportable and verifiable now applied to what developed nations were to do for developing ones, de-emphasizing the accountability of developing countries in meeting their mitigation commitments. 'Mr President, this is our preference,' the Indian delegate's politeness indicating the strength of the hand India and its allies were playing.

Not so the Chinese delegation, whose intervention let the cat out of the bag. It was wrong to ask the meeting to adopt a text when the other meeting, convened by the Indonesian foreign minister and other members of the G77 plus China, was discussing this very matter. 'At this moment, we cannot adopt this decision.'

Witoelar suspended the session.

When it resumed, the Chinese delegation intervened again. The Indonesian foreign minister was still in consultations. It was therefore still inappropriate for the matter to be discussed. A second member of the delegation took the microphone. In English, he accused the secretariat of intentionally holding the session in the hall when he knew that the G77 was still meeting the Indonesian foreign minister. To applause, he demanded apologies from the secretariat.

'Yes, I have been offended,' the microphone picked up de Boer telling Witoelar. 'This is a process I was not aware of.'

Pakistan, on behalf of the G77, asked for a further suspension while negotiations continued outside the hall.

Witoelar began the resumed session with an apology; as chair of the conference, he had not been without faults. But he had brought heavy reinforcements. 'I come before you very reluctantly,' Ban Ki-moon told the delegates. 'Frankly I'm disappointed at the lack of progress.' Everyone (i.e., the US) should be ready to make compromises. 'No one leaves this chamber fully satisfied.'

Witoelar made another apology, asking for delegates' understanding and forgiveness for any unintentional mistakes (i.e., that the G77 plus China stitch-up had become public in such an embarrassing manner). The apology wasn't enough for the Chinese delegation, who wanted to know from the executive secretary why China had needed to make two speeches on a point of order earlier in the day.

Yvo de Boer switched on his microphone. The words didn't come easily; two or three at a time before trailing off. Then a complete sentence. 'The secretariat was not aware that parallel meetings were taking place and was not aware that text was being negotiated elsewhere.' He switched off his microphone, closed his light blue UN folder and left the platform. China had made its point.

Take two. India took the floor. The G77 plus China had accepted 'somewhat different' language and read out the text again.* 'This is our preference.'

---

* It was adopted as 1 (b) (ii) of the Bali Action Plan. The placement of the word 'enabled' differed from the version read out in the morning session, but the two versions were functionally identical.

Portugal expressed the EU's support. Cheering and applause filled the hall.

Now it was the turn of the US. Dobriansky explained that the US had come to Bali with the hope of agreeing a strong statement about common global responsibility to address climate change, recognizing differences among national circumstances. However, the formulation proposed by India was not one the US could accept as it represented a change in the balance that the parties had been working towards.

She was met with boos and catcalls.

Japan, in a series of circumlocutions, avoided explicitly accepting or rejecting the proposal.

To applause and cheers, South Africa's Marthinus van Schalkwyk said that the suggestion that developing countries were not willing to assume their full responsibilities was 'most unwelcome and without any basis'. They were saying voluntarily they were willing to commit to measurable, verifiable mitigation actions. 'It has not happened before.' The US should reconsider its position.

Brazil said the text was a balanced and fair basis.

More cheering.

US-born Kevin Conrad for Papua New Guinea unleashed more. The US should either lead or get out of the way, conference-speak for the US to do the opposite and fall into line and accede to the position of China, India and the rest of the G77.

The US was isolated.

Dobriansky indicated she wished to speak again. 'We've listened very closely to many of our colleagues here,' she said. The US had come to Bali 'very committed' to developing a long-term, global greenhouse gas emission goal. It was also committed to giving very serious consideration to the views of Japan, Canada and the EU, to lead to a halving of global emissions by 2050. It had sought agreement on the principle that commitments should be measurable, reportable and verifiable, including emission reduction or limitation objectives in a way that ensures comparability between countries' different circumstances. 'We have all come a long way here,' Dobriansky continued. The US just wanted to ensure that everyone acted together – 'we will go forward and join consensus'.

Later that day, the White House released a statement welcoming the outcome of the talks. Many features of the decision were 'quite positive'. But it also had 'serious concerns'. The negotiations needed to proceed on the basis that emissions cuts solely by developed countries would be insufficient.

> It is essential that the major developed and developing countries be prepared
> to negotiate commitments, consistent with their national circumstances,
> that will make a due contribution to the reduction of global emissions.[5]

Had Dobriansky been right to reverse her position and join the consensus? The result, as she indicated, fell short of what the US was seeking. In particular, there was no explicit recognition of the need to limit the rise or to reduce developing counties' emissions. The effect of the Berlin Mandate had been to preclude any discussion of emissions limitations for non-Annex I countries. In this respect, Bali was a breakthrough. The principle of developing countries taking on commitments to limit or reduce their emissions was on the table.

No power on Earth was strong enough to compel China, India and the other major developing economies to accept commitments on emission reductions against their will. What Bali achieved was keeping the option on the table that they might do so voluntarily, avoiding a second Kyoto and an almost certain Senate rejection. If Dobriansky had not acceded, the US would have been blamed a second time for derailing the climate change negotiations. It was a gutsy call and a class performance.

The Bush administration would bequeath its successor a viable negotiating framework. At Copenhagen two years later, the world would discover just how far leading developing countries were willing to go.

On September 15th 2008 Lehman Brothers filed for bankruptcy protection.

The impact on the Second Environmental Wave was similar to the Egyptian army smashing through the Bar Lev Line on the First. After Lehman, the language and the rhetoric were the same, but the intensity had gone. Saving the planet became less important than rescuing the banking system and staving off global economic collapse.

There was an additional effect. The West, specifically its governments, which seemed, or presumed, to have the answers to the world's problems, were suddenly exposed.

They didn't.

SHOWDOWN IN COPENHAGEN

If we are willing to work for it, and fight for it, and believe in it, then I am absolutely certain that, generations from now, we will be able to look back and tell our children … this was the moment when the rise of the oceans began to slow and our planet began to heal…

Barack Obama, June 3rd 2008

President Obama, acting the way he did, definitely eliminated any differences between him and the Bush presidency.

Lumumba Di-Aping of Sudan on behalf of the G77,
December 19th 2009

According to the Met Office, 2008 was the tenth warmest of the last one hundred and fifty-eight years. As recently as the 1970s or 1980s, globally 2008 would have been considered warm, observed climate scientist Myles Allen of Oxford University, 'but a scorcher for our Victorian ancestors'.[1]

Evidently the Victorians were made of sterner stuff. Without global warming, many parts of the world would have experienced arctic conditions. 2008 began with China's worst winter for half a century. Heavy snow closed the Chinese steel industry and killed one hundred and twenty-nine people. For the first time in living memory, snow settled in Baghdad.

For Britain and other parts of Northern Europe, the summer was marked by the lack of direct sunshine and South America was experiencing a particularly cold winter. Australian skiers had one of their best seasons, with snow depths around twice the previous ski season. In the spring, it snowed in sub-tropical southern Brazil – 'If snow is rare, to get accumulation is astonishing,' the Met-sul Brazilian weather centre reported – and in October, Sydney had early summer snow.[2] As Parliament debated the Climate Change Act, London had its first October snowfall since 1934. The tenth warmest year on record closed with freak snow storms in Southern California and up to eight inches of snow fell in Las Vegas, a record for the most snow in the month of December since official records began in 1937.[3]

Disbelief about the exceptional warmth of 2008 extended to American consumers, who bought record numbers of snow blowers. Sales of the machines were up 'high double digits' over the previous year, one chain reported, particularly among the heavier-duty big-ticket models, spurred by December weather that broke more than two thousand snowfall records.[4]

'When climate scientists like me explain to people what we do for a living we are increasingly asked whether we "believe in climate change,"' Vicky Pope, the Met Office's head of climate change advice, wrote in February 2009.[5] To Pope's dismay, a November 2008 poll for *The Times* found that only forty-one per cent of those surveyed accepted as an established fact that global warming was taking place and was largely man-made. Only twenty-eight per cent believed that global warming was happening and that it was 'far and away the most serious problem we face as a country and internationally'. Awareness of the scale of the problem resulted in people taking refuge in denial, Pope explained.[6]

On a pre-inaugural whistle-stop tour in January 2009, Barack Obama spoke of the dangers of a planet 'warming from our unsustainable dependence on oil'. A poll suggested the view was held by a minority of his fellow citizens. Forty-one per cent of Americans blamed global warming on human activity, compared to forty-four per cent who thought long-term planetary trends were the cause. The numbers were a sharp reversal from a similar Rasmussen poll taken in July 2006, when forty-seven per cent blamed global warming on human activity compared to thirty-four per cent who viewed long-term planetary trends as the culprit. A Pew poll suggested that Americans did not view global warming as a priority, the issue coming twentieth out of twenty (down from eighteenth in a January 2007 poll).

In July 2009, President Obama joined other leaders for his first G8 summit. The venue had been switched to L'Aquila in central Italy after it had been struck by a severe earthquake. Since the previous G8, a financial earthquake had hit the global economy. Now the financial crisis was yoked together with climate change and the elimination of poverty in an all-encompassing mega-crisis, the G8 leaders stating their determination to tackle these 'interlinked challenges' with what they hopefully called a 'green recovery'. 'A shift towards green growth will provide an important contribution to the economic and financial crisis recovery,' the G8 claimed.

Recognizing the 'broad scientific view' that the average global temperature should not rise more than 2°C above pre-industrial levels, the G8 wanted 'to share' the goal of cutting global emissions by at least fifty per cent by 2050. Global emissions would have to peak 'as soon as possible'. It would imply that developed countries would have to cut their emissions by eighty per cent or more.

On the summit's second day, a meeting of the Major Economies Forum indicated limited willingness to share the G8's self-imposed burden. Affirming the 2°C goal, the Major Economies leaders – which included Brazil, China, India and South Africa – agreed on work to 'identify a plan for substantially reducing global emissions' by 2050. However, its declaration avoided reference to any emissions reduction target.

Speaking on behalf of the G8, Silvio Berlusconi stated that 'the active agreement of all major emitting countries through quantitative mitigation action was regarded by the G8 as an indispensable condition to tackling climate change'. Nonetheless, the G8 praised the Forum's 'constructive contribution' and looked forward to a global, wide-ranging and ambitious post-2012 agreement in Copenhagen.

As in 2007 and the lead up to Bali, there was a steady drum-beat to the COP in December. In September, Ban Ki-moon hosted a climate change summit at the UN attended by more than a hundred world leaders. None was as dedicated as Britain's Gordon Brown, one of the first leaders to say he would be attending the COP.

In October, Brown hosted a meeting of leaders' representatives of the Major Economies Forum at Lancaster House. There were less than fifty days to reach an agreement and avoid catastrophe, Brown declared. 'In just twenty-five years the glaciers in the Himalayas, which provide water for three quarters of a billion people could disappear entirely,' Brown told them, recycling a discredited IPCC claim.[7] 'Failure to avoid the worst effects of climate change could lead to global GDP being up to twenty per cent lower than it would otherwise be,' Brown said, repeating the most alarmist claim of the Stern Review, one that had collapsed under critical scrutiny, '[a]nd that is an economic cost greater than the losses caused by two world wars and the Great Depression.' Developed countries had to come forward with offers of finance, Brown said. He had been working on a $100 billion per year package in 'predictable public and private funding by 2020'.

There was a fragility that hadn't been apparent two years earlier. It was evidenced in the shrill reaction to the fall-out from the release of the Climategate emails in November. Writing in *The Times*, former chancellor Nigel Lawson slammed the integrity of the scientific evidence deployed by the IPCC to base far-reaching and hugely expensive policy decisions. 'The reputation of British science has been seriously tarnished,' Lawson wrote.[8]

Climate secretary Ed Miliband branded Lawson and other sceptics 'climate saboteurs'. He accused them of being 'dangerous and deceitful' for misusing data and misleading people in an attempt to derail the Copenhagen conference.[9] Brown then weighed in. 'With only days to go before Copenhagen we mustn't be distracted by the behind-times, anti-science, flat-earth climate sceptics,' Brown told the *Guardian*. 'We know the science. We know what we must do.'[10]

It was delusory to believe a small group of climate skeptics might sway governments' assessment of their national interest and swing the outcome at Copenhagen. The bluster betrayed insecurity. The air was coming out of the balloon.

Still more dramatic were political developments in Australia. At Bali, Kevin Rudd had been the COP hero. By a single vote in a December 1st party caucus ballot, the opposition Liberals ditched Malcolm Turnbull and his policy of cooperating with Labor to pass Rudd's emissions trading scheme. In came Tony Abbott. Two months before, Abbott had described climate change as 'absolute crap' – something the newly elected leader now described as 'a bit of hyperbole'.[11] The next day, the Australian Senate voted down the ETS for the second time. If Rudd had the courage of his convictions on the ETS, he could have called a double dissolution and fought an election on climate change. He blinked. Rudd's hold on power was slipping, a victim of climate change.

Copenhagen provoked millennial expectations among some of the committed. Tom Burke, NGO leader, government adviser, corporate environmental guru and academic extraordinaire, boldly declared 2009 the most important year in human history.[12] The World Council of Churches asked churches around the world to ring their bells on the Sunday midway through the conference. Bill McKibben, a leading environmental activist and Sunday school teacher, spoke of the special role of churches. 'Where I live, in the United States, before we had radio when somebody's house caught fire, we rang the church bells so that everybody would know and come out to do something about it,' McKibben explained. 'Well, something's on fire now.'[13]

In Britain, Christian, Jewish, Muslim, Hindu, Sikh, Buddhist, Bahá'í, Jain and Zoroastrian faith leaders put their names to a joint statement overseen by Ed Miliband and filmed by the Foreign Office for worldwide distribution. Tackling the causes of global warming was an unequivocal moral imperative, the statement declared.

Two days before the conference, more than twenty of Britain's church leaders painted their hands blue and called for an ambitious deal in Copenhagen. Addressing an ecumenical service in Westminster City Hall, the Archbishop of Canterbury provided some perspective. 'It looks in the last few decades particularly and perhaps the last few millennia as if the human race has on the whole not been very good news for the rest of creation,' Rowan Williams told the congregation.[14]

From the spiritual to the temporal, Copenhagen would test the willingness of developing countries to accept the idea that cutting greenhouse gas emissions would boost their economies. If the notion of a green recovery was widely believed, countries would be falling over themselves to outbid each other with offers to cut their emissions. In the event, although President Obama talked the talk, only the EU walked the walk, offering to up their twenty per cent emissions cut to thirty per cent if others followed. None did.

The environmentalist agenda made inroads in developing nations. Unlike the West, especially Western Europe, economic considerations were paramount. The refusal of developing countries to subscribe claims about the benefits of 'green growth' set the scene for a confrontation between Western environmentalism and the Third World's growth ambitions. Copenhagen thus brought the series of UN environmental conferences full circle.

Separated by thirty-seven years, Stockholm and Copenhagen shared similar symbolic actions. Nixon's EPA announced its DDT ban and Obama's EPA issued its finding that carbon dioxide was dangerous. There is the role of environmental NGOs; introduced by Maurice Strong at Stockholm as messaging propagators, but drastically curtailed in Copenhagen's Bella Center as the conference teetered on the brink of chaos.

Then there are the roles of India and China. For Strong, both were crucial. He had courted Indira Gandhi, who delivered the conference keynote address.

Did her warning that the welfare of men came before the preservation of beasts still reflect developing country attitudes? Copenhagen would provide the answer.

In 1972, China was groping to find its place in the world. Strong had coached its delegation so the Stockholm Declaration could be adopted by consensus. In 2009, China was the world's second largest economy. Its diplomats were confident and highly proficient in navigating their way around the climate change negotiations. At Copenhagen's climactic moment, China fielded a mid-level official in face-to-face negotiations with the president and other world leaders which determined the outcome.

The accommodation between environmentalism and the developing world – hammered out in the Founex formula by Maurice Strong and Barbara Ward in 1971 – was based on a non-binding aspiration by developing countries not to emulate the developed world's path of industrialization. If their development path deviated from the one preferred by environmentalists – Barbara Ward and Fritz Schumacher in the 1970s and their successors such as Al Gore and Prince Charles today – each developing country could decide for itself its trade-off between economic growth and environmental objectives.*

The Founex formulation was not written into the international climate change agreements, but it permeated their every pore. There are innumerable references to 'sustainable development' from the preamble of the 1992 convention to the critical paragraph 1 (b) (ii) of the Bali Action Plan on nationally appropriate mitigation actions by developing countries 'in the context of sustainable development.'

Although the use of the words 'sustainable' and 'sustainability' became *de rigueur* in the West by companies and governments, for developing countries, the meaning of sustainable development is about environmental policies not constraining human needs and aspirations and the Brundtland Report giving overriding priority to the world's poor. The texts do not define what is meant by sustainable development, but it is clear from them that the concept chiefly applies to developing countries. There is no mention, for example, of sustainable development in the corresponding paragraph 1 (b) (i) of the Bali Action Plan with respect to developed country parties.

For the West to prevail at Copenhagen, the large emerging economies, principally China and India, but also Brazil, Indonesia, Mexico, South Africa and South Korea, would have to agree to at least one of two propositions: first, that they no longer considered themselves to be poor or developing nations; second, that the threat posed to them by global warming was so grave that it overrode the condition accepted by Strong to gain Third World participation at the Stockholm conference – that environmental protection would not fetter their development ambitions.

---

*   The third part of the Founex formula related to money. The principle was uncontroversial, but there were always disputes over the amount, who would get it and who would control it.

This made the Copenhagen conference unlike its predecessors and something international conferences try and avoid. The logic of global warming – the harm caused by an extra tonne of carbon dioxide in the atmosphere was the same irrespective of how it got there or who put it there – and the economic priorities of developing countries made confrontation inescapable. It permitted only a binary outcome. One side would win; the other would lose.

The alignment of the main blocs at Copenhagen was also different. In Rio, Kyoto and Bali, the US and the EU had clashed. For the first time, they pitched camp on the same ground. There was EU grumbling about the scale of the US cut – four per cent below their 1990 levels by 2020 compared to the EU's twenty per cent. But both believed a credible agreement had to involve the major emitting nations.

American negotiators came to Copenhagen determined to apply the lessons from the failure of Kyoto. They didn't want an agreement they couldn't get through the Senate and warned that Obama wasn't going to arrive in Copenhagen and act as a *deus ex machina*, as Gore had done. 'We don't want to promise something we don't have,' chief negotiator Todd Stern told reporters.

Congressional Democrats were as supportive to the American position in Copenhagen as they had been hostile at Bali, despite it being identical. 'The concerns that kept us out of Kyoto back in 1997 are still with us today,' Senator Kerry told the conference. 'To pass a bill, we must be able to assure a senator from Ohio that steelworkers in his state won't lose their jobs to India and China because those countries are not participating in a way that is measurable, reportable and verifiable.'[15]

Gore made the same point. Obligations applied to one part of the world but not to another part could heighten people's fears about their economic circumstances. 'I would ask for an understanding of the difficulty that goes for elected officials.'

On the other hand, divisions within the EU were more visible. Britain and France wanted to raise the EU's offer to a thirty per cent cut. 'Because of the economic recession, a thirty per cent cut is much more like a twenty per cent cut two years ago,' Britain's foreign secretary, David Miliband, explained from Brussels. At least the recession was helping save the planet. Speaking for coal-dependent central and eastern European states, Poland's Mikolaj Dowgielewicz, minister for European affairs, was having nothing of it. 'The conditions for that are non-existent.'

Arrayed on the other side were the G77 plus China, along with the usual assortment of NGOs. Sudan, China's principal ally in Africa, spoke for the bloc. Its core was formed by Brazil, South Africa, India and China – the BASIC nations, the last two constituting the bloc's inner core.

Before the conference, China announced that it would cut its carbon-intensity by forty to forty-five per cent in 2020 compared to 2005, a move which unsettled India. At a November meeting in Beijing, the BASIC group reiterated its non-negotiable position. A relieved Jairam Ramesh, India's environment minister, expressed confidence that China would not 'ditch us'. Should industrialized countries seek to override them, the BASIC four would stage a collective walk-out.

Both sides knew the other's strategy. The aim of the US was to peel off as many members of the G77 as possible and leave China isolated. Its big bazooka was the third part of the Founex formula: money, huge amounts of it, targeted at the African nations.

Halfway through the conference, China's vice foreign minister, He Yafei, told the *Financial Times* that China would not be the fall guy if there were a fiasco. 'I know people will say if there is no deal that China is to blame,' he said. 'This is a trick played by the developed countries.'

The rest of what went on at Copenhagen was noise. It was fractious, at times farcical – the largest gathering of world leaders outside the UN in New York – but such were the tensions between them, no team photo.

There was the usual PR hoopla of climate conferences, but more so. The Nepalese Cabinet was helicoptered to a remote plateau in the Himalayas. They took part in a traditional Sherpa prayer ceremony before approving the speech that Nepal's PM would deliver in Copenhagen. Sonoma County, California, despatched a seven-person delegation at a cost of $225,000. 'This is a Disneyland for policy wonks,' exclaimed Gary Gero of the Los Angeles-based Climate Action Reserve.

Coca Cola was the most visible brand in Copenhagen, its logo and 'Hopenhagen' splashed across city billboards. More than forty thousand people came, but the lines to enter the conference were a lot longer than at Disney. On the first day, a snag in credentialing left more than a thousand delegates shivering outside the fifteen-thousand-capacity Bella Center for nine hours. Less remarked was the bitter cold. Climate change conferences were irony-free zones when it came to adverse weather events like heavy snowfall.

'Maybe I'm naïve, but I'm feeling optimistic about the climate talks starting in Copenhagen,' economist Paul Krugman wrote in the *New York Times*. So was the White House. Pointing to signs of progress toward a 'meaningful' agreement, the president's trip was rescheduled to the end of the conference. Gordon Brown wrote that Copenhagen was poised to achieve 'a profound historical transformation: reversing the road we have traveled for 200 years'. Quite why it should make sense to turn the clock back to the beginning of the nineteenth century, Brown didn't say.

It was different on the ground. The COP's opening was pushed off balance by the furore of the Climategate emails. 'It's an eleventh-hour smear campaign,' Hockey Stick author Michael Mann said, adding, 'I've done nothing wrong; I have nothing to hide.'[16] The National Oceanic and Atmospheric Administration's James Overland complained: 'It has sucked up all the oxygen.' IPCC vice-chairman Jean-Pascal van Ypersele agreed: 'We are spending lots of useless time discussing this.'

Saudi Arabia's Mohammad al-Sabban called for an independent inquiry. The 'level of trust' had been shaken, 'especially now that we are about to conclude an agreement that ... is going to mean sacrifices for our economies'. The IPCC's Pachauri went on the offensive. 'The only debate is who is behind it, I think we should catch the culprits.' Climategate was an attempt to tarnish the IPCC. 'The Fourth Assessment Report is completely objective, totally unbiased and solid in its scientific assessment,' Pachauri claimed.[17]

Striking a markedly less belligerent note, the UN's Yvo de Boer conceded the emails looked 'very bad' and were fuelling skepticism, but media scrutiny was not unwelcome:

> It's very good that what is happening is being scrutinized in the media because this process has to be based on solid science. If quality and integrity is being questioned, that has to be examined.[18]

Pachauri's bravado didn't last. As well as the inquiry mandated by the Dutch Parliament, Pachauri and Ban Ki-moon subsequently asked the InterAcademy Council to investigate the IPCC.

CEOs of Western corporations flocked to Copenhagen. 'Not one single Indian CEO is here, and do you know why? Because they do not consider Copenhagen to be the most important event in one hundred years,' a participant observed. De Boer was blunt. 'Basically you're not playing any role in a serious way.'[19]

Five thousand miles away, Exxon Mobil demonstrated its seriousness – a $41 billion deal to acquire shale gas and fracking corporation XTO Energy. It was the largest energy sector transaction in four years and Exxon's largest since it bought Mobil Corp in 1999. 'This is not a near-term decision,' said CEO Rex Tillerson. 'This is about the next ten, twenty, thirty years.' Combined with XTO's holdings, Exxon Mobil would control about eight million acres of land on top of unconventional natural gas.

The quantum expansion of natural gas reserves unlocked by fracking gas shale presents the prospect of cheap, abundant energy. No one in Copenhagen seemed to notice. 'As politicians dither and debate,' the *Daily Telegraph*'s Damian Reece wrote, 'the market has taken another decisive step in dictating where the world's energy dollars are invested, whether campaigners like it or not.' For those professionally engaged in worrying about global warming, shale gas was not part of the plan. It wasn't 'clean' energy. Of the three paths to large-scale emission reductions, the collapse in Soviet communism was unrepeatable and the 2008–2009 global recession undesirable. Only Britain's dash-to-gas – replacing coal with gas-fired power stations – required no government subsidies or artificial price support.

No big climate conference could do without Al Gore or the Prince of Wales. Both were in Copenhagen. Prince Charles garnered the better press with his plea to slow down tropical deforestation. 'The quickest and most cost-effective way to buy time in the battle against catastrophic climate change is to find a way to make the trees worth more alive than dead,' he told the conference.[20]

Although not entirely eschewing the histrionics of his Bali performance ('the future of our civilization is threatened as never before'), Gore was low key. His role was subsidiary, more of a John the Baptist preparing the way for He who would be flying in on Air Force One. Even so, Gore managed to land himself in hot water over his Arctic ice prediction. The date for the 'possibility' of an ice-free Arctic ocean 'for a short period in summer' had been pushed back to 'perhaps as early as 2015' (at Bali, the forecast had been as early as 2012 to 2014). This time, Gore was swiftly rebutted. Wieslaw Maslowski, a climatologist at the US Naval Postgraduate School in California, said that it misrepresented the information he'd given Gore's office. 'Why would you take anything Al Gore said seriously?' MIT's Richard Lindzen asked. All Gore had done was extrapolate from 2007, when there was a big retreat in the sea ice, and got zero, Lindzen explained.

Kyoto took place towards the end of a decade of rising temperatures. Coming near the end of a trendless temperature chart, Copenhagen was more of a challenge. The Met Office did the next best thing. 2010 was 'more likely than not' to be the world's warmest year on record and man-made climate change would be a factor. 'If 2010 turns out to be the hottest year on record it might go some way to exploding the myth, spread by the climate conspiracy theorists, that we're experiencing global cooling,' Greenpeace's Ben Stewart said. 'In reality the world is getting hotter, possibly a lot hotter, and humans are causing it.'[21]

The trio of Venezuela's Hugo Chavez, Zimbabwe's Robert Mugabe and Bolivia's Evo Morales entertained the conference with their denunciations of capitalism as the cause of climate change. 'When these capitalist gods of carbon burp and belch their dangerous emissions, it's we, the lesser mortals of the developing sphere who gasp and sink and eventually die,' declared Mugabe, the octogenarian Marxist who had destroyed Zimbabwe's economy. 'A ghost is stalking the streets of Copenhagen' Chavez told the conference. 'Capitalism is that ghost,' provoking wild applause from the representatives of civil society, aka the environmental NGOs, who seemed not to know that the collapse of Soviet communism accomplished more for the environment than any other event in history.[22]

The violence of some of the language in the conference hall was reflected outside. Three hundred youths shrouded in black threw bricks and smashed windows as around thirty thousand people demonstrated in central Copenhagen. Police made nine hundred and sixty-eight arrests.

NGOs stoked the anger with inflammatory rhetoric. 'Each year three hundred thousand people are dying because of climate change,' Kumi Naidoo of Greenpeace said at a rally.[23] But they had little purchase over the governments of India, China and others in the developing world. So the story they spun at the end of Copenhagen – widely taken up by the media – was to blame the rich nations for blocking progress. 'Rich countries have condemned millions of the world's poorest people to hunger, suffering and loss of life,' stormed Nnimmo Basey of Friends of the Earth. 'The blame for this disastrous outcome is squarely on the developed nations.'[24]

As a description of the showdown at Copenhagen, it was pure NGO fantasy. Nor was Copenhagen a confrontation between capitalism and socialism. It was a battle between the West that had signed on to the environmentalist agenda and the largest emerging economies asserting their right to a prosperous future.

It mostly took place away from the media behind closed doors. From time to time, the clash spilled out into the open. It didn't need a code-breaker to decipher the positions of the main protagonists, which were clearly signalled in speeches in the conference hall.

There was a clear victor. Equally clearly, there was a side that lost more comprehensively than at any international conference in modern history where the outcome had not been decided beforehand by force of arms.

32
NEVER AGAIN

If by the end of next week we have not got an ambitious agreement, it will be an indictment of our generation.

Gordon Brown, December 7th 2009

How are we going to look on Friday or Saturday if there are more than a hundred heads of state and government from all over the world and that what we say to the world is it was not possible to come to an agreement?

José Manuel Barroso, December 14th 2009

Battle was joined on the second day. The *Guardian* leaked what quickly became known as the 'Danish text', a negotiating draft circulated by the Danes as COP president.

Sudan's Lumumba Di-Aping claimed the draft destroyed the Kyoto Protocol and the UN. 'It sets new obligations for developing countries,' Di-Aping charged. 'It does away with two years of negotiations.'[1] Oxfam's Antonio Hill said the draft was a backward step. 'It tries to put constraints on [emissions in] developing countries when none were negotiated in earlier UN climate talks.' Andy Atkins for Friends of the Earth called the text profoundly destructive: 'It violates the principles of UN negotiations.'[2]

In part, the vehemence was synthetic. 'Some changes, but nothing earth-shattering,' a veteran developing country negotiator told the *Earth News Bulletin*. For India and China, the thirteen-page document contained two highly objectionable features.

The first was substantive. The parties were to agree that global emissions should peak 'as soon as possible,' with 2020 in square brackets as the backstop year and that *global* emissions should be cut by fifty per cent by 2050 (compared to 1990) and by an unspecified X per cent by 2020.[3] It set out a vision of long-term cooperative action, under which all parties, except the least developed nations, would have been obliged to take some form of mitigation action, i.e., be subject to limit or reductions in their greenhouse gas emissions.

Developing countries would undertake to reduce their collective emissions compared to business as usual (here the text provided a Y) to peak at a specified date and decline thereafter. On Attachment B, each developing country was to list its proposed actions together with the quantified emissions outcome expected from each action. Attachment B would morph into Appendix II of the Copenhagen Accord, in the process losing its essential feature – quantification.

Their second objection was procedural. The Danish text attempted to bring together in one agreement the two post-Bali streams, the first under the convention and the second under the Kyoto Protocol. The latter enshrined the Berlin

Mandate and provided the best defence against the threat of quantified mitigation targets.

It also had the benefit of keeping the eyes of the world on the performance of Annex I parties in meeting their Kyoto obligations. 'You will find a huge gap if you make a comparison between their pledges and the actions they have taken so far,' China's envoy Yu Qintai observed on the conference's third day. 'We have no lack of legal documents, but a lack of sincerity for taking action.'[4]

It was in the interests of India and China to frustrate post-Bali discussions. Even the EU, wedded as it was to Kyoto, recognized its insufficiency. On the conference's first day, Sweden, speaking for the bloc, stated that Kyoto alone could not achieve the goal of having emissions peak by 2020 and halve them by 2050 to meet the 2°C limit. When discussed two days later, India, China and Saudi Arabia opposed a new protocol, with China urging focus on countries implementing their existing Kyoto commitments.

In an attempt to close down this avenue, the formal discussions on future cooperation under the convention produced an unworkable outcome. 'The negotiating text evolved into the most complex document in the history of the UNFCCC, with nearly two hundred pages reflecting various proposals by all UNFCCC parties and thousands of brackets indicating areas of disagreement,' the authoritative *Earth News Bulletin* reported.

Seemingly arcane arguments over whether Copenhagen should produce a unified document or two separate ones in reality were a fight to determine the future of the climate change regime. A unified approach, reflected in the Danish text, would have demolished the Berlin Wall. Two documents meant primacy for the Kyoto Protocol. Because the US was not a party to the Protocol, the first agreement would probably be the last as it would most likely be dead-on-arrival in the United States Senate.

So the G77 plus China lined up foursquare behind Kyoto. 'The death of the Kyoto Protocol would be the death of Africa,' declared Ethiopia's PM Meles Zenawi, speaking for the fifty-three-nation African Union. 'I have been assured of China's support and India will probably take the same position.'[5] Surprisingly, Ban Ki-moon lent the neutrality of his office to the two documents camp.* Until there was a legally binding treaty, 'The Kyoto Protocol remains the only legally-binding instrument that captures reduction commitments,' Ban told the conference. 'As such it must be maintained.'[6]

On the conference's fourth day, a second front opened, with a skirmish between the US and China over money. China argued that countries like the US had a duty to pay out billions of dollars in compensation to poorer, developing countries. US negotiator Jonathan Pershing rejected China's demand. 'If you

---

* Before becoming UN Secretary-General, Ban Ki-moon had been South Korea's foreign minister, a country which was a major beneficiary of Kyoto in being a highly industrialized non-Annex I nation.

think about what will be prioritized in terms of the needs of the ... poorest countries, the countries that are hardest hit, I wouldn't start with China.'[7]

The spat got personal. Responding to a comment by Todd Stern that China was too prosperous to be a recipient of US climate funding, He Yafei expressed shock. 'I don't want to say the gentleman [Stern] is ignorant ... but I think he lacked common sense ... or he's extremely irresponsible,' He told a press conference. (Stern later described his comment as 'a bit unfortunate', but the position remained the same).[8]

At stake was one of the key principles of the climate change negotiations. 'Common but differentiated responsibilities' runs through the texts as frequently as 'sustainable development'. Indeed, 'common but differentiated responsibilities' is mentioned on the first page of the text of the convention and 'sustainable social and economic development' on the third. Yet the West and the developing world had a profoundly different understanding as to what it implied.

Western negotiators emphasized the 'common' part of the formulation and interpreted responsibilities to mean forward-looking obligations, an interpretation supported in the convention which goes on immediately to speak of countries' 're-spective capabilities and their social and economic conditions.' As the West is richer, it is more capable of taking action. As developing countries become wealthier, they too can assume greater responsibilities.

Developing nations emphasized 'differentiated'. For them, responsibility is historic – the nations who had been responsible for increasing the amount of carbon dioxide in the atmosphere are the ones obliged to take action. Developing nations could also point to the convention. Its third paragraph states:

> the largest share of historical and current global emissions of greenhouse gases has originated in developed countries, that per capita emissions in developing countries are still relatively low and that the share of global emissions originating in developing countries will grow to meet their social and development needs.[9]

At a meeting with journalists in Copenhagen, China's He Yafei argued that history was the basis on which the negotiations could move forward:

> For developed countries, they have to face the history squarely. The obligations for developed countries to live up to their commitments in emission reduction and the provision of funds and technology transfer, is an obligation they have undertaken ... The key, the prerequisite, for a successful Copenhagen conference, is that developed countries need to live up to their responsibilities.[10]

On America's denial of China's eligibility for climate change funding, it was as if

> all of us are sitting for dinner, we finish the main course, and then comes the dessert. The poor man walks in and sits down and has dessert. And we say right, you have to pay for the meal.[11]

Was China like the poor guy being slapped with the bill?

> We still have over one hundred and fifty million people under the poverty line, according to UN standards. If you care to go into the interior parts of China, the south-western parts of China, you will see lots of poverty. So poverty reduction – to provide a better life for Chinese people – is and will be the priority for the Chinese government.[12]

Whatever the outcome of Copenhagen, He warned, it should not be done at the expense of the rights to development by developing countries.

Yet the US and the EU and other Western nations were adamant that in some way or another, developing countries would have to limit and then reverse the growth of their greenhouse gas emissions. Was there any way of bridging the divide, perhaps with 'green growth'? Takers for that proposition were hard to come by.

As each day passed, expectations were downgraded. Before the start of the COP, the possibility of Copenhagen producing a treaty text was replaced with the aim of a strong political declaration with a firm timetable to a treaty before 2010 was out.

On December 15th, Pope Benedict XVI sent a World Peace Day plea. 'It is indeed important to recognize, among the causes of the current ecological crisis, the historic responsibility of the industrialized countries.'

The next day, Connie Hedegaard, Denmark's environment minister, resigned as COP president. Her handling of the conference had not been a success. In 2010, she was appointed the EU's first climate change commissioner, where she endeavored to start a trade war with China by extending the EU Emissions Trading Scheme to international airlines.

Danish Prime Minister Lars Løkke Rasmussen took over. His debut as COP president was peppered with hostile interventions challenging the status of the texts proposed by the Danish presidency. It was an issue of 'trust between the host country and parties', China said, noting that the process had not been transparent.

That afternoon, the working group on long-term cooperative action under the convention held its closing plenary. It was chaired by the convention's first and longest serving executive secretary, Michael Zammit Cutajar. So many changes had been suggested that it would not be possible to prepare texts in time for the COP plenary two days later, Zammit Cutajar said. His proposal that the entire package be adopted as 'unfinished business' was accepted.

If this group wasn't going to produce a text, who was?

A day later, with world leaders gathering in Copenhagen, Angela Merkel voiced her fears. 'The news reaching us is not good.' A sense of foreboding spread through the delegations. 'There are more than one hundred and thirty leaders here. If they cannot seal a deal, who can?' asked Ban Ki-moon. 'I believe in God. I believe in miracles,' Brazil's President Lula declared the next day.

The outside world had little inkling how badly things were going. The Danes ensured that the country delegations were hermetically sealed from the NGOs and the media. According to ITN's Jon Snow, covering the conference for Channel 4 News, tight security made it very difficult to doorstop conference participants. So the media relied on NGOs, who didn't know either, and Yvo de Boer, whose press briefings seemed perpetually upbeat.

The West had one more card to play. 'It was unforgettable political theater,' reported the *Independent*'s Michael McCarthy. 'Like a poker player with a sudden new bet, the power-dressed Mrs Clinton changed the game instantly as she pulled her gigantic sum out of the US back pocket.' The United States, announced Clinton, was willing to work with other countries towards a goal of mobilizing $100 billion a year by 2020 – conditional on all major economies standing behind 'meaningful mitigation actions'.

The message was clear – and it wasn't subtle: an extra $100 billion a year for Africa if China capitulated; not a cent if it didn't. In any case, it was made up of funny money – 'public and private, bilateral and multilateral, including alternative sources of finance', Clinton said in her speech. As an administration official explained, 'The private sector is going to be the engine that drives all this.' In other words, flows which in all likelihood would be happening anyway, so much of it wouldn't be additional money. 'A lot of this is not aid in the traditional sense of aid,' the official said. How could it be otherwise, given that the West's public finances had been shot to pieces? Republican congressional leaders said they would introduce a 'disapproval resolution'. Republican House leader John Boehner commented, 'The administration wants to give billions in US taxpayer dollars we don't have to other countries.'

On the last Thursday of the COP, there was a high-level segment for speeches by world leaders. Gordon Brown, the son of a Presbyterian minister, was biblical. 'Hurricanes, floods, typhoons and droughts that were once all regarded as the acts of an invisible god are now revealed to be also the visible acts of man,' Brown said. His talks during the week had convinced him that while reaching agreement was difficult, there was, 'no insurmountable [sic] wall of division' that prevented agreement.

Where Brown was declamatory, Angela Merkel, the daughter of a Protestant pastor, was earnest, more of a 'let's all hold hands' homily than a sermon. 'We need to help each other. But we also need to stand ready to change our way of living, our lifestyle.' Renewable energy was 'so important'. Germany and the European Union were ready and willing to 'open their arms' to take the negotiations forward and reach agreement so that world leaders could face the world on Friday and be able to say they had got the message: 'Life cannot go on as it was. The world needs to change.'

Jabbing the air, French President Nicolas Sarkozy harangued delegates. 'A failure in Copenhagen would be a catastrophe for each and every one of us,' he said. Much of it was directed against China. 'Who could dare to say they should be against giving money to the poorest countries?' He was crude. If there wasn't an agreement, 'Let me say to my African friends, you'll be the first to suffer from it.' Without naming them, he dared the Chinese to come to the podium. 'Who would dare say that the poor countries of Asia should be treated the same way as Brazil and China, the giants of tomorrow?' He railed: '*Mes chers amis*, time is short. Let's stop posturing.' It was pretty desperate.

Even as the Europeans spoke of the perils and dangers of climate change, a rumble from Europe's periphery betokened a real crisis. During the conference's first week, credit rating agency Fitch had cut Greece's rating to BBB+ with negative outlook the day after Standard & Poor's said it was considering downgrading Greece.

The conference's Big Three – the leaders of the US, China and India – were slated to speak the following day. Before then, Queen Margrethe II hosted a dinner. Sarkozy had called for the leaders to negotiate the agreement after the gala dinner, so the outcome could be presented to the world on Friday, the COP's last day.

During the dinner, Wen Jiabao picked up a rumor that he hadn't been invited. Annoyed at the perceived snub, Wen repaired to his hotel, sending He Yafei in his place. It meant that China was represented by someone who knew the negotiating texts better than any of the leaders in the room.

As daylight broke on Friday morning, Air Force One touched down at Copenhagen's Kastrup airport. It was cold and windy and started to snow as Obama stepped into his limo – his second trip to Scandinavia in a week. The first was to Oslo to pick up his Nobel Peace Prize. It was also his second trip to Copenhagen in three months, having flown there to support Chicago's losing bid for the 2016 Olympics. This time Obama was prepared for failure and armed with a draft statement prepared by Jonathan Pershing for that eventuality.

To Jon Snow, it seemed as if the drama that tumultuous day centred on the schedule of the president's plane. Obama hadn't planned to stay long. A snowstorm was forecast to hit the eastern seaboard, bringing forward his return flight.

The next twelve hours were not only to be the end point of two years of negotiations since Bali to agree a successor to the Kyoto Protocol. For the first time, the international community was to come together to agree a comprehensive, global approach. 'Kyoto was a treaty that aimed at making a point, but less successful at making a policy,' Blair told a meeting at the conference, 'Copenhagen is where we need to make a policy.'[13] December 18th 2009 was thus the culmination of nineteen years of climate change negotiations since the UN General Assembly adopted the resolution establishing the Intergovernmental Negotiating Committee for a Framework Convention on Climate Change nineteen years before – on December 21st 1990.

The speeches of the Big Three delineated the gulf separating the West and the world's major emerging economies. President Obama used his not to persuade, but to lecture. In similar language used to justify the Iraq War, climate change posed a 'grave and growing danger'.[14] America was taking action to cut its emissions not only to meet its global responsibilities, but also because it was in America's self-interest. America would proceed to a 'clean energy' economy, no matter what happened in Copenhagen. It was essential to America's national security to reduce its dependence on foreign oil, an argument used to domestic audiences, which jarred with an international one comprising America's trade partners and oil exporters.

Any agreement must have a mechanism to assess compliance. 'Without such accountability, any agreement would be empty words on a page,' he said. 'It would be a hollow victory.' This was aimed at China, which had been resisting pressure to accept some form of verification. By then, it had signaled it was ready to concede – in a way that made the entire issue moot.

With time running short, 'The question is whether we will move forward together or split apart, whether we prefer posturing to action.' He left a sting in the tail. 'We have charted our course. We have made our commitment,' he concluded. The message was stark: take it or leave it.

To those who hoped Obama would rescue the conference by making some grand offer, it was a terrible letdown. At no point in his remarks had the president mentioned the goal – so dear to the Europeans – of limiting the increase in global temperatures to 2°C, which Obama and other members of the G8 had affirmed at L'Aquila in July.

The reaction in the Tycho Brahe conference hall was as chilly as the weather outside. Suddenly the realization dawned on the NGOs – and from them to the world beyond – that Copenhagen was heading for disaster and Obama was doing nothing to stop it. In truth, there was nothing he could have done to have turned the conference around.

Every inch undisputable Third World authenticity in a dark Nehru jacket and blue turban, India's Manmohan Singh recognised the sensibilities of Western environmentalism. He praised the 'valiant efforts' to build a global consensus on 'highly complex issues'.[15] The vast majority of countries did not support any renegotiation or dilution of the principles and provisions of the climate change convention and in particular the principle of equity enshrined in 'common but differentiated responsibilities and respective capabilities'.

More action should be expected than at the time of the Rio treaty and the Kyoto Protocol. That meant adhering to the Bali mandate for future discussions

– a clear signal that agreement wasn't going to be reached at Copenhagen. Kyoto should stand and the parties to it should deliver on their 'solemn commitments'. India, entirely voluntarily, had made a pledge to reduce the emissions intensity of its economic growth by around twenty per cent by 2020. India would deliver this goal regardless of the outcome of Copenhagen.

Speaking deliberately, Singh told the conference that any agreement on climate change must respect the need for development and growth in developing countries. Every citizen had an equal right of entitlement to the global atmospheric space, Singh concluded.

The growth of developing country emissions was now the single largest driver of global emissions. India's position was incompatible with Western demands for a global limit on carbon dioxide emissions and setting a date when emissions would peak and then decline. If the computer models relied on by the IPCC were to be believed, the 2°C aim could be consigned to history.

Like Obama and Singh, Wen Jiabao did not mention the 2°C goal in his speech, half of which was taken up with telling the conference what China had already done. In modernizing its economy, China had scrapped heavily polluting steel, cement and coke-making plants. It was investing in renewable energy and had planted the world's largest acreage of man-made forests. It had reduced its carbon dioxide intensity by forty-six per cent in fifteen years and aimed to do so again.

'Dear colleagues,' Wen continued, the principle of 'common but differentiated responsibilities' was the 'core and bedrock' of international cooperation on climate change. 'It must not be compromised,' Wen stated. Developed countries had accounted for eighty per cent of total carbon dioxide emissions since the Industrial Revolution:

> If we all agree that carbon dioxide emissions are the direct cause of climate change, then it is all too clear who should take the primary responsibility. Developing countries only started industrialization a few decades ago and many of their people still live in abject poverty today. It is totally unjustified to ask them to undertake emission reduction targets beyond their due obligations and capabilities in disregard of historical responsibilities, per capita emissions and different levels of development.[16]

Wen criticized the West for its failure to honor its commitments. A one-thousand-mile journey starts with the first step, yet emissions from many developed countries had increased. It was more important to focus attention on achieving promised reductions in the short-term than setting long-term goals, Wen argued.

Like India, China's carbon-intensity reduction was its own affair and not linked to any other country's target. The formulation used by both leaders conveyed a clear message. India and China were not going be parties to an agreement which specified global emissions caps or dates by when emissions should peak.

Gordon Brown called Copenhagen 'the most important conference since the Second World War'.[17] The decisive confrontation took place at a four-and-a-half-hour meeting that afternoon in the Bella Center's Arne Jacobsen room.

Present were the conference Big Three – Obama, Singh and He Yafei for China; the European Big Three – Merkel, Sarkozy and Brown. The other two BASIC countries were there – South Africa's Zuma and Brazil's Lula, plus an assortment of other leaders, including Ethiopia's Meles Zenawi representing Africa, the Maldives' Mohamed Nasheed for the small island states, Mexico's Felipe Calderón (host of the next COP), Australia's Kevin Rudd (climate change's golden boy), and Norway's Jens Stoltenberg (after Canada's loss of interest, representing the original conscience of global warming) – twenty-five countries in a summit chaired by Denmark's Rasmussen. With them were Ban Ki-moon, other senior UN officials and leaders' advisers. Altogether there were fifty to sixty people in the room.

The leaders spent their time working on a draft that resembles an evolved version of the controversial Danish text based on the framework of a global cap – a reduction of 'X' by 2020 and 'Y' by 2050. A 1.2-gigabyte sound recording of the meeting found its way to *Der Spiegel*, which in May 2010 published an account of the meeting.*

Rasmussen asked if there were any major objections. 'We just have to go,' Merkel responded. 'And if we do not agree today then we have to say within four weeks because we cannot go over and say nice things but X and Y please wait one year or so.'

He Yafei said the meeting needed to go through the draft line by line, comment by comment (death by square bracket, the fate that had befallen Zammit Cutajar's text).

An Indian representative spoke in support of China: 'We have all along been saying "don't prejudge options,"' provoking Merkel to retort, 'Then you don't want legally binding.' To which the Indian responded: 'Why do you prejudge options? All along you have said don't prejudge options. This is not fair.'

He Yafei expressed surprise the numbers were still there: 'We have made our objections very clear this morning.'

Gordon Brown intervened. 'We are trying to cut emissions by 2020 and 2050. That is the only way we can justify being here. It is the only way we can justify the public money that is being spent to do so' – an observation that must have struck the Chinese and Indians as otiose.

It was also a misreading of their goal. The duo's strategic objective consisted of destroying any prospect that their economies would be subject to international emissions controls – now and at any point in the future.

---

* Excerpts from the recording can be streamed at http://www.spiegel.de/international/ world/0,1518,692861,00.html

That was clear when China vetoed the inclusion of a target by industrialized countries of eighty per cent cut by 2050. 'Why can't we even mention our targets?' demanded Merkel, recorded Mark Lynas, an environmental activist advising Nasheed of the Maldives.[18] At this point, according to Lynas, Rudd started banging his microphone and a Brazilian representative asked why rich countries should not even announce a unilateral cut. When the Chinese delegate said 'no', Merkel threw up her hands.

Merkel had one more go. 'Let us suppose a one hundred per cent reduction, that is, no CO_2 in the developed countries any more. Even then, with the two degrees, you have to reduce carbon emissions in the developing countries. That is the truth.'

China fully understood the logic of the two degree target. 'Thank you for all these suggestions,' He responded. 'We have said very clearly that we must not accept the fifty per cent reductions. We cannot accept it.'

A lit fuse reached the gunpowder. Sarkozy exploded. Speaking in French, he angrily declared China's position 'absolutely unacceptable' before accusing China of hypocrisy. Poor countries would not be getting the money promised them because of China's refusal to shoulder responsibility for the fifty per cent.

Then President Obama spoke.

Unlike his French colleague, the American was clinical, almost detached. The politics meant that the financing commitments had to be linked to adoption of the targets. 'In order for us to mobilize the political will within each of our own countries to not only engage in substantial mitigation efforts ourselves, which are very difficult, but also then to channel some of the resources from our countries, is a very heavy lift,' Obama said. 'If there is no sense of mutuality in this process, it is going to be difficult for us ever to move forward in a significant way.'

In the conference hall earlier in the day, Obama had been a unilateralist. Now, behind closed doors, Obama was a multilateralist. He could read the polls as well as anyone. A Gallup survey two days earlier suggested that a majority of Americans would support him on a climate pact, but that eighty-five per cent of Americans wanted the economy, not global warming, to be the focus. By linking economic issues to tackling climate change, poll-driven talk of green growth was for domestic consumption, not an argument to be used with India and China.

An AP Stanford University poll suggested how much or, rather, how little Americans were willing to pay. While three quarters of respondents said they would support action to address climate change, fifty-nine per cent said they wouldn't support any action if it increased their electricity bills by $10 a month.

Obama then expressed irritation at Wen Jiabao's absence. 'I am very respectful of the Chinese here,' Obama said, addressing He Yafei, 'but I also know there is a premier here who is making a series of political decisions. I know he is giving you instructions.'

That rebuke was as nothing as to what the president said next. Obama turned to Sarkozy. 'Nicolas, we are not staying until tomorrow,' he said. 'I'm just letting you know, because all of us obviously have extraordinarily important other business to attend to.'

He Yafei reacted coolly, as well he might, for Obama had just told him all he needed to know. Obama wanted an agreement. 'I am saying that, confident that, I think China is as desirous of an agreement as we are,' Obama had told the meeting. Saving face was Obama's objective. Any agreement would do.

Now the Europeans were not only isolated. They had been crushed. For them, climate change was an existential issue. There could be no more important business than saving the planet. Obama wanted a deal with China and India more than he was willing to use the collapse of the talks to try to isolate the duo and force them to concede.

'I heard President Sarkozy talk about hypocrisy,' He Yafei said in his perfect English. 'I think I'm trying to avoid such words myself.'

He then asked for the meeting to be suspended for a few minutes for some consultations. It was around 4pm.

The meeting did not reconvene.

According to the conference schedule, the closing plenary was meant to have started at 3pm and be over by 6pm. An hour after it was due to have started, there was still no agreement. The explanation was simple: there was no agreement because there was no agreement.

Events then took a bizarre turn. In an account given to journalists traveling back with the president on Air Force One, Obama wanted one more try. He decided to meet directly with the leaders of the BASIC group. The president's advance team was told that Singh was already at the airport; the Brazilians didn't want to meet without the Indians there. Zuma, according to this account, also cried off. 'If they're not coming, I can't do this.'[19] Wen's team then came on the line to say the Chinese were ready to meet Obama.

Obama's team then went to find a room for the meeting, only to find they could not get in. 'We've now figured out why we can't get into that room: because that room has Wen, Lula, Singh and Zuma,' the official said. 'They're all having a meeting.'[20]

Obama decided to go in.

'Mr Premier, are you ready to see me? Are you ready? Mr Premier, are you ready to see me? Are you ready?' Obama cried.[21]

'I'm going to sit by my friend Lula,' as an aide to the Brazilian president gave up his chair.[22]

They spent eighty minutes working on a text. The outcome was the Copenhagen Accord.

'I don't think anything like this has ever happened, and I'm not sure whether something like this will ever happen again,' Yvo de Boer commented in what must be the understatement of the conference.[23] Speaking to journalists in Berlin, Merkel swore she would not risk the same humiliation again.

She was as good as her word. Of the European leaders, only Norway's Jens Stoltenberg attended the next COP in Cancún.

33
AFTERMATH

Nothing except a battle lost can be half so melancholy as a battle won.

Duke of Wellington, 1815

The first requirement of statesmanship is a sense of reality – Disraeli rather than Gladstone; Bismarck rather than Napoleon III; Nixon rather than Carter. At Copenhagen, the Europeans did not comprehend the entrenched positions of China and India. Barack Obama did. Having concluded an agreement with the BASIC Four, Obama did what many advise American presidents finding themselves in intractable overseas entanglements. He declared victory and left.

At 10.30pm, Obama held a press conference to announce a 'meaningful and unprecedented breakthrough'.[1] It was, the president claimed, the first time in history all major economies had come together to take action on climate change. 'OK, thank you very much everybody, we'll see some of you on the plane.'

It was over. Obama couldn't get out of Copenhagen fast enough. He had, as he'd told the Europeans, more important business to attend to.

On the flight back, the president secured the sixtieth Senate vote for his healthcare legislation. 'After a nearly century-long struggle, we are on the cusp of making healthcare reform a reality in the United States of America,' Obama exulted.[1]

Climate change was a priority for Obama. It just wasn't the most important. In his June 2008 speech on winning the Democratic nomination, the first thing people would be telling their children wasn't about the slowing of the rise of the oceans and the healing of the planet. It would be about when America started providing care for the sick.

There was one breakthrough. Malta became the first non-Annex I party to attain Annex I status in the history of the convention.* If it was a miracle, it wasn't a divine agent but Malta's Zammit Cutajar to whom the credit belonged.

China spoke of a 'positive' result, though not of a breakthrough. 'All should be happy.' Happiness was in short supply in the Bella Center. Jon Snow saw eager young men and women in NGO T-shirts in tears. Copenhagen ended 'more dramatically badly than any conference I've ever been to', Snow – who had covered every major summit from the Reagan-Gorbachev meetings – recalled. It was meant to have been the great moment. 'This was going to be it.' Instead it turned out to be the moment when all the wind went out of the climate change negotiations.

Denmark's Rasmussen had the thankless task of trying to get the Accord approved by the COP plenary. The session started at 3am on Saturday. It was a rocky

---

* To date, Malta is the only developing country to have graduated to Annex I. Croatia, the Czech Republic, Slovakia and Slovenia were the products of parties to the convention that had split. The other addition to the original list was the Principality of Liechtenstein.

ride, lasting nearly thirteen hours. Tuvalu, one of the most vocal small island states, could not accept the Accord, despite the offer of financing. To do so would 'betray our people and sell our future, our future is not for sale'.[2] Saudi Arabia's delegate declared Copenhagen 'the worst COP plenary' he had ever attended. It was evident there was no consensus and the parties were simply restating their positions.

Speaking for the G77, Lumumba Di-Aping compared the proposed deal to the Holocaust. '[This] is asking Africa to sign a suicide pact, an incineration pact in order to maintain the economic dependence of a few countries. It's a solution based on values that funnelled six million people in Europe into furnaces.'[3]

Di-Aping's remarks were immediately condemned by the UK, Norway, the EU and others. They hadn't objected when Gore made the comparison at Bali two years earlier or when the world's most famous climate scientist, James Hansen, called coal trains 'death trains'.

The NGOs went into over-drive, demonstrating how they had lost the plot. 'By delaying action, rich countries have condemned millions of the world's poorest people to hunger, suffering and loss of life,' said Nnimmo Basey, chair of Friends of the Earth. 'The blame for this disastrous outcome is squarely on the developed nations.'[4] Kumi Naidoo for Friends of the Earth denounced the Accord as a 'betrayal of the poor' and blamed the conference failure on racism. Why was there such a lack of urgency? 'Is it because of the colour of their skin' of those in the front line of climate change, Naidoo asked?[5] It was a question better addressed to Beijing and New Delhi rather than Berlin and Paris.

As the plenary session wore on, Ethiopia, on behalf of the African Union, supported the Accord. Tuvalu then backtracked and decided to betray its future, suggesting that it be adopted as a 'miscellaneous document' of the COP. The UK proposed adopting the document as a COP decision. Opposition from Bolivia, Cuba and Venezuela and two or three others prevented consensus. At 10.35 on Saturday morning, it was proposed that the COP 'take note' of the Copenhagen Accord. Parties could signify their (non-legally binding) support for the Accord by being listed in an appendix.

Hours of procedural wrangling followed. Late on Saturday morning, the COP finally decided to 'take note' of the Accord. 'You sealed a deal,' Ban Ki-moon told exhausted delegates – only they hadn't.[6] 'We will try to have a legally binding treaty as soon as possible in 2010,' Ban promised reporters.[7]

There wasn't a treaty in 2010. Or in 2011.

The institutional gulf in the Accord separating Annex I nations from the rest was laid bare in the respective appendices for both groups. Appendix I listed each Annex I party's quantified emissions target for 2020. In Appendix II, there was a list for *unquantified* mitigation actions by developing country parties. In Attachment B of the Danish Text, there had been an additional column for the quantified impact of each action in units of millions of tons of CO_2 equivalent.

In Appendix II of the Accord, it had disappeared. Obama's warning that it would be a hollow victory unless developing country commitments were 'measurable, reportable and verifiable' turned out to be exactly that. The actions might be reportable and even verifiable, but they were no longer measurable.

Out too went any specified year when global and national emissions should peak. The Accord's aim was to achieve a peak as soon as possible

> recognizing that the time frame for peaking will be longer in developing countries and bearing in mind that social and economic development and poverty eradication are the first and overriding priorities of developing countries and that a low-emission development strategy is indispensable for sustainable development.[8]

There was just enough in the Accord to keep the whole negotiating process going indefinitely and provide cover for European governments to continue with their global warming policies. Everything could go on much as before. And, if the scientists and the IPCC were right, it would mean unabated global warming.

Angela Merkel put on a brave face. 'It is a first step toward a new world climate order, nothing more but also nothing less,' she told the *Bild am Sonntag*. What was needed, Merkel thought, was 'a UN environment organization that could control the implementation of the climate process.' Quite how that could come about, she didn't say.

Nicolas Sarkozy was closer to the mark. The deal was the only one that could be reached after the summit had revealed deep rifts. The UN process of moving forward by consensus of one hundred countries was not workable. Although China was a permanent member of the UN Security Council, if India were too, 'it would be far easier to get her to shoulder a greater proportion of her responsibilities', Sarkozy said. 'Seeing a system like this makes it blindingly obvious.'[9]

More likely, Sarkozy's explanation had it the wrong way round. Given the opprobrium heaped on those viewed as being on the wrong side of the argument about global warming, the UN negotiating forum was ideal for China and India. Fundamental disagreements could be masked as procedural objections and proxies could make arguments on their behalf. If the world's major economies were genuinely agreed on a way forward, it is hard to believe that the rest of the world would not have followed, as happened with ozone-depleting substances and the Montreal Protocol. The fundamental reality at Copenhagen was the failure of the West to get its way.

Some found it hard to come to terms with. Gordon Brown and Ed Miliband decided to pick a quarrel with the world's second largest economy. Writing in the *Guardian* on the Monday after the conference, Miliband accused China of vetoing quantified caps. 'We will make clear to those countries holding out against a binding treaty that we will not allow them to block global progress,' Miliband threatened. The process had been hijacked, presenting a 'farcical picture'.[10] There needed to be major reform in the way climate change negotiations were conducted.

'Such an attack was made in order to shirk the obligations of developed countries to their developing counterparts and foment discord among developing countries,' a spokeswoman for China's foreign ministry said.[11] The next day, Brown took up the cudgels. 'Never again should we face the deadlock that threatened to pull down those talks,' Brown thundered in sub-Palmerstonian vein. 'Never again should we let a global deal to move towards a greener future be held to ransom by only a handful of countries.'[12] How was Britain going to take on China? Send a gunboat up the Yangtze? Brown didn't say.

Meanwhile in Beijing, French premier François Fillon arrived with a delegation of ministers and business leaders. 'Even though it is our first meeting in person, I feel like we are friends,' Wen told Fillon. 'Our two countries' partnership is unmatched,' Fillon replied. The two signed agreements on aviation and nuclear cooperation.[13] A source said the French would leave China with €6.3 billion of signed contracts.

Nonetheless, China remained sensitive about its role at Copenhagen. The same day, Wen declared that China had played an 'important constructive' part at the climate conference. Maintaining China's climate alliance with India was vital. At a meeting with Manmohan Singh, the Chinese premier gave an assurance that China 'would like to pursue relations with India on the basis of equality', according to India's foreign minister Nirupama Rao.[14] Global warming had brought together these two longstanding Asian rivals into an alliance to protect their common interests against the West.

'It is easy to discern China's fingerprints all over the international climate change fiasco,' wrote John Tkacik, a former chief China analyst at the State Department's Bureau of Intelligence and Research.[15] If his intelligence briefers had been reading the Chinese press, Obama would have seen it coming. In a September article in Beijing's *Science Times*, Ding Zhongli, China's top paleo-climatologist and VP of the Chinese Academy of Sciences, had written that 'the idea that there is significant correlation between temperature increases and concentrations of atmospheric carbon dioxide lacks reliable evidence in science'. Given the deep scepticism of the Chinese Academy of Science's senior climatologist, Tkacik wrote, 'It is clear that no one in the Chinese politburo is truly anxious about the climatic consequences of global warming.'

In New Delhi, the opposition Communist Party of India (Marxist), known by its initials CPM, accused the Singh government of being too soft and making pre-summit concessions which enabled the US to undermine the Kyoto Protocol. 'The CPM had warned the government that unilateral concessions before the negotiations, and without conditional linkages to deep cuts by developed countries, will not yield results. This is indeed what has happened.'[16]

As for the Indian government, in a March 2010 interview in the *Wall Street Journal*, environment minister Jairam Ramesh described himself as a climate agnostic. Ramesh, who headed India's delegation at Bali and addressed the Copenhagen conference, said the climate negotiations were in a 'complete quagmire' and heading nowhere.[17] 'In many parts of India people are dying because of excess pesticides in the water, or arsenic in the water,' according to Ramesh. 'That's more important and more urgent than climate change.' The other BASIC nations shared the same perspective. They had bonded 'very well' at Copenhagen. 'We are united in our desire not to have a binding agreement thrust upon us which will constrict our developmental options,' Ramesh said.

China could even claim the support of the weather. The first week of 2010 saw Beijing temperatures plunge to -18°C as a blizzard dumped the heaviest snow fall in a single January day since 1951. For two months beforehand, China had experienced widespread gas shortages as demand rose because of the unusually cold weather. In Seoul, more than ten inches of snow fell on the Korean capital – the greatest amount since records began in 1937. In Mongolia, temperatures fell to -50°C. Ferocious winter conditions killed almost ten million sheep, cattle, goats, horses and camel – a fifth of the country's total.

In Washington, Senator Inhofe's grandchildren built an igloo with a sign saying 'Al Gore's New Home'. 'This isn't a good old-fashioned winter for the District of Columbia, not unless you're remembering the last ice age. And it doesn't disprove global warming,' wrote Bill McKibben, the Copenhagen bell-ringer, in the *Washington Post* in February. 'Instead, the weird and disruptive weather patterns around the world are pretty much exactly what you'd expect as the planet warms,' McKibben rationalized.

Whether or not global warming caused the heavy snowfall, McKibben's claim that more snow was exactly what was expected contradicted one of the most famous predictions in the history of global warming. Within a few years, winter snowfall will become 'a very rare and exciting event', David Viner, of the University of East Anglia's Climatic Research Unit told the *Independent* in March 2000. 'Children just aren't going to know what snow is,' Viner forecast.

Mongolian nomadic herdsmen had a name for the harsh winter – *dzud*, meaning 'white death'. Writing three days after McKibben, the *New York Times'* Thomas Friedman coined his own. Avoid the term 'global warming', Friedman advised. 'I prefer the term global weirding.' As global temperatures rise, the weather gets weird, Friedman asserted.

Global weirding even found its way onto the website of the National Oceanic and Atmospheric Administration (NOAA). 'Global warming or global weirding?' Don't draw long-term, large-scale conclusions from short-term local weather patterns, was NOAA's careful answer.

The scientists at NOAA had good reason to be cautious. As a scientific concept, global weirding is, well, weird. Following Popper's criterion of falsifiability, the more a theory states that certain things cannot happen, the stronger the theory is. The problem with global weirding is that it doesn't preclude anything – rather it suggests the opposite. Anything could happen.

And how is weirdness measured? In Britain in 1975, near the start of a three-decade rise in global average temperatures in a record-breaking hot summer, the weather suddenly turned cold in the first week of June. Derbyshire was playing Lancashire in a county cricket championship match at Buxton. It started to rain, then snow. 'When I went out to inspect the wicket, the snow was level with the top of my boots. I'd never seen anything like it,' recalled umpire Dickie Bird. Is summer snow in Buxton more or less 'weird' than winter snow in Washington?

Despite this, global weirding captures a reality that changes in global average temperature do not. We only experience local temperature and local weather. No one experiences a global average temperature – a statistical artefact created by climate scientists. As measured by global average temperature, 2010 was one of the warmest years on record. Yet what was experienced by Mongolian nomads, the inhabitants of Beijing or Senator Inhofe's grandchildren in their igloo was quite different. If climate scientists had not been pre-disposed to worry about global warming, would anyone have noticed that 2010 was statistically one of the warmest years on record?

The lack of observed warming during the first decade of the new century began to create a stir among climate scientists. In May 2009, Phil Jones, Britain's best known climate scientist, reassured a colleague in a government funding agency. 'Bottom line – the no upward trend has to continue for a total of fifteen years before we get worried,' Jones wrote – giving himself more time by dating the fifteen years from 2004/05, and not 1998, as that was an El Niño year.[18] It was as if a doctor was dismayed rather than pleased to find a patient's disease hadn't progressed as fast as he had anticipated. In such a situation, most people would find themselves another doctor.

While climate scientists might be over-committed to the idea of climate warming, political leaders most over-invested in global warming turned out to be Copenhagen's biggest political casualties. No politician had toiled as hard as Gordon Brown. Trailing in the polls, a breakthrough at Copenhagen was Brown's last chance of pulling around his electoral fortunes, even though, when talking climate change, it often sounded like a foreign language Brown hadn't mastered. In the May 2010 election, Brown led Labour to its worst defeat since 1983.

For Australia's Kevin Rudd, Copenhagen was, in the words of opposition leader Tony Abbott, 'an unmitigated disaster'. Before the 2007 election, Rudd had described climate change as 'the greatest moral, economic and environmental challenge of our generation'. He then upped the stakes. Rudd dismissed opposition arguments to delay introducing cap-and-trade legislation until the outcome of Copenhagen was clear. 'The argument that we must not act until others do is an argument that has been used by political cowards since time immemorial,' Rudd said in November 2009. 'There are two stark choices – action or inaction. The resolve of the Australian Government is clear – we choose action, and we do so because Australia's fundamental economic and environmental interests lie in action,' Rudd declared.[19]

An RAAF plane was put on standby to fly Rudd to Copenhagen at short notice. A provisional delegate list numbered one hundred and fourteen Australians, including Rudd's official photographer, compared to a seventy-one-strong UK delegation.

In April 2010, Rudd did an about-turn. Blaming the opposition and the pace of international negotiations, he announced a delay of his Carbon Pollution Reduction Scheme until the end of 2012. Two months later, the Labor caucus heaved Rudd out of office and replaced him with his deputy Julia Gillard, whom Rudd also blamed for his decision to postpone the plan.

Global warming played a part in Rudd's downfall, according to his predecessor John Howard. It hurt him personally: one moment he had been saying it was the greatest moral challenge, then he was putting it off for two years, precipitating the sharp fall in his poll ratings.[20]

The biggest casualty of Copenhagen was the West's standing in the world. It had declared global warming the most serious issue of the age. The US and the EU had come together and pushed hard for a comprehensive agreement covering all the major economies. The build-up of pressure going into the conference was immense.

However, the West misunderstood the contingent nature of the Third World's participation in international environmental negotiations. Western politicians talked a story that the world's poor would be hardest hit by global warming. But the nations with the largest numbers of poor people had priorities that conflicted with what the West wanted. Its leaders understood better than their colleagues in the West what Bastiat had found in the nineteenth century – the best defence against capricious nature is wealth.

In this regard, it is striking how little had changed since 1972 and the Stockholm conference. There was some movement. The language of environmentalism made inroads. China committed to huge investments in renewable energy. Brazil moved from the outright hostility of its military rulers. Even so, Brazil stuck by China and India. When he got home from Copenhagen, Lula pointed the finger: 'The United States is proposing a reduction of four per cent from the date fixed by the Kyoto Protocol. That is too little,' Lula said on his weekly radio show *Coffee with the President*.[21]

At the same time, the power of the idea of global warming meant China and India couldn't simply say 'no'. So they fought the West by other means and played for time: fielding high caliber negotiators with a mastery of the conference texts and procedures; holding on to ground taken in previous rounds for as long as possible, especially the Berlin Mandate; using proxies to articulate their arguments and, most of all, maintaining the cohesiveness of the G77, despite the disparate interests of the small islands states and Africa from those of China and India. Throughout, they were aided by NGOs for whom blaming the West was encoded in their DNA.

In the end China and India succeeded. If there was going to be an agreement, it would have happened at Copenhagen. The Durban COP in December 2011 demonstrated the enormity of their victory at Copenhagen. Canada, Russia and Japan had already announced they would not enter into a second commitment period under the Kyoto Protocol. In setting the goal of reaching agreement to cover the period from 2020, the Durban Platform confirmed the failure of the Bali Road Map to reach its intended destination. There would be a gap between 2012 – the end of the first commitment period under the Kyoto Protocol – and the start of a first commitment period of a yet to be negotiated treaty.

The Durban Platform had a glaringly obvious credibility problem. If negotiating a legally binding agreement to cover the period after 2012 was too difficult, why would it be any easier to negotiate one for the period after 2020?

The scale of the challenge was made apparent immediately after the COP when Canada announced its withdrawal from the Kyoto Protocol. Environment minister Peter Kent said quitting the Protocol would save Canadians $13.6 billion it would otherwise have to spend on buying emissions credits from other countries.

Canada's repudiation of Kyoto – inconceivable two years before at Copenhagen – made the prospect of a new treaty even less plausible. It strengthened China and India's ability to resist pressure to cap their emissions. In public, they could hardly welcome Canada's move, so they did the next best thing. 'Any attempt by developed countries to casually set aside their existing legal commitments while calling for a new legally-binding agreement seriously questions their credibility and sincerity in responding to the climate crisis,' a February 2012 statement on behalf of the BASIC Four said.[22]

The long-standing position of China, India and the rest of the G77 was that developed nations must first demonstrate their commitment by actually cutting their emissions, the only interpretation of Kyoto that makes any sense. The US had not ratified Kyoto and now Canada had repudiated it, ostensibly because it did not cover developing nations. The result was to leave the climate negotiations in an unconsummated equilibrium with the potential to last indefinitely.

It left the EU to negotiate with itself and with Norway and Switzerland over the emissions cuts for a second Kyoto commitment period. Outside this hard-core, there remained a handful of unilateralists. At enormous political cost, the Australians under Julia Gillard passed a carbon tax. In the US, California is a staunch climate change unilateralist. The three branches of the federal government are split between multilateralists and unilateralists. Cap-and-trade legislation was unable to make it through the Senate and had no chance after the 2010 mid-term elections, when Republicans took control of the House of Representatives. In its five to four decision on *Massachusetts v. EPA*, the Supreme Court endorsed unilateralism. Within the executive branch, the EPA under its administrator Lisa Jackson has adopted a hard-line unilateralist position. Obama's position depended on the audience. In public, he was a unilateralist. Negotiating with India and China, he morphed into a multilateralist.

At the conclusion of the Copenhagen conference, Ban Ki-moon hailed the Copenhagen Accord as an 'essential beginning'.[23] At Bali two years earlier, Copenhagen was to be the final destination on the Bali Road Map.

'Pathwalker, there is no path,' Gore told the delegates at Bali in words that were more prophetic about the climate change negotiations than he'd hoped. Like the Voyager spacecraft after a multi-decade planetary tour, perhaps the fate of the COPs and the MOPs and their subsidiary bodies is to leave the solar system and journey into outer space.

34
REFLECTIONS

It is by solidity of criticism more than by the plenitude of erudition, that the study of history strengthens, and straightens, and extends the mind ... Be more severe to ideas than to actions; do not overlook the strength of the bad cause or the weakness of the good.

Lord Acton, 1895 Inaugural Lecture

For most cities which were great once are small today; and those which used to be small were great in my own time. Knowing, therefore, that human prosperity never abides long in the same place, I shall pay attention to both alike.

Herodotus

On February 4th 2010, Stephen Schneider gave one of his last lectures at Stanford, where he was professor of interdisciplinary environmental studies and of biology. Six months later, he died at the age of sixty-five. 'No one, and I mean no one, had a broader and deeper understanding of the climate issue than Stephen,' Michael Oppenheimer, Princeton professor of geosciences, told the *Washington Post*. 'More than anyone else, he helped shape the way the public and experts thought about this problem – from the basic physics of the problem, to the impact of human beings on nature's ecosystems, to developing policy.'[1]

Schneider's lecture showed why. He mixed a conversational style with an easy authority. No one could doubt who was the smartest person in the room. In ninety minutes, Schneider – who had been involved with the IPCC since 1988 – addressed issues rarely, if ever, ventured by other leading proponents of global warming. For this reason, Schneider's talk can claim to be the most important presentation by a climate scientist since James Hansen's congressional testimony.

Thomas Kuhn thought scientists were little better than laymen at characterizing the established bases of their field.[2] Schneider made a similar point. 'Very few people learn about the basic philosophy of science and how it works.'[3] Universities were handing out a Ph.D. in science with no 'Ph' in it. 'OK guys, now we're doing epistemology,' Schneider would tell his freshmen class – and spoke of his frustration that many of his colleagues did not study it.

Other climate scientists showed how they could have benefited from attending Schneider's class. One such was Kevin Trenberth, a senior physicist at Boulder, Colorado's National Centre for Atmospheric Research, IPCC lead chapter author and one of the world's most cited geophysicists. Trenberth's foray had been provoked by a Climategate email of his that had gone viral.

Responding to Schneider's request for help in rebutting a BBC report suggesting that there had been no warming since 1998, Trenberth emailed Schneider and Michael Mann. 'The fact is that we can't account for the lack of warming at the moment and it is a travesty that we can't.' People in Boulder were asking where the heck global warming was. The previous two days had smashed all previous cold records. 'This is January weather,' Trenberth wrote on October 12th 2009.[4]

In a January 2011 paper to the American Meteorological Society in tribute to Schneider, Trenberth argued that the null hypothesis test should be stood on its head. For there to be a relationship between two variables, the null hypothesis – that there is no relationship – should first be shown to be false.

Trenberth took the opinion expressed by the IPCC as foundational truth. Because the Fourth Assessment Report had declared global warming 'unequivocal' and 'very likely' due to human activities, 'the null hypothesis should now be reversed, thereby placing the burden of proof on showing that there is no human influence'.[5]

It is an axiom of the null hypothesis that it cannot be proven. By reversing the test, Trenberth set critics an impossible task. Even so, he mis-specified the problem. It is not whether humans might influence the climate, but by how much. Here Trenberth's viral email pointed to the real difficulty. It wasn't so much a travesty that Trenberth and other climate scientists were unable to reconcile rising carbon dioxide concentrations and flat global temperatures, for Trenberth might have been on a fool's errand. 'For small changes in climate associated with tenths of a degree, there is no need for any external cause,' Richard Lindzen wrote in 2009. 'The earth is never exactly in equilibrium.'[6]

Schneider was too smart to make such elementary mistakes on the null hypothesis and knew the obstacles he had to navigate around. Science subject to test by falsification should be distinguished from system science. 'Is the science of anthropogenic climate change settled?' he asked. It was a dumb question, which the class fell for. Why? 'Climate science is not like test tube science. You don't falsify. Eventually you do, but not right away,' Schneider explained. 'It's system science.'

Schneider characterized system science as built on a base of well-established components, then a layer of competing explanations and finally a layer of what he called speculative components. 'Every single complicated system science, whether we're talking climate science, healthcare, security, education, always is going to be in this category,' Schneider said, illustrating the convergence of natural science with economics and social sciences.

The convergence enables scientists to stake their claim to formulating government policy, traditionally the province of the social sciences, enhanced by its reputation as a hard science, even though system science was diluting it. Now natural scientists were trading the rigor of knowledge derived from experimentation and falsifiability for a lead role in determining public policy.

'Opinion' perhaps better describes its output. 'Knowledge' implies what is known whereas opinion indicates a statement of belief. In the physical sciences, what is determinative is not what scientists think or believe, but what can be demonstrated by formulating hypotheses and testing them against nature.

In forsaking falsifiability, climate scientists kept a problematic feature of scientific practice – the strong collective tendency to operate within an unquestioned, dominant paradigm. Whereas arguments between economists of different schools of economic thought (and often within them) created skepticism about any claim made by an economist, the adherence of scientists to a dominant paradigm creates the opposite impression.

This makes even more problematic the form of system science falsification permitted by Schneider. 'We do not falsify by single experiments. We falsify on the basis of accumulated numbers of papers and numbers of bits of information.' Determining the relative credibility of each was not simple. It took assessment groups like the US National Academy of Sciences, with their multiple disciplines arguing out the relative merits of various competing or speculative components. Yet the process described by Schneider could be applied to any field of learning. The process of institutional oversight, peer review of papers and so forth, sometimes mistakenly described as the scientific method, is not unique to science.*

Schneider returned to the subject of falsifiability towards the end of his lecture. 'There are still some people who think [climate science] operates on the basis of falsification.' In the case of system science it does – by 'community action over decades'. Thus the 'scientific community' is accorded the determinative role formerly given to experiments conducted on nature.

What defined this community? By scientific community, 'I'm talking about those people who actually do the work,' not non-climate scientists who drop in opinions from the outside. The release of the Climategate emails had made scientists 'very, very angry about their critics', Schneider said. The critics almost never showed up at scientific meetings.

> They just write blogs and screeds and do 'audits' without really being members of the community. So they're not welcome. That's absolutely true because they're not part of the debate. That's cultural. That's not a matter of who's right and who's wrong.

Climate science, Schneider continued, gave plenty of scope for those with ideological agendas to fasten on to particular elements to prosecute their case. Their arguments might be technically correct, but they would neglect others in what Schneider

---

* An example is provided by Naomi Oreskes, historian of science and co-author of *Merchants of Doubt: How a Handful of Scientists Obscured the Truth on Issues from Tobacco Smoke to Global Warming* (2010). In an interview on Australian radio, Oreskes said: 'Well, of course, no one can ever say, a hundred per cent, that any particular individual piece of science is correct, but that's what the scientific process is all about. And over the past four hundred years, scientists have developed mechanisms for evaluating scientific work, and the most important mechanism is peer review.' *The World Today*, October 19th 2010.

called 'selective inattention to inconvenient components'. Schneider dubbed this 'courtroom epistemology'; in other words: 'It's not my job to make my opponent's case.' Scientists, Schneider went on, would view that as an 'immoral philosophy'. Schneider was more robust. 'I don't agree with that because if I were accused of something, I don't want my lawyer dwelling on abstractions of truth. I want him to get me off.'

Schneider was particularly exercised by the role of the media. In reporting both sides of the debate, the media presented a spurious balance of the extreme ends of the bell curve of the possible outcomes of global warming.

> It is inconceivable that debate polarized between 'end of the world' and 'good for you' – which for me are the lowest probability outcomes – can possibly be properly communicated through that kind of advocacy dichotomy.

The job of climate scientists was to evaluate risk and the role of 'the community' was to winnow out the relative probabilities. For the media not to report the middle of the bell curve was to 'miscommunicate' the nature of science.

This, Schneider suggested, raised a further question: can democracy survive complexity? The media focus on points of contention and its alleged neglect of the mainstream view had created public confusion about the science. This, Schneider argued, led to policy paralysis.

The view that public confusion about the science – sowed by malign fossil fuel interests – stalled global action is only plausible if the history of global warming is ignored. The 1992 climate change convention had written the scientific consensus into a treaty signed by over one hundred and sixty-five nations. It had been swiftly and unanimously ratified by the United States Senate. The most consistent finding of opinion surveys is not skepticism about the science, but that tackling climate change came way down the list of voters' concerns. It was a convenient community myth to blame the West, when the true block on global action was the refusal of India and China. But then, what pull do climate scientists – or NGOs for that matter – have in New Delhi and Beijing?

Schneider's second problematic claim was to suppose that nature would conform to a bell curve of climate scientists' expectations of the future. In an unusually candid description, Schneider characterized scientific judgment as an objective set of issues with subjectivity buried when it's about the future, because there's no data about the future. In projecting the future, 'It's always subjective by definition.'*

---

*   'Buried' is the right word to use about the IPCC's handling of subjectivity. It does not appear in any of the summaries for policymakers in the IPCC's 2007 Fourth Assessment Report. Neither does it appear in the crucial Chapter Nine on understanding and attributing climate change, but is buried in the chapter's supplementary material.

In 1957, the scientists Hans Suess and Roger Revelle wrote of human beings carrying out a large-scale geophysical experiment by increasing the amount of carbon dioxide in the atmosphere. It is bad scientific practice to prejudge the outcome of a unique experiment. The laws of physics can't be used beforehand to determine the climate sensitivity of carbon dioxide, so no one knows if the climate sensitivity assumed by the IPCC will turn out to have been correct.[†]

Global warming involved a second experiment – not a geophysical one, but a political experiment of a scale and ambition surpassing anything before it. Because fossil fuels are the principal source of energy of industrial civilization, attempts to decarbonize society affect virtually every facet of government policy: energy policy, but also economic policy, land use planning and housing, transportation, agriculture, industrial policy and international relations – successive G8 summits and the largest gatherings of world leaders were devoted to solving global warming. Never has the impact of scientists on how societies are governed been as great. During the age of global warming, the West came closest to realizing Francis Bacon's ideal of a republic governed by a body of scientists that he made nearly five centuries ago.

For leading scientists, the political elevation of science was its due. In April 2010, the presidents of the Royal Society and the National Academy of Science, Martin Rees and Ralph Cicerone, wrote: 'Our academies will provide the scientific backdrop for the political and business leaders who must create effective policies to steer the world toward a low-carbon economy,' seemingly oblivious to the fact that at Copenhagen four months earlier, the world had failed to agree steps to decarbonize their economies.[7]

No one questioned the role scientists had acquired. Based on their collective record, natural scientists would be among the very last people to involve in government policy. During the First Environmental Wave of the late 1960s and early 1970s, they had predicted the imminent collapse of civilization and called for the abandonment of economic policies designed to satisfy humanity's material needs and generate rising prosperity. Their dire predictions turned out to be completely wrong and their recommendations disastrous – had any government been foolish enough to have acted on them.

[†] In its 2007 Fourth Assessment Report, the IPCC acknowledged that observational uncertainty, uncertainty about external influences on the climate ('forcing uncertainty') and internal climate variability affect the width of probable outcomes ('the prior distribution'). The prior distribution used in the calculations, the IPCC authors wrote, 'indicates that little is known, a priori, about the parameters of interest except that they are bounded below and above. Even so, the choice of prior bounds can be subjective'. Hegerl, G.C., F.W. Zwiers, P. Braconnot, N.P. Gillett, Y. Luo, J.A. Marengo Orsini, N. Nicholls, J.E. Penner and P.A. Stott, '2007: Understanding and Attributing Climate Change' in *Climate Change 2007: The Physical Science Basis. Contribution of Working Group I to the Fourth Assessment Report of the Intergovernmental Panel on Climate Change*, Appendix 9.B, SM.9–3.

And so it proved in the era of global warming. Across every dimension, global warming policy has been a costly fiasco. Unsustainable commitments to solar and wind energy in Germany and Spain; the morally abhorrent diversion by rich countries of resources from growing food into making biofuels; the collapse of the EU's carbon market; the transformation of the UK's liberalized energy market producing some of the cheapest electricity in Europe to become Europe's most expensive electricity producer; the scandals associated with the Clean Development Mechanism; the destruction of tropical rainforests to make way for palm oil plantations – all provide material for students of policy folly.

After the West pushed its demands to breaking point, global warming has been a story of defeat and retreat. Its failure at Copenhagen is a milestone in the deterioration of the West's prestige and the ascendance of China and India.

The implications of Copenhagen on the efficacy of global warming policies are nothing short of disastrous. Whatever the predictive merits of the science, the absence of a regime capping global greenhouse gas emissions rendered the West's global warming policies completely pointless. The results of global warming's political experiment already provide a definitive verdict: global warming policies have made the world unambiguously worse off, a conclusion which holds irrespective of the outcome of the geophysical experiment.

In addition to the higher cost of producing electricity from renewables (their capital costs alone are estimated to total $160 billion a year before the 2008 recession) over the most efficient conventional means – there are hidden costs. Energy accounts for a higher proportion of poorer household spending, so those on lower incomes are disproportionately hit by higher energy prices. For social democrats such as Tony Crosland in the 1970s, concerns about the welfare of the working class and poverty reduction trumped the claims of environmentalism, the opposite of the situation today.

Then there are the opportunity costs of global warming policies – the valuable activities that the world has foregone. These include the innovation and productivity boost from Silicon Valley venture capital dollars diverted into green tech investments. Because alternative energy projects depend on government support, entrepreneurs and energy utility executives are turned into government lobbyists maximizing their take from global warming policies.

Perhaps the biggest casualty is science. The rapid growth of climate science was not for the sake of pure knowledge, but because it is the leading branch of global therapeutics. Scientific knowledge can be of immense therapeutic benefit. Medicine is the outstanding example. The desire on the part of the physician or medical researcher to cure the sick is one of the finest of all human qualities. A similar motive drove Schneider, whose decision to become a climate scientist was, he once said, 'a marriage of convenience and deep conviction' – a decision he had made on Earth Day 1970 at the age of twenty-five to devote himself to the environment.

In medicine what matters is not the motive of the practitioner but the efficacy of the therapy. A century ago, Professor Lawrence Henderson of Harvard drew attention to the remarkable advances in medical science, technology and therapy. 1912, Henderson claimed, marked a 'Great Divide' when 'for the first time in human history, a random patient with a random disease consulting a doctor chosen at random stands a better than fifty-fifty chance of benefiting from the encounter'.[8]

In his book *The Role of Medicine: Dream, Mirage or Nemesis?* medical historian and demographer Thomas McKeown wrote that 'the notion that treatment of disease may be useless, unpleasant, and even dangerous has been expressed frequently and vehemently, particularly in French literature', notably by Montaigne, Molière and Proust.[9] 'Remarkably, considering the eminence of the critics, such views had little effect on medicine or the public's estimate of it,' McKeown observed.

'The fashionable doctors ... stood as they do now, in admiration of their own science. As now, they talked as if illness and death were mastered ... The learned, magic, meaningless words, the grave looks at each other, the artful hesitation between one worthless formula and another – all are there,' Nancy Mitford wrote in her biography of Louis XIV.

McKeown argued that

> patients have been and continue to be exposed to pain and injury from misguided attempts to do them good. Suffering is marginally more tolerable when inflicted with the best of intentions, and the death of Charles II under treatment by his doctors was much more cruel than that of his father at the hands of his executioner.[10]

When did reputable doctors retrospectively become quacks and when did clinical interventions, based on the medical science of the day, become of net benefit to patients? Even with the benefit of hindsight, it's hard to know. One answer is certainly wrong – at doctors' evaluation of their own abilities. Good intentions and strength of belief are highly misleading indicators of the quality of scientific knowledge. Physicians had been swearing the Hippocratic Oath to do no harm for twenty-three centuries prior to Henderson's Great Divide.

A tiny minority of climate scientists scrupulously avoided lending their voices to policy pronouncements. The vast majority saw it as part of their vocation. In the Fourth Assessment Report's chapter on climate change and sustainable development, the Working Group II authors said that the 'real message' from the IPCC's estimates of future global average temperatures was that

> no threshold associated with any subjective judgment of what might constitute 'dangerous' climate change can be guaranteed by anything but the most stringent mitigation interventions, at least not on the basis of current knowledge.[11]

Despite uncertainties acknowledged by the authors, there can be no mistaking the therapy these doctors believed necessary – 'on the basis of current knowledge'.

How can we know whether current knowledge on climate science is correct? In a video in the run up to the Copenhagen climate conference, Schneider made the comparison with tobacco smoking and lung cancer.

> We to this day do not know the precise mechanism whereby smoking causes lung cancer, but the statistics are so overwhelming that it would be irresponsible not to act. It's the same in climate.[12]

Al Gore also made the link. It's human nature to take time to connect the dots, Gore says in *An Inconvenient Truth* of his sister's death from lung cancer.

Leading climate scientists have invoked the American tobacco industry's denial of the link between smoking and lung cancer.* The attacks on climate science ahead of the Copenhagen conference mirrored the earlier tactics of the tobacco industry, according to IPCC vice-chair Jean-Pascal van Yperselethe. Five years before, Robert May, president of the Royal Society at the time and previously the British government's chief scientific adviser, accused those questioning the science of using tactics 'reminiscent of the tobacco lobby's attempts to persuade us that smoking does not cause lung cancer'.[13]

David King, May's successor as chief scientific adviser, mounted a similar argument after the collapse of the Copenhagen conference.

> When paid lobbyists try to discredit the scientific theory that smoking causes lung cancer, they used the argument that it wasn't a proven fact. Well it wasn't then, and nor will it ever be, but would you now bet against it? ... And in the case of climate change, the scientific probability that the world is warming, and that humans are the chief cause, is overwhelming.[14]

King made the comment in 2010 – at the end of a decade which showed no statistical trend in average global temperatures.

The clear implication is that the relationship between rising levels of carbon dioxide in the atmosphere and changes in global temperature is as solidly founded as that between tobacco smoking and lung cancer. There is little dispute that tobacco smoking is, in the words of Eric Feldman and Ronald Bayer, authors of *Un-*

---

* Notoriously the US tobacco industry ran an advertising campaign in the 1950s claiming that its customers' health was the industry's overriding concern. The response of the UK industry was different. After the industry's chief statistician was sacked for saying he would resign unless the tobacco industry accepted that smoking caused lung cancer, he was reinstated six weeks later and the industry agreed not to say anything to imply that smoking did not cause lung cancer. Conrad Keating, *Smoking Kills: The Revolutionary Life of Richard Doll* (2009), pp. 183–6.

filtered: Conflicts over Tobacco Policy and Public Health, 'the single most important preventable source of morbidity and mortality in advanced industrial societies'.

The parallel with smoking and lung cancer is a potent advocacy tool to discredit opponents. Global warming was an idea in search of evidence, reversing Popper's formulation that a theory must allow us to *explain* the observations that created the problem. Here, the theory preceded the observations: speculation that rising levels of carbon dioxide would lead to rising global temperatures led concerned scientists – many of whom were worried about the possibly deleterious consequences on the environment if such a rise were to occur – to seek evidence for it.

Superficially, the most compelling evidence for global warming was derived from temperature reconstructions that purported to show that the rise in temperatures during the twentieth century was unprecedented and way outside the bounds of natural variability. After the Hockey Stick was discredited, the IPCC changed tack. Now the principal evidence was the ability of computer models to replicate global temperature trends over the second half of the twentieth century. The rise in average global temperature could only be explained by the rise in carbon dioxide (although the computer models used assumptions about cooling induced by sulphate aerosols that were little more than guesswork).* Secondary evidence was also adduced, including glacier retreat, arctic sea ice extent, the number of polar bears, extreme weather events and sometimes lurid forecasts of what the world was going to look like, amplified and often distorted by NGOs who shared scientists' concern about the environment.

In terms of Schneider's 'courtroom epistemology', carbon dioxide was in the dock from the start. To change the metaphor, the disease pathway had been found before evidence of disease – the opposite of the case with lung cancer.

Until the 1920s, cancer of the lung was considered rare. Death rates from infectious diseases such as tuberculosis, typhus and cholera had fallen steadily during the nineteenth century. As deaths from infectious epidemics fell, lung cancer rates climbed. In Britain, lung cancer rose by one thousand, five hundred per cent between 1922 and 1947 and overtook the declining death rate from tuberculosis. 'By the late 1940s,' according to the medical historian Conrad Keating, 'Britain had the highest lung cancer rates in the world, and the reasons for this were completely unknown.'[15]

In 1947, the Ministry of Health wrote to the Medical Research Council. At a meeting of thirteen leading researchers, various hypotheses were put forward.

---

* According to Richard Lindzen, 'The models focus on aerosols and solar variability, and generally assume that natural internal variability is accurately included and accounted for. That models each use different assumptions for aerosols and solar variability makes clear that these are simply adjustable parameters.' Richard S. Lindzen, 'Response to the critique of my lecture in the House of Commons on February 22nd 2012', April 12th 2012 http://thegwpf.org/the-climate-record/5437-richard-lindzen-response-to-the-critique-of-my-house-of-commons-lecture.html

These included proximity to gas works, tar on roads, vehicle exhaust fumes and cigarette smoking. 'There was no consensus of opinion.'[16]

The task of finding the cause was given to Austin Bradford Hill (a pipe smoker) and Richard Doll (who smoked a pipe and non-tipped cigarettes). 'I was not antagonistic to tobacco when, in 1947, I began to study its effects,' Doll wrote in 1999.[17] 'Originally Doll thought the increase in motor cars and the tarring of roads were likely to be responsible for the epidemic,' Keating wrote in his biography of Doll. 'Hill, typically, was reported to have entered the study "with an open mind."'

The survey questionnaire designed by Hill and Doll was 'no frontal attack on smoking, which formed one section out of nine – eleven questions out of nearly fifty', others being on social class, diet, electric or gas cookers, whether those surveyed lived near a gas works. After the study had been extended for a year, Hill and Doll's analysis started to reveal evidence neither had anticipated. In October 1949, Doll wrote to the Medical Research Council of a 'real association between smoking and cancer of the lung'.[18] It was only then that Doll quit smoking. 'That so many diseases – major and minor – should be related to smoking is one of the most astonishing findings of medical research in this century,' Doll wrote fifty years later.[19]

Coincidentally in 1948 Ernst Wynder, a summer student at New York University, decided to conduct a case-control study on smoking and lung cancer. In February 1949, Wynder presented the results of his study to the national meeting of the American Cancer Society. They aroused little interest. In May 1950 Wynder and his supervisor, Evert Graham, published a paper in the *Journal of the American Medical Association*. It concluded: 'Smoking, especially in the form of cigarettes, plays an important role in the aetiology of lung cancer.'[20]

Two studies – completely independent of one another – came to the same conclusion. The two teams were not part of a community, comparing notes as part of a collective endeavor. In the case of the Englishmen, both were rigorous in separating scientific research from policy. Doll, a committed communist for much of his life, never permitted his personal political views to cross over into his work as a scientist. Doll had an ideological immune system and scrupulously followed what Hill had taught him: 'If you were going to contribute to a subject it was very important to separate the presentation of evidence from the discussion of what should be done on the basis of that evidence.'[21] A scientist was an expert witness, not a policy advocate.

Doll was acutely conscious of the susceptibility of epidemiological evidence to distortion: 'The only safeguard is always to suspect the influence of bias, consider every way it could have entered the study and then test to see if it has.'[22] According to Keating, 'In nearly every way Doll embodied Charles Darwin's definition of a value-free experimenter. "A scientific man ought to have no wishes, no affections – a mere heart of stone."'[23] Doll was thus an exemplar of the scientific method that Gore vehemently denounced in *Earth in the Balance*, guilty of man's breach with nature and contributing to the evils perpetrated by Hitler and Stalin.

For his part, Doll was critical of claims – inspired by Rachel Carson and *Silent Spring* – made by the US environmental movement that the widespread use of chemicals and other forms of industrial pollution were causing a cancer epidemic. A 1981 paper written with Richard Peto concluded: 'Were it not for the effects of tobacco, total US death rates would be decreasing substantially more rapidly than they already are.'[24] Indeed Keating makes the case that US environmentalists helped Big Tobacco off the hook:

> The claim that twenty per cent of cancers were a result of the actions of a rapacious chemical industry created an apologia for Big Tobacco … In 1980-81 the American cigarette industry had a record year … Commenting on this, the Chairman of the largest American cigarette manufacturer was reported as saying that he thought the 'cancer problem' was no longer hitting sales as hard as before because 'so many things have been linked to cancer' that people might be getting skeptical.[25]

The implication is inescapable. Environmentalism cost lives.

With global warming, there is an additional layer of opaque uncertainty – the extent to which climate science is subject to a paradigm in the sense Kuhn wrote about in *The Structure of Scientific Revolutions*. 'Current scientific knowledge' views large-scale changes in the Earth's climate, whether natural or man-made, through the prism of a paradigm of CO_2-induced climate change. Accordingly changes in concentrations of carbon dioxide are assumed by climate scientists to play a major role in the succession of glacial cycles.*

'The proponents of competing paradigms are always to some extents at cross-purposes,' Kuhn wrote. 'They are bound partly to talk through each other in a battle that cannot be resolved by proofs.'[26] Thus debates between scientists are not going to settle the issue. Neither would counting the number of scientists for or against the prevailing consensus resolve the matter. The existence of consensus might only constitute evidence of a paradigm's hold on the minds of contemporary scientists.

---

* For example, Brian Hoskins, J. Mitchell, T. Palmer, K. Shine & E. Wolff wrote that the termination of the most recent glaciation, with a small initial rise in temperatures followed by several thousand years in which temperatures and CO_2 rose together, 'is entirely consistent with the role of CO_2 as an amplifier of an otherwise small external forcing' (*A critique of the scientific content of Richard Lindzen's Seminar in London, February 22nd 2012*). In response, Lindzen wrote: 'The notion that the small changes in globally and annually averaged insolation are the crucial driver is implausible to say the least, but it stems from the current simplistic view of climate consisting in a single variable (globally averaged temperature anomaly) forced by some globally averaged radiative forcing – an idea that permeates the critics' discussion.' Richard S. Lindzen, 'Response to the critique of my lecture in the House of Commons on February 22nd 2012' April 12th 2012 http://thegwpf.org/the-climate-record/5437-richard-lindzen-response-to-the-critique-of-my-house-of-commons-lecture.html

However there is an indicator that provides a rough and ready litmus test. Popper's search for a principle to distinguish science from pseudo-science was sparked by the contrast between Einstein's theory of relativity and Marx's theory of history, Freud's psycho-analysis and Alfred Adler's 'individual psychology'. Einstein's theory was supported by passing the severe test conducted by Eddington, a test it could have failed. By contrast, subscribers to the theories of Marx, Freud and Adler found confirmatory evidence wherever they looked. Their theories seemed to explain practically everything within the fields to which they referred. Whatever happened always confirmed it. 'It was rather that I felt that these three other theories, though posing as science, had in fact more in common with primitive myths than with science; that they resembled astrology rather than astronomy,' Popper wrote.[27]

Common to the three, Popper noted, was their treatment of unbelievers – or, to use the terminology of global warming, sceptics and deniers. Their defiance in the face of manifest truth had a ready explanation, for, as we've already seen,

> unbelievers were clearly people who did not want to see the manifest truth; who refused to see it, either because it was against their class interest, or because of their repressions which were still 'un-analyzed' and crying out for treatment.

Attempts to settle a dispute by pointing to the other side's motives are the stock in-trade of the political campaigner. In framing the argument on global warming in terms of scientific truth versus the false consciousness promoted by special interests, proponents of the idea of global warming inadvertently proclaim their adherence to a pseudo-science.

The argument that the public became confused about the science because of the activities of malign special interests (i.e., fossil fuel companies) is so ubiquitous as to be an intrinsic component in the morphology of global warming. It has been articulated by leading scientists (Michael Mann has written an entire book blaming the shortcomings of the Hockey Stick on special interests), politicians (Gordon Brown and Kevin Rudd, as well as Al Gore), and environmental NGOs. In 2012, William Nordhaus compared the 'distortion' of climate science to the activities of tobacco companies. 'Scientists, citizens, and our leaders will need to be extremely vigilant to prevent pollution of the scientific process by the merchants of doubt,' Nordhaus wrote, criticizing global warming skeptics.[28]

It is also unhistorical. Despite the opposition of tobacco farmers and cigarette manufacturers, governments ran extensive anti-smoking campaigns and adopted other preventive measures, including vertiginously high taxes on tobacco. 'By the close of the twentieth century, anti-tobacco advocates, public health officials, physicians' groups, and international organizations, separately and in concert, had succeeded in putting tobacco control on the policy agenda of every industrialized democracy,' Feldman and Bayer wrote.[29] Thanks to them, cigarette smoking has fallen in the Western world. As predicted by Doll, after a lag, lung cancer rates fell too.

Even the most ingenious maker of conspiracy theories would be hard pressed to forge a chain of causality linking the activities of Western fossil fuel companies to the refusal of India and China to accede to the West's demands for a global cap on greenhouse gas emissions, especially as the Third World's resistance to the claims of Western environmentalism had been adopted in the early 1970s – a decade and a half before the arrival of global warming in world politics in 1988. So we come to the final paradox of global warming. The science is weak, but the idea is strong. The science is inherently weak because it is not capable of being falsified in the here and now. Voluminous evidence is itself testament to its weakness, for the same reasons Popper noted with respect to the trio of pseudo-sciences prevalent in Vienna in the 1920s.

Predictions that global warming will be harmful or dangerous are future statements which cannot be verified. By default, computer models have been used to guide governments, but they too depend on unverified assumptions about the cooling effect of aerosols.* But these models have failed to capture the amplitude of natural variability or to predict the absence of a clear trend in average global temperatures during the twenty-first century in what could be called Trenberth's Travesty.†

Bias in the IPCC is endemic, from language of the expected signal of global warming emerging in the future, the cover-up of flaws in the Hockey Stick, its politicization through government selection of lead authors and their control over what the IPCC says, to its infiltration by NGOs. 'Despite the numbers of persons involved, and the lengthy formal review procedures, the preparation of the IPCC Assessment Reports is far from being a model of rigor, inclusiveness and impartiality,' David Henderson wrote in a critique of the IPCC.[30]

After the Fourth Assessment Report was found to be full of errors, Working Group II co-chair Martin Parry wrote to colleagues characterizing it as a 'clamor without substance' fed by critics and skeptics, at the same time defending the inclusion of NGO literature in IPCC output. Bob Watson, the IPCC's second chairman, thought otherwise. 'The mistakes all appear to have gone in the direction of making it seem like climate change is more serious by overstating the impact. That is worrying,' Watson said. 'The IPCC needs to look at this trend in the errors and ask why it happened.'[31]

---

*   'There are considerable uncertainties in estimating the impact of aerosols on climate,' according to Hoskins et al., who go on to say that 'uncertain' does not mean 'unknown' or that the cooling effect of aerosols is zero. Brian Hoskins, J. Mitchell, T. Palmer, K. Shine & E. Wolff, *A critique of the scientific content of Richard Lindzen's Seminar in London, February 22nd 2012* (undated).

†   According to Lindzen, 'The assumption that the models adequately represent natural internal variability is seriously mistaken.' Richard S. Lindzen, 'Response to the critique of my lecture in the House of Commons on February 22nd 2012'.

Yet global warming's success in colonising the Western mind and in changing government policies has no precedent. It dominated the international agenda from the 1992 Rio Earth Summit until the 2008 banking crisis and the West's defeat at Copenhagen a year later. The prospect of planetary salvation inflated the science so it became too big to fail, justifying the anti-scientific practice of withholding data and methods from potential critics and de-legitimizing critical argument.* 'What is called objectivity consists solely in the critical approach,' Popper wrote.[32]

The credibility of global warming rested on the prestige of science – the branch of knowledge that had advanced further and faster than any other in human history. Underneath though, the nature of science was changing. It is hard to see a scientist such as Percy W. Bridgman in the middle of the twentieth century accepting with equanimity unverifiable future statements as science. Without the possibility of verification, Bridgman wrote, truth becomes meaningless. Apply his criterion to the output of the IPCC, and there wouldn't be much science left to furnish the basis of an immensely costly call to action.

This change was accompanied by a highly expansive view of the social role of scientists as uniquely capable of identifying and devising solutions to the problems threatening the survival of humanity and the planet. To this, scientists brought their cultural aversion to learning from the past. For them, history is not so much a closed book as irrelevant to the problems of the future. So they didn't ask why previous predictions of imminent catastrophes, from Malthus and Jevons in the nineteenth century to the limits to growth debates of the early 1970s, were all wrong.

True to form, in 2012 the Royal Society produced the latest in the series of doomsday predictions. Unsurprisingly *People and the Planet* argued there were too many of the former for the latter.

> Over the next thirty to forty years the confluence of the challenges described in this report provides the opportunity to move towards a sustainable economy and a better world for the majority of humanity, or alternatively the risk of social, economic and environmental failures and catastrophes on a scale never imagined.[33]

The last claim was untrue. Such catastrophes had been imagined. Exactly forty years earlier, Julian Huxley, winner of the Royal Society's Darwin Medal, together with four other Fellows of the Royal Society and twenty-nine eminent scientists

---

* Hill and Doll's findings on tobacco and lung cancer were fiercely attacked by R.A. Fisher, the leading geneticist and statistician of the period (Fisher formulated the null hypothesis test), who demanded to see the underlying data. Hill and Doll passed two data sets to Fisher. Conrad Keating, *Smoking Kills: The Revolutionary Life of Richard Doll* (2009), p. 191.

and experts, predicted the collapse of industrial civilization in famine, epidemic and war within the lifetimes of people then living. 'On a planet with finite resources there are limits to growth,' the Royal Society stated, repeating the message of the Club of Rome's *The Limits to Growth*, also from 1972, and making the same error.[34]

When it comes to learning from their mistakes, collectively scientists vie with the Bourbons.

In believing scientists and politicians can solve the problems of a far distant future, the tangible needs of the present are neglected. We cannot know what the future holds. It is like walking into a dense fog armed with a flashlight and one's wits, picking out objects which we only know for sure what they are when nearly upon them.

No one knows what the Earth's climate will be at the end of the century. Based on history, it is possible to hazard a prediction of a different kind. Before the end of the century, the Western mind will conceive another environmental crisis necessitating the ending of the modern industrial economy, the only form of economic arrangements that has lifted mankind to undreamt of prosperity.

The big question is whether the Western mind will be sovereign at the century's end and the West remain the core of the world economy or relegated to its periphery – something only the passing of the present century can answer.

NOTES

1
THE IDEA

1. From two hundred and eighty parts per million (that is 0.028 %) before industrialisation to three hundred and seventy-nine ppm in 2005 (0.0379 %) according to the Intergovernmental Panel on Climate Change (IPCC), *Climate Change 2007: Synthesis Report Summary for Policymakers* (2007), p. 37.
2. IPCC, *Climate Change 2007: Synthesis Report Summary for Policymakers* (2007), p. 39.
3. Barbara Ward and René Dubos, *Only One Earth* (1974), p. 66.
4. Speech by HRH The Prince of Wales at The Prince's Foundation for the Built Environment conference 2009 titled 'Globalisation from the Bottom Up', February 5th 2009, http://princeofwales.gov.uk/speechesandarticles/
5. Peter Raven, 2002 AAAS Presidential Address, Science, *Sustainability, and the Human Prospect* http://www.sciencemag.org/cgi/content/full/297/5583/954
6. 'Human consumption: Flying in the face of logic', *Guardian*, July 16th 2008.
7. Bjorn Lomborg, *The Skeptical Environmentalist* (2001), p. xix.
8. Nina Mazar & Chen-Bo Zhong, 'Do Green Products Make Us Better People,' *Psychological Science*, April 2010 Vol. 21 No. 4, pp. 494–8.

2
PROMETHEAN REVOLUTION

1. Ronald H. Coase, 'The Problem of Social Cost', *Journal of Law and Economics* (1960).
2. William Rawley, *The Life of the Right Honourable Francis Bacon*, Vol. 1 of the Complete Works (1730), p. 5.
3. Quoted in Benjamin Farrington, *Philosopher of Industrial Science* (1951), p. 64.
4. Quoted in Farrington op. cit., p. 54.
5. Karl Popper, *The Myth of the Framework* (1994), p. 198 (emphasis in the original).
6. ibid., p. 193.
7. A.N. Whitehead, *Science and the Modern World* (1967), p. 12.
8. Popper, *The Myth of the Framework* (1994), p. 203.
9. ibid., p. xiii.
10. ibid., p. 110.
11. Martin Wolf, *Why Globalisation Works* (2004), p. 41.
12. Thomas Malthus, *An Essay on the Principle of Population* (1798).
13. Maurice Strong, *Where on Earth are We Going?* (2001), p. 7.
14. Charles Darwin, *Autobiography* (2010), p. 82.

3
ANTECEDENTS

1. Joseph A Schumpeter, *History of Economic Analysis* (1994), p. 826.
2. Harro Maas, *William Stanley Jevons and the Making of Modern Economics* (2005), p. 33.
3. W. Stanley Jevons, *The Coal Question* (1865), p. 149.
4. Jevons, *The Coal Question* (1865), p. 117.
5. Meek, *Marx and Engels on Malthus* (1953), p. 82.
6. Gilbert (ed.), *TR Malthus Critical Responses* (1998), Vol. 3, p. 191.
7. ibid., p. 191.
8. Meek, *Marx and Engels on Malthus* (1953), p. 110.
9. Wood (ed.), *William Stanley Jevons: Critical Assessments* (1988), Vol. 1, p. 65.
10. ibid.
11. Thomas S. Kuhn, *The Structure of Scientific Revolutions* (1996), p. 139.
12. ibid., p. 96.
13. Brock, *The Fontana History of Chemistry* (1992), pp. 111–12.
14. James Rodger Fleming, *The Callendar Effect* (2007), p. 66.
15. Quoted in Mike Hulme, 'On the origin of "the greenhouse effect": John Tyndall's 1859 interrogation of nature', *Weather*, Royal Meteorological Society (May 2009), Vol. 64, No. 5, p. 121.

4
FIRST STIRRINGS

1. 'The Changing Arctic,' *Washington Post*, November 2nd 1922.
2. Theodore Roosevelt, *An Autobiography* (1913), p. 94.
3. ibid., p. 47.
4. Roosevelt, *An Autobiography* (1913), p. 435.
5. ibid., p. 424.
6. Annual Message to Congress, December 6th, 1923, http://www.presidency.ucsb.edu/ws/index.php?pid=29564
7. Anna Bramwell, *Ecology in the 20th Century* (1989), pp. 162, 173.
8. T.S. Eliot, *Christianity and Culture* (1988), p. 26.
9. Peter Hall, Harry Gracey, Roy Drewett & Ray Thomas, *The Containment of Urban England* (1973), p. 628.
10. George Orwell, *The Lion and Unicorn: Socialism and the English Genius* (1941), included in *Essays* (2002), p. 292.
11. Cecil Palmer, 'Do We Agree? A Debate Between G. K. Chesterton And Bernard Shaw with Hilaire Belloc in the Chair' (1928) http://www.cse.dmu.ac.uk/~mward/gkc/books/debate.txt
12. Rolf Gardiner & Heinz Rocholl (ed.), *Britain and Germany: A Frank Discussion Instigated by Members of the Younger Generation* (1928), p. 19.

13. ibid., p. 124.
14. Quoted from Henry Williamson, *Phoenix Generation*, cited in Anna Bramwell, *Ecology in the 20th Century* (1989), p. 142.
15. ibid., p. 110.
16. ibid., p. 110.
17. Gardiner & Rocholl (ed.), *Britain and Germany: A Frank Discussion Instigated by Members of the Younger Generation* (1928), p. 123.
18. ibid., p. 129.

5
TURNING POINT

1. Fleming, *The Callendar Effect* (2007), p. 72.
2. *Resources for Freedom*, A Report to the President by The President's Materials Policy Commission, Washington (1952), Vol. 1, *Foundations for Growth*, p. 1.
3. *Resources for Freedom* (1952), Vol. 1, *Foundations for Growth*, p. 21.
4. ibid.
5. ibid., p. 20.
6. ibid., p. 3.
7. ibid., p. 77.
8. Statement of Charles G. Groat, Director USGS, before the subcommittee on Energy and Mineral Resources Committee on Resources US House of Representatives 'The Role of Strategic and Critical Minerals in Our National and Economic Security', 17th July 2003 http://www.usgs.gov/congressional/hearings/testimony_17july03.asp
9. *Resources for Freedom* (1952), Vol. 4, *The Promise of Technology*, p. 23.
10. Gary Kroll, 'Rachel Carson's *The Sea Around Us*, Ocean-Centrism, and a Nascent Ocean Ethic' in Lisa H. Sideris & Kathleen Dean Moore, *Rachel Carson: Legacy and Challenge* (2008), p. 119.
11. Paul Brooks, *Speaking for Nature* (1980), pp. 276–7.
12. Brooks, *Speaking for Nature* (1980), p. 278; Linda Lear, *Rachel Carson Witness for Nature* (1999), p. 90.
13. Brooks, *The House of Life – Rachel Carson at Work* (1972), p. 299.
14. ibid.
15. Henry David Thoreau, *Walden and Civil Disobedience* (1986), pp. 362–3.
16. Rachel Carson, *Silent Spring* (2000), p. 22.
17. ibid., p. 195.
18. ibid., p. 212.
19. ibid., p. 193.
20. ibid., p. 24.
21. Linda Lear, *Rachel Carson Witness for Nature* (1999), p. 429.
22. Chris J. Magoc, *Environmental Issues in American History* (2006), p. 227.
23. Carson, *Silent Spring* (2000), p. 121.
24. ibid., p. 161.
25. ibid., p. 164.
26. Brooks, *The House of Life – Rachel Carson at Work* (1972), p. 316.

27. Sideris, 'The Ecological Body: Rachel Carson, *Silent Spring*, and Breast Cancer' in Sideris & Moore, *Rachel Carson: Legacy and Challenge* (2008), p. 137.

28. Lear, *Rachel Carson Witness for Nature* (1999), p. 450.

29. Brooks, *The House of Life – Rachel Carson at Work* (1972), p. 325.

30. http://motherjones.com/politics/1997/01/stephen-jay-gould?page=1

31. Brooks, *The House of Life – Rachel Carson at Work* (1972), p. 319.

32. Carson, *The Sea Around Us* (1989), p. 15.

33. Wilfred Beckerman, *In Defence of Economic Growth* (1976), p. 265.

6
SPACESHIP EARTH

1. Raúl Prebisch, *Power, Principle and the Ethics of Development* (2006), p. 54.

2. Edgar J. Dosman, *The Life and Times of Raúl Prebisch* (2008), pp. 248–9.

3. ibid., p. 249.

4. J.T. Cuddington, 'Long-Run Trends In 26 Primary Commodity Prices – A Disaggregated Look at the Prebisch-Singer Hypothesis' in *Journal of Development Economics* 39 (2) (1992), pp. 207–27.

5. Kenneth Boulding, 'The Economics of the coming Spaceship Earth' in Henry Jarret (ed.), *Environmental Quality In a Growing Economy – Essays from the Sixth RFF Forum* (1966), p. 3.

6. Barbara Ward, *Space Ship Earth* (1966), p. 22.

7. ibid., p. 24.

8. Barbara Wood, *Alias Papa: A Life of Fritz Schumacher* (1984), p. 241.

9. ibid., p. 222.

10. ibid., p. 248.

11. Schumacher, *This I Believe* (2004), p. 8.

12. ibid., p. 131.

13. ibid., p. 42.

14. ibid., p. 44.

15. ibid., p. 43.

16. ibid., p. 44.

17. ibid., p. 129.

18. Wood, *Alias Papa: A Life of Fritz Schumacher* (1984), p. 353.

19. ibid., p. 320.

20. Schumacher, *Small is Beautiful* (1993), p. 131.

21. 'Norman Borlaug, scientist who "saved 245m lives", dies aged 95', *The Times*, September 14th 2009.

22. Harry Cleaver, *American Economic Review*, Vol. 62, Issue 2 (1972), pp. 177–86.

23. Douglas Kellner (ed.), *Collected Papers of Herbert Marcuse*, Vol. 3, *The New Left and the 1960s* (2005), p. 173.

7
LIMITS TO GROWTH

1. Richard Reeves, *President Nixon: Alone in the White House* (2001), p. 163.
2. ibid.
3. ibid., p. 238.
4. Lou Cannon, *Governor Reagan: His Rise to Power* (2003), p. 300.
5. Wilfred Beckerman, *In Defence of Economic Growth* (1976), p. 247.
6. 'Pollution: Nuisance or Nemesis? A Report on the Control of Pollution' (1972), p. 7.
7. Meadows et al, *The Limits to Growth: A Report for the Club of Rome's Project on the Predicament of Humanity* (1972), pp. 9–10.
8. ibid., p. 23.
9. ibid.
10. Edward Goldsmith el al, *A Blueprint for Survival* (1972), p. 28.
11. ibid., p. 14.
12. 'Pollution: Nuisance or Nemesis? A Report on the Control of Pollution' (1972), p. 3.
13. ibid., p. 9.
14. ibid., p. 80.
15. Beckerman, *In Defence of Economic Growth* (1976), p. 14
16. ibid., p. 35.
17. Anthony Crosland, *A Social Democratic Britain* (1971).
18. Beckerman, *In Defence of Economic Growth* (1976), p. 110.

8
STOCKHOLM

1. Maurice Strong, *Where on Earth are We Going?* (2001), p. 51.
2. Strong, 'Our common future – 15 years after the Stockholm conference' in World Media Institute, *TRIBUTE…to Barbara Ward: Lady of Global Concern* (1987), p. 94.
3. Strong, *Where on Earth are We Going?* (2001), p. 156.
4. ibid., p. 125.
5. Harold Wilson, *The Labour Government 1964–1970: A Personal Record* (1971), p. 499.
6. Satterthwaite, *Barbara Ward and the Origins of Sustainable Development* (2006), p. 46.
7. ibid., p. 15.
8. Barbara Ward, *A New Creation? Reflections on the Environmental Issue*, first published by the Pontifical Commission Justice and Peace in Vatican City in 1973, in World Media Institute, *TRIBUTE…to Barbara Ward: Lady of Global Concern* (1987), p. 16.
9. ibid., p. 15.
10. ibid., p. 31.
11. Barbara Ward, 'The End of an Epoch?' in *Economist*, May 27th 1972.
12. Barbara Ward, *Space Ship Earth* (1966), p. 3.
13. Barbara Ward and René Dubos, *Only One Earth – The Care and Maintenance of a Small Planet* (1974), p. 10.

14. ibid., p. 266.

15. ibid., p. 145.

16. ibid., p. 84.

17. ibid., p. 85.

18. 'UN secretary-general calls on all nations to meet crisis of a polluted planet' in *The Times*, June 6th 1972.

19. 'Delegates' bicycles vanish in Stockholm scramble' in *The Times*, June 8th 1972.

20. http://www.mauricestrong.net/2008091028/video/video/unche.html

21. ibid.

22. ibid.

23. 'Mrs Gandhi blames profits race for crisis' in *The Times*, June 15th 1972.

24. McCormick, *The Global Environmental Movement: Reclaiming Paradise* (1989), p. 99.

25. UNEP, *In Defence of the Earth: The basic texts on environment: Founex. Stockholm* (1981), pp. 42–7.

26. ibid., p. 82.

9
BREAKING WAVE

1. Henry Kissinger, *Years of Upheaval* (2000), p. 854.

2. E.F. Schumacher, *This I Believe* (2004), p. 21.

3. John Dumbrell, *The Carter Presidency: A re-evaluation* (1993), p. 172.

4. Jimmy Carter, *Keeping Faith: Memoirs of a President* (1995), pp. 117–18.

5. Ray Maghroori & Stephen Gorman, *The Yom Kippur War: A Case Study in Crisis Decision-Making in American Foreign Policy* (1981), p. 61.

6. Address to the Nation About Policies To Deal With the Energy Shortages, November 7th 1973 http://www.presidency.ucsb.edu/ws/print.php?pid=4034

7. US Energy Information Administration, MTTIMUS2, http://tonto.eia.doe.gov/dnav/pet/hist/LeafHandler.ashx?n=PET&s=MTTIMUS2&f=A

8. Kissinger, *Years of Upheaval* (2000), p. 885.

9. Prices are for Saudi Light from 1971-1974 and Imported Refiner Acquisition Cost (IRAC) from 1975 expressed in March 2009 dollars.

10. Kissinger, *Years of Upheaval* (2000), p. 897.

11. President Carter address on Energy Policy, April 18th 1977 http://www.pbs.org/wgbh/amex/carter/filmmore/ps_energy.html

12. George Melloan & Joan Melloan, *The Carter Economy* (1978), p. 118

13. Environment Message to the Congress, May 23rd 1977 http://www.presidency.ucsb.edu/ws/index.php?pid=7561

14. Melloan & Melloan, *The Carter Economy* (1978), p. 143.

15. BP 'Statistical Review of World Energy 2012'

16. Don Richardson (ed.), *Conversations with Carter* (1998), p. 173.

17. ibid., pp. 203–4.

18. US Energy Information Administration, MTTIMUS2

19. Thornton, *The Carter Years: Towards a New Global Order* (1991), p. 421.

20. Biven, *Jimmy Carter's Economy: Policy in an Age of Limits* (2002), p. 259.

21. Gerald O. Barney (study director), *The Global 2000 Report to the President: Entering the Twenty First Century* (1980), Vol. 1, p. 42.

22. Angus Maddison, *The World Economy: A Millennial Perspective* (2001), p. 125.

23. United Nations, Department of Economic and Social Affairs, *World Population Prospects: The 2006 Revision Highlights* (2007), Fig. 4.

24. UN Food and Agriculture Organisation http://.fao.org/fileadmin/templates/ess/img/chartroom/72.gif

25. BP 'Statistical Review of World Energy 2012'

26. Barney, *The Global 2000 Report to the President: Entering the Twenty First Century* (1980), Vol. 1, p. 36.

27. James Rodger Fleming, *Historical Perspectives on Climate Change* (1998), p. 132.

28. Kiron Skinner, Annelise Anderson & Martin Anderson, *Reagan In His Own Hand* (2001), p. 329.

29. Skinner, Anderson & Anderson, *Reagan In His Own Hand* (2001), p. 326 & p. 339.

10

PUPATION

1. Mihajlo Mesarovic & Eduard Pestel, *Mankind at the Turning Point* (1975), p. ix.

2. ibid., p. 123.

3. Leonard Silk, 'Scholars Favor Global Growth' in *New York Times*, April 13th 1976.

4. UNEP, *In Defence of the Earth: The basic texts on environment: Founex, Stockholm, Cocoyoc* (1981), p. 115.

5. Dinesh D'Souza, *Ronald Reagan: How an Ordinary Man Became an Extraordinary Leader* (1997), p. 230.

6. Independent Commission on International Development, *North-South: A Programme for Survival* (1980), pp. 19–20.

7. ibid., p. 33.

8. Edgar J. Dosman (ed.), *Raúl Prebisch: Power, Principle, and the Ethics of Development* (2006), p. 57.

9. Julian L. Simon & Herman Kahn (ed.), *The Resourceful Earth: A Response to Global 2000* (1984), p. 1.

10. David Henderson, 'Survival, Development and the Report of the Brandt Commission' in *The World Economy*, Vol. 3 No. 1 (June 1980), pp. 87–117.

11. World Commission on Environment and Development, *Our Common Future* (1990), p. 4.

12. Gro Harlem Brundtland, Speech to UNEP, Nairobi, June 8th 1987 http://www.regjeringen.no/upload/SMK/Vedlegg/Taler%20og%20artikler%20av%20tidligere%20statsministre/Gro%20Harlem%20Brundtland/1987/Presentation_of_Our_Common_Future_to_UNEP.pdf

13. World Commission on Environment and Development, *Our Common Future* (1990), p. 43.

14. Henderson, 'Survival, Development and the Report of the Brandt Commission' in *The World Economy*, Vol. 3 No. 1 (June 1980), pp. 87–117.

15. World Commission on Environment and Development, *Our Common Future* (1990), p. 44.
16. ibid., p. 46.
17. World Commission on Environment and Development, *Our Common Future* (1990), p. 76.
18. ibid., p. 44.
19. Brundtland, Speech to UNEP, Nairobi.
20. World Commission on Environment and Development, *Our Common Future* (1990), p. 8, p. 45 & p. 9.
21. Mikhail Gorbachev, *Memoirs* (1996), p. 205.
22. Statement by the UNEP/WMO/ICSU International Conference on the Assessment of the Role of Carbon Dioxide and of other Greenhouse Gases in Climate Variations And Associated Impacts Villach, Austria, October 9–15th 1985

11
ANNUS MIRABILIS

1. Brian Mulroney, *Memoirs 1939–1993* (2007), p. 591.
2. Speech to the Royal Society, September 27th 1988 http://www.margaretthatcher.org/speeches/displaydocument.asp?docid=107346
3. R.K. Pachauri and A. Reisinger (ed.), *Climate Change 2007: Synthesis Report* (2007), Table 3.1.
4. Bert Bolin, *A History of the Science and Politics of Climate Change: The Role of the Intergovernmental Panel on Climate Change* (2007), pp. 48–9 & p. 57.
5. ibid., p. 118.
6. ibid.
7. James Hansen, 'Global Warming Twenty Years Later: Tipping Points Near' www.columbia.edu/~jeh1/2008/TwentyYearsLater_20080623.pdf
8. Author interview with Sir Crispin Tickell, January 14th 2010.
9. George Guise email to author, March 31st 2010.
10. Author interview with Sir Crispin Tickell, January 14th 2010.
11. Speech to the Royal Society, September 27th 1988
12. Karl Popper, *Conjectures and Refutations* (2002), p. 47.
13. Karl Popper, 'Science: Problems, Aims, Responsibilities' in *The Myth of the Framework*, ed. M.A. Notturno (2006), p. 94.

12
TWO SCIENTISTS

1. P.W. Bridgman, *The Way Things Are* (1959), p. 69.
2. ibid., p. 239.
3. ibid., p. 70.
4. H.H. Lamb, *Climate History and the Modern World* (1982), p. 6.

5. ibid., p. 12.
6. ibid., p. 14.
7. ibid., p. 330.
8. Bridgman, *The Way Things Are* (1959), p. 56.
9. Karl Popper, 'Science: Problems, Aims, Responsibilities' in *The Myth of the Framework* (1997), p. 110.
10. Bridgman, *The Way Things Are* (1959), p. 55.
11. ibid., p. 56.
12. Mike Hulme, *Why We Disagree About Climate Change* (2009), pp. 51–2.
13. ibid., p. 95.
14. Bridgman, *The Way Things Are* (1959), p. 129.
15. Hulme, *Why We Disagree About Climate Change* (2009), p. 77.
16. Popper, 'Science: Problems, Aims, Responsibilities', in *The Myth of the Framework*, ed. M.A. Notturno (2006), p. 93.

13
GREEN WARRIOR

1. Declaration of The Hague.
2. Al Gore, *Earth in the Balance* (1993), p. 8.
3. Michael Dukakis, 'A New Era of Greatness for America': Address Accepting the Presidential Nomination at the Democratic National Convention in Atlanta, July 21st 1988.
4. George H.W. Bush, Address Accepting the Presidential Nomination at the Republican National Convention in New Orleans, August 18th 1988.
5. Larry B. Stammer, 'Forty-Nation Environmental Panel Act Quickly on Global Warming, Baker Asks' in the *Los Angeles Times*, January 31st 1989.
6. Gore, *Earth in the Balance* (1993), p. 172.
7. Judy Leep email to author on behalf of George Shultz, October 26th 2010.
8. Scott Barrett, *Environment and Statecraft: The Strategy of Environmental Treaty-Making* (2003), p. 239.
9. House of Commons debates, February 23rd 1989, *Hansard*, Col. 1143.
10. UNFCCC figures extracted from Time series – Annex I, Total CO_2 Emissions without Land Use, Land-Use Change and Forestry.
11. Margaret Thatcher, Speech to United Nations General Assembly (Global Environment), November 8th 1989.
12. Paul Lewis, 'Thatcher urges pact on climate' in the *New York Times*, November 9th 1989.
13. Michael McCarthy, 'US fears over cost of global warming' in *The Times*, May 16th 1990.

14
RUSH TO JUDGMENT

1. Bert Bolin, *A History of the Science and Politics of Climate Change: The Role of the Intergovernmental Panel on Climate Change* (2007), pp. 61–2.

2. J.T. Houghton, G.J. Jenkins, J.J. Ephraums, *Climate Change: The IPCC Scientific Assessment* (1990), p. xii.

3. ibid., p. 199.

4. ibid., p. 233.

5. ibid.

6. ibid., p. 254.

7. ibid., p. xi.

8. ibid., p. xxxix.

9. John Houghton, 'World climate needs concerted action' in the *Financial Times*, November 10th 2010.

10. Houghton, Jenkins, Ephraums, *Climate Change: The IPCC Scientific Assessment* (1990), p. xvii.

11. ibid, p. 79.

12. ibid., p. 80.

13. ibid.

14. ibid., p. 80.

15. ibid., p. 73.

16. J.T. Houghton, *Global Warming: The Complete Briefing* (1994), p. 68.

17. Houghton, Jenkins, Ephraums, *Climate Change: The IPCC Scientific Assessment* (1990), p. 73.

18. ibid., p. 17.

19. Margaret Thatcher, Speech opening Hadley Centre for Climate Prediction and Research, May 25th 1990 http://www.margaretthatcher.org/document/108102

20. Marlise Simons, 'Scientists urging gas emissions cuts' in the *New York Times*, November 5th 1990.

21. Margaret Thatcher, Speech at Second World Climate Conference, November 6th 1990.

22. Marlise Simons, 'US view prevails at climate parley' in the *New York Times*, November 8th 1990.

23. Maurice Strong address to the Second World Climate Conference, November 1990, in J. Jäger and H.L. Ferguson (ed.), *Climate Change: Science, Impacts and Policy: Proceedings of the Second World Climate Conference* (1991), p. 434.

24. ibid., p. 519.

25. ibid., p. 522.

26. Nigel Hawkes, 'Is this really a scientist speaking?' in *The Times*, November 8th 1990.

27. Thatcher's speech at Second World Climate Conference, November 6th 1990 http://www.margaretthatcher.org/document/108237

28. ibid.

29. Bolin, *A History of the Science and Politics of Climate Change: The Role of the Intergovernmental Panel on Climate Change* (2007), p. 68.

30. Daniel Bodansky, 'Prologue to the Climate Change Convention' in Irving M. Mintzer & J. Amber Leonard (ed.), *Negotiating Climate Change: The Inside Story of the Rio Convention* (1994), p. 67.

15
A HOUSE DIVIDED

1. Chandrashekhar Dasgupta, 'The Climate Change Negotiations' in Irving M. Mintzer & J. Amber Leonard (ed.), *Negotiating Climate Change: The Inside Story of the Rio Convention*, Cambridge (1994), p. 141.
2. William Reilly interview with author, December 21st 2010.
3. Michael Boskin interview with author, December 20th 2010.
4. Robert Watson interview with author, December 6th 2010.
5. D. Allan Bromley, *The President's Scientists: Reminiscences of a White House Science Adviser* (1994), p. 21 and p. 148.
6. John H. Sununu, interview with author, November 11th 2010.
7. Reilly interview with author.
8. Bromley, *The President's Scientists: Reminiscences of a White House Science Adviser* (1994), pp. 144–5.
9. Boskin, Schmalensee & Taylor, *The Annual Report of the Council of Economic Advisers* (1990), pp. 214–15.
10. ibid., p. 222.
11. Michael Weisskopf, 'Bush says More Data on Warming Needed' in the *Washington Post*, April 18th 1990.
12. Bob Hepburn, 'Bouchard rebukes Bush for stalled pollution fight' in the *Toronto Star*, April 18th 1990.
13. Bert Bolin, *A History of the Science and Politics of Climate Change: The Role of the Intergovernmental Panel on Climate Change* (2007), p. 60.
14. Jerome Idaszak, 'Bush, critics clash on steps to combat climate change' in the *Chicago Sun-Times*, April 18th 1990.
15. Richard Benedetto, 'Europeans press US to act to prevent global warming' in *USA Today*, April 18th 1990.
16. Reilly interview with author.
17. Richard Benedetto, 'Bush does about-face at warming conference' in *USA Today*, April 18th 1990.
18. Philip Shabecoff, 'Bush Denies Putting off Action on Averting Global Climate Shift' in the *New York Times*, April 19th 1990.
19. J. Jäger and H.L. Ferguson (ed.), *Climate Change: Science, Impacts and Policy: Proceedings of the Second World Climate Conference* (1991), p. 431.
20. 'North and South Hold Environment Hostage' in the *Seattle Times*, June 3rd 1992.
21. Dasgupta, 'The Climate Change Negotiations' in Irving M. Mintzer & J. Amber Leonard (ed.), *Negotiating Climate Change: The Inside Story of the Rio Convention* (1994), p. 141.
22. Goldemberg interview with author.
23. Jonathan Dimbleby, *The Prince of Wales: A Biography* (1994), p. 605.
24. Reilly interview with author.
25. Clayton Yeutter memorandum to author, December 31st 2010.
26. Robert Reinstein email to author, April 29th 2011.
27. Michael Howard interview with author, April 12th 2010.
28. Christopher Marquis and Sam Dillon, 'Bush's US Seen as Environmental Outlaw' in *Miami Herald*, June 3rd 1992.

16
PRESIDENT BUSH GOES TO RIO

1. Steve Fainaru, 'Rio becomes a City under siege as leaders meet at Summit' in the *Boston Globe*, June 11th 1992.
2. Dianne Dumanowski and John Mashek, 'US is Isolated in Opposing Biodiversity Treaty' in the *Boston Globe*, June 9th 1992.
3. Michael Weisskopf, '"Outsider" EPS Chief being Tested' in the *Washington Post*, June 8th 1992.
4. Paul Hoversten, 'Treaty for Rio eludes EPA chief' in *USA Today*, June 8th 1992.
5. Susan Benesch, 'Senators: Anti-US sentiment strong in Rio' in the *St Petersburg Times*, June 7th 1992.
6. 'Wirth: Bad PR taints US image' in the *Denver Post*, June 10th 1992.
7. Ann Devroy, 'White House scorns summit critics' in the *Washington Post*, June 10th 1992.
8. 'As Population Grows, People will live "Like Rats", Cousteau says' in the *Los Angeles Times*, June 6th 1992.
9. Vaclav Havel, 'Rio and the New Millennium' in the *New York Times*, June 3rd 1992.
10. Fiona Godlee, 'Rio Diary: a fortnight at the earth summit' in *British Medical Journal* Vol. 305 (July 11th 1992), p. 103.
11. 'North and South Hold Environment Hostage' in the *Seattle Times*, June 3rd 1992.
12. Strong, *Where on Earth are We Going?* (2001), p. 210.
13. Ed Meese interview with author, March 11th 2010.
14. John Holusha, 'The Earth Summit Poll finds Scepticism in US about Earth Summit' in the *New York Times*, June 11th 1992.

17
TWO PROTOCOLS

1. Richard Benedick, *Morals and Myths: A Commentary on Global Climate Policy*, 109 WZB-Mitteilungen (2005).
2. Scott Barrett, *Environment and Statecraft: The Strategy of Environmental Treaty-Making* (2003), p. 360.
3. J.T. Houghton, L.G. Meira Filho, B.A. Callander, N. Harris, A. Kattenberg & K. Maskell (ed.), *Climate Change 1995: The Science of Climate Change: Contribution of WG1 to the Second Assessment Report of the Intergovernmental Panel on Climate Change* (1996), p. 5.
4. Cass R. Sunstein, 'Of Montreal and Kyoto: A Tale of Two Protocols' in *Harvard Environmental Law Review*, Vol. 31 (2007), p. 35.
5. ibid., p. 45.
6. ibid., pp. 45–6.
7. UNFCCC figures extracted from Time series – Annex I, Total CO_2 Emissions without Land Use, Land-Use Change and Forestry.
8. Dieter Helm, *Energy, the State, and the Market: British Energy Policy since 1979* (2003), p. 169.

9. Farley, 'Gore vows flexibility in climate talks' in the *Los Angeles Times*, December 8th 1997.
10. Timothy E. Wirth email to author, March 25th 2011.

18
CHINA SYNDROME

1. Al Gore, *Earth in the Balance* (1993), p. 213.
2. ibid., p. 232 & p. 228.
3. ibid., p. 230.
4. ibid., p. 253.
5. ibid., pp. 256–7.
6. John Cushman & David Sanger, 'Global Warming No Simple Fight' in the *New York Times*, December 1st 1997.
7. ibid.
8. Dawn Erlandson, 'The Btu Tax Experience: What Happened and Why It Happened' in *Pace Environmental Law Review*, Vol. 12 (1994) pp. 175–6.
9. Bill Clinton, *My Life* (2004), p. 522.
10. John Gummer interview with author, April 8th 2011.
11. Depledge, *The Organisation of Global Negotiations: Constructing the Climate Change Regime* (2005), p. 47.
12. ibid., p. 65.
13. ibid., Box 8.1.
14. Taylor Branch, *The Clinton Tapes: Wrestling History with the President* (2009), p. 456.
15. Chuck Hagel interview with author, February 25th 2011.
16. Indira Lakshmanan, 'Kerry says cuts would benefit Mass' in the *Boston Globe*, December 8th 1997.
17. William Stevens, 'Gore, in Japan, signals that US may make some compromises' in the *New York Times*, December 8th 1997.
18. Harlan L. Watson interview with author.
19. Raúl Estrada-Oyuela, 'First Approaches and Unanswered Questions' in José Goldemberg (ed.), *Issues & Options: The Clean Development Mechanism* (1998), p. 25.
20. Watson interview with author.

19
THE MORNING AFTER

1. Willis Witter, 'Gore dares Congress to resist pact' in the *Washington Times*, December 9th 1997.
2. John Godfrey, 'White House to hold off sending climate pact to Hill' in the *Washington Times*, December 12th 1997.
3. Andrew Turnbull interview with author, April 7th 2001.
4. William K. Stevens, 'Argentina Takes a Lead in Setting Goals on Greenhouse Gases'

in the *New York Times*, November 12th 1998.

5. John H. Cushman, 'Washington Skirmishes over Treaty on Warming' in the *New York Times*, November 11th 1998.

6. Jacques Chirac, 'Speech By Mr. Jacques Chirac French President To The VIth Conference of the Parties to the United Nations Framework Convention on Climate Change The Hague' November 20th 2000 http://sovereignty.net/center/chirac.html

7. Mary H. Cooper, 'Global Warming Treaty' in *CQ Researcher*, Vol. 11, No. 3 (January 26th 2001).

8. Andy McSmith, 'French anger at "macho" Prescott' in the *Daily Telegraph*, November 28th 2000.

9. Bill Clinton, 'Remarks During the White House Conference on Climate Change' October 6th 1997.

10. Stuart Eizenstat, Prepared Testimony On Kyoto Protocol (Delivered before Senate Foreign Relations Committee, February 11th 1998).

11. Frank Loy interview with author.

12. Clayton Yeutter email to author, March 28th 2011.

13. Erskine Bowles email to author, March 22nd 2011.

14. George W. Bush, Text of a Letter from the President to Senators Hagel, Helms, Craig, and Roberts, March 13th 2001.

15. Andrew C. Revkin, 'Bush's Shift Could Doom Air Pact, Some Say' in the *New York Times*, March 17th 2001.

16. David E. Sanger, 'Bush Will Continue to Oppose Kyoto Pact on Global Warming' in the *New York Times*, June 12th 2001.

17. Scott Barrett, *Environment and Statecraft: The Strategy of Environmental Treaty-Making* (2003), p. 371.

18. ibid., p. 374.

19. T.M.L. Wigley, 'The Kyoto Protocol: CO_2, CH4 and climate implications' in *Geophysical Research Letters*, Vol. 25 (1998), p. 2287.

20. Numbers derived from World Resources Institute, Climate Analysis Indicators Tool (CAIT UNFCCC) Version 4.0 (2011).

21. IEA, *CO2 Emissions From Fuel Combustion Highlights* (2010), p. 8.

20
TURNING UP THE HEAT

1. Henry D. Jacoby, Ronald G. Prinn & Richard Schmalensee, 'Kyoto's Unfinished Business' in *Foreign Affairs*, July/August 1998 Vol. 77, No. 4, p. 57.

2. Christopher Essex, 'What do climate models tell us about global warming?' in *Pure and Applied Geophysics* Vol. 135, Issue: 1 (1991), pp. 125–6.

3. Houghton, Meira Filho, Callander, Harris, Kattenberg & Maskell (ed.), Climate Change 1995: The Science of Climate Change: Contribution of WG1 to the Second Assessment Report of the Intergovernmental Panel on Climate Change (1996), p. 418.

4. Bert Bolin, *A History of the Science and Politics of Climate Change: The Role of the Intergovernmental Panel on Climate Change* (2007), p. 197.

5. http://climateaudit.org/2010/06/22/kellys-comments/

6. Houghton, Meira Filho, Callander, Harris, Kattenberg & Maskell (ed.), Climate Change 1995: The Science of Climate Change: Contribution of WG1 to the Second Assessment Report of the Intergovernmental Panel on Climate Change (1996), p. 413.
7. ibid., p. 438.
8. Bolin, *A History of the Science and Politics of Climate Change: The Role of the Intergovernmental Panel on Climate Change* (2007), p. 113.
9. Associated Press, 'Chief scientist responsible for global warming' December 1st 1997.
10. Frederick Seitz, 'A Major Deception on Global Warming' in the *Wall Street Journal*, June 12th 1996.
11. S. Fred Singer, Letter to IPCC (Working Group 1) Scientists, undated http://www.his.com/~sepp/Archive/controv/ipcccont/ipccflap.htm
12. IPCC, Report of the Ninth Session of the Intergovernmental Panel on Climate Change, June 29–30th 1993, Appendix G.
13. Paul N. Edwards & Stephen H. Schneider, 'Broad Consensus or "Scientific Cleansing"?' in *Ecofable/Ecoscience* 1:1 (1997), pp. 3–9.
14. ibid.
15. Houghton, Meira Filho, Callander, Harris, Kattenberg & Maskell (ed.), Climate Change 1995: The Science of Climate Change: Contribution of WG1 to the Second Assessment Report of the Intergovernmental Panel on Climate Change (1996), p. 28.
16. ibid., p. 62.
17. ibid., p. 61.
18. Fred Pearce, *The Climate Files: The Battle for the Truth about Global Warming* (2010), p. 44.
19. John R. Christy, Testimony to a House Science, Space and Technology Committee, March 31st 2011, p. 5.
20. Chris Folland email to Michael E. Mann and others, September 22nd 1999.
21. Keith Briffa email to Michael E. Mann and others, September 22nd 1999.
22. Briffa email to Mann and others.
23. Michael E. Mann email to Keith Briffa, September 22nd 1999.
24. Christy, Testimony to a House Science, Space and Technology Committee, March 31st 2011, p. 6 http://science.house.gov/hearing/full-committee-hearing-climate-change
25. Houghton, Ding, Griggs, Noguer, van der Linden, Dai, Maskell, Johnson (eds), *Climate Change 2001: The Scientific Basis* (2001), p. 2.

21
QUIS CUSTODIET?

1. Thomas S. Kuhn, *The Structure of Scientific Revolutions* (1996), p. 140.
2. ibid., p. 96.
3. ibid., p. 64.
4. ibid., p. 47.
5. ibid., p. 68.
6. ibid., p. 94.
7. Stephen McIntyre & Ross McKitrick, 'Corrections to the Mann et al. (1998) Proxy Data Base and Northern Hemispheric Average Temperature Series' in *Energy & Environment* Vol. 14. No. 6 (2003), p. 766.

8. Richard A. Muller, 'Global Warming Bombshell' in *Technology Review*, October 15th 2004.
9. Bert Bolin, *A History of the Science and Politics of Climate Change: The Role of the Intergovernmental Panel on Climate Change* (2007) p. 167.
10. Montford, *The Hockey Stick Illusion: Climategate and the Corruption of Science* (2010), p. 434.
11. David Verado email to Stephen McIntyre, December 17th 2003, reproduced in Michael E. Mann letter to Joe Barton, July 15th 2005.
12. Alan Greenspan, *The Age of Turbulence* (2007), p. 495.
13. P.W. Bridgman, *The Way Things Ares* (1959), p. 56.
14. Roger Scruton, *The Uses of Pessimism* (2010), p. 170.
15. Bridgman, *The Way Things Are* (1959), p. 56.
16. Stephen H. Schneider, *Science As A Contact Sport* (2009), pp. 147–8.
17. Richard A Kerr, 'Draft Report Affirms Human Influence' in *Science*, Vol. 288, 28th April 2000.
18. Phil Jones email to Tom Wigley, October 21st 2004.
19. National Research Council, *Surface Temperature Reconstructions for the Last 2,000 Years* (2006), p. 113.
20. ibid., p. 99.
21. Brumfiel, 'Academy affirms hockey-stick graph' in *Nature*, 441, 1032-1033 & Revkin, 'Panel Supports a Controversial Report on Global Warming' in the *New York Times*.
22. Edward J. Wegman, David W. Scott & Yasmin H. Said, 'Ad Hoc Committee Report on the "Hockey Stick" Global Climate Reconstruction' July 2006, p. 4.
23. Wigley email to Jones.
24. 'Climate change: Is the US Congress bullying experts?' in *Nature*, Vol. 436, No. 7047, July 7th 2005.

22
CLIMATEGATE

1. Bert Bolin, *A History of the Science and Politics of Climate Change: The Role of the Intergovernmental Panel on Climate Change* (2007), p. 208
2. Eric Steig, 'Al Gore's movie' May 10th 2006.
3. J.T. Houghton, 'An Overview of the Intergovernmental Panel on Climate Change (IPCC) and Its Process of Science Assessment' in *Issues in Environmental Science and Technology*, No. 17, Royal Society of Chemistry (2002), p. 6.
4. Phil Jones email to Michael Mann (Mann's response dated May 29th 2008.
5. Todd J. Zinser letter to Senator James Inhofe, February 18th 2011.
6. George Monbiot, 'Pretending the climate email leak isn't a crisis won't make it go away' in the *Guardian*, November 25th 2011.
7. James Rodger Fleming, *The Callendar Effect* (2007), p. 94.
8. Ben Webster, 'Top scientists rally to the defence of the Met Office' in *The Times*, December 10th 2009.
9. House of Commons Science and Technology Committee, *The disclosure of climate data from the Climatic Research Unit at the University of East Anglia Eighth Report of Session 2009–10*, Volume II (2010), Ev 171.

10. Hoggart, 'The sight of another scientist being skewered makes for painful viewing' in the *Guardian*, March 2nd 2010.
11. Quentin Letts, 'Lord Lawson labelled them climate alarmists' in the *Daily Mail*, March 2nd 2010.
12. House of Commons Science and Technology Committee, *The disclosure of climate data from the Climatic Research Unit at the University of East Anglia Eighth Report of Session 2009–10*, Volume II (2010), Q136, Ev 32.

23
STATE OF DENIAL

1. House of Commons Science and Technology Committee, *The disclosure of climate data from the Climatic Research Unit at the University of East Anglia Eighth Report of Session 2009–10*, Volume II (2010), Q 207, Ev 60.
2. Ronald Oxburgh et al., *Report of the International Panel set up by the University of East Anglia to examine the research of the Climatic Research Unit*, (2010), p. 3.
3. *The disclosure of climate data from the Climatic Research Unit at the University of East Anglia Eighth Report of Session 2009–10, Report, together with formal minutes* (2010), para 137.
4. ibid., para 54.
5. Ben Webster, 'Climate-row professor Phil Jones should return to work, say MPs' in *The Times*, March 31st 2010.
6. Michael E. Mann email to Tim Osborn, July 31st 2003.
7. Harold T. Shapiro et al., *Climate change assessments: Review of the processes and procedures of the IPCC* (2010), p. viii.
8. Clive Crook, 'Climategate and the Big Green Lie' in the *Atlantic*, July 14th 2010.

24
TIME'S WINGÈD CHARIOT

1. Joseph A. Schumpeter, *History of Economic Analysis* (1994), p. 53.
2. Kenneth E. Boulding 'The Economics of the coming Spaceship Earth' in Henry Jarrett (ed.), *Environmental Quality in a Growing Economy – Essays from the Sixth RFF Forum* (1966), pp. 12–13.
3. William D. Nordhaus, 'World Dynamics: Measurement without Data' in *The Economic Journal*, Vol. 83 332 (1973), p. 1183.
4. William D. Nordhaus, *The Efficient Use of Energy Resources* (1979), pp. xviii–xix.
5. Nordhaus, *The Efficient Use of Energy Resources* (1979), p. 131.
6. ibid., p. 142.
7. William Cline, 'Scientific Basis for the Greenhouse Effect' in *The Economic Journal*, Vol. 101 407 (1991), p. 913.
8. William D. Nordhaus, 'To Slow or not to Slow: The Economics of the Greenhouse Effect' in *The Economic Journal*, Vol. 101 407 (1991), p. 933.

9.   *Yale Symposium on the Stern Review* (2007), p. 131.
10. David Henderson, 'Economists and Climate Science: A Critique' in *World Economics*, Vol. 10, No. 1 (2009), p. 67.

25
STERN REVIEW

1.   P.T. Bauer, *Reality and Rhetoric – Studies in the Economics of Development* (1984), p. 147 .
2.   Decca Aitkenhead, '"We're the first generation that has had the power to destroy the planet. Ignoring the risk can only be described as reckless"' in the *Guardian*, March 30ᵗʰ 2009.
3.   ibid.
4.   Yale Center for the Study of Globalization, *Yale Symposium on the Stern Review* (2007), p. 8.
5.   Ian Byatt, Ian Castles, David Henderson, Nigel Lawson, Ross McKitrick, Julian Morris, Alan Peacock, Colin Robinson and Robert Skidelsky, 'The Stern Review "Oxonia Papers": A Critique', p. 5.
6.   Nicholas Stern, 'Reply to Byatt et al' in *World Economics*, Vol. 7, No. 2 (April–June 2006), p. 155.
7.   Stern, *The Economics of Climate Change: The Stern Review* (2007), p. xv.
8.   ibid., p. 35.
9.   ibid., p. 54.
10. Partha Dasgupta, 'Commentary: The Stern Review's Economics of Climate Change' in *National Institute Economic Review*, No. 199, January 2007, p. 6.
11. William Nordhaus, *The Stern Review on the Economics of Climate Change* (2007), pp. 14–15.
12. ibid., p. 25.
13. William Nordhaus, *A Question of Balance: Weighing the Options on Global Warming Policies* (2008), p. 167.
14. Stern, *The Economics of Climate Change: The Stern Review* (2007), p. 448.
15. ibid., p. 523.
16. ibid., p. 468.
17. Nigel Lawson, *An Appeal to Reason: A Cool Look at Global Warming* (2008), p. 44.
18. Stern, *The Economics of Climate Change: The Stern Review* (2007), p. 191.
19. ibid., p. 320.
20. ibid., Figs 13.1 & 13.2.

26
SELLING SALVATION

1.   Martin Wolf, 'A compelling case for action to avoid a climatic catastrophe' in the *Financial Times*, November 1ˢᵗ 2006.
2.   Scheherazade Daneshku, 'Change that is costing the Earth' in the *Financial Times*, October 31ˢᵗ 2006.

3.   'What planet are they on?' in the *Daily Mail*, October 31st 2006.
4.   Robert J. Samuelson, 'Greenhouse Guessing' in the *Washington Post*, November 10th 2006.
5.   ibid.
6.   Yale Center for the Study of Globalization, *Yale Symposium on the Stern Review* (2007), p. 83.
7.   *Yale Symposium on the Stern Review* (2007), p. 100.
8.   ibid., p. 73.
9.   ibid., p. 124.
10. Nicholas Stern, *The Economics of Climate Change: The Stern Review* (2007), p. 664.
11. ibid., Table PA. 3.
12. Martin Weitzman, 'On Modelling and Interpreting the Economics of Catastrophic Climate Change' in *The Review of Economics and Statistics*, Vol. XCI, No. 1 (2009), p. 5.
13. Martin Wolf, 'Why obstacles to a deal on climate are mountainous' in the *Financial Times*, July 8th 2008.
14. Nigel Lawson, *An Appeal to Reason: A Cool Look at Global Warming* (2008), p. 90.
15. Bert Bolin, *A History of the Science and Politics of Climate Change: The Role of the Intergovernmental Panel on Climate Change* (2007), p. 96.

27
CUCUMBERS INTO SUNBEAMS

1.   Martin Wolf, *Why Globalization Works* (2004), p. 74.
2.   Donald Mitchell, *A Note on Rising Food Prices*, World Bank Policy Research Working Paper 4682 (July 2008), p. 2.
3.   Paul-Frederik Bach, *The Variability of Wind Power* (2010), p. 32.
4.   Manuel Frondel, Nolan Ritter, Christoph M Schmidt, Colin Vance, *Economic Impacts from the Promotion of Renewable Technologies: The German Experience* (2009), p. 20.
5.   John Constable, *The Green Mirage* (2011), p. 93.
6.   Kathryn Harrison, *The Road Not Taken: Climate Change Policy in Canada and the United States* (August 2006), p. 18.
7.   James M. Inhofe, 'The Science of Climate Change' July 28th 2003.
8.   Maurice Strong, *Where on Earth are We Going?* (2001), p. 219.
9.   Schmidheiny, *Changing Course: A Global Perspective on Development and the Environment* (1992), p. 2.
10. Supreme Court of the United States, Opinion of the Court, 549 US (2007), p. 22.

28
HUGGING HUSKIES

1.   *Hansard*, House of Commons, June 9th 2008, Col. 38.
2.   Department of Energy and Climate Change, *Climate Change Act 2008 Impact Assessment* (March 2009), Box 5.

3. Browne, April 22nd 1999.
4. *Hansard*, House of Lords, January 12th 2012, Col. 280.

29
DANGEROUS CLIMATE CHANGE

1. UNFCCC, 'Time series Annex I – Total CO_2 Emissions without Land Use, Land-Use Change and Forestry (2011).
2. Dieter Helm, 'Forget the Huhne hype about wind power' in *The Times*, February 6th 2012.
3. Peter Schwartz and Doug Randall, *An Abrupt Climate Change Scenario and Its Implications for United States National Security* (October 2003), p. 7.
4. Tony Blair, *A Journey* (2010), p. 557.
5. George W. Bush, 'President Bush Discusses Global Climate Change' June 11th 2001 http://georgewbush-whitehouse.archives.gov/news/releases/2001/06/20010611-2.html
6. Bush, 'President Bush Discusses Global Climate Change'.
7. Swaminathan S. Anklesaria Aiyar, 'Global warming or global cooling' in the *Times of India*, February 27th 2005.
8. Joint science academies' statement: Global response to climate change (June 2005).
9. Václav Klaus, *Blue Planet in Green Shackles* (2008), pp. 108–9.
10. George W. Bush, 'President Bush Participates in Major Economies Meeting on Energy Security and Climate Change' September 28th 2007.

30
BALI

1. Elizabeth Rosenthal & Andrew Revkin, 'Science Panel Calls Global Warming "Unequivocal"' in the *New York Times*, February 3rd 2007.
2. David Adam, 'Kerry blasts Bush for resisting Bali climate goals' guardian.co.uk, December 10th 2007.
3. Catherine Brahic, 'Al Gore tells Bali the inconvenient truth on US', in the *New Scientist*, December 13th 2007.
4. The account and quotes presented here are drawn from a four part video posted on YouTube starting with 'Bali climate summit final plenary / part1'.
5. Dana Perino, Office of the Press Secretary, 'Statement by the Press Secretary' December 15th 2007.

31
SHOWDOWN IN COPENHAGEN

1. Jack Lefley, 'Last 10 years have been warmest on record because of man-made climate change' in the *Daily Mail*, December 16th 2008.
2. http://icecap.us/images/uploads/RarelatewintersnowfallinBrazil.pdf
3. http://www.lasvegassun.com/news/2008/dec/17/rain-snow-moving-las-vegas-valley/
4. *Wall Street Journal*, January 15th 2009.
5. Vicky Pope, 'Scientists must rein in misleading climate change claims' in the *Guardian*, February 11th 2009
6. Ben Webster and Peter Riddell, 'Global warming is not our fault, say most voters in Times poll' in *The Times*, November 14th 2009
7. Gordon Brown, 'PM's speech to the Major Economies Forum' October 19th 2009.
8. Nigel Lawson, 'Copenhagen will fail – and quite right too' in *The Times*, November 23rd 2009.
9. John Vidal and Damian Carrington, 'Ed Miliband attacks Tory climate "saboteurs"' in the *Guardian*, December 3rd 2009.
10. Damian and Suzanne Goldenberg, 'Gordon Brown attacks "flat-earth" climate change sceptics' in the *Guardian*, December 4th 2009.
11. Michelle Grattan, 'We will have climate policy, Abbott says' in *The Age*, December 2nd 2009.
12. Tom Burke, 'The Future of Climate Policy' June 18th 2009.
13. World Council of Churches, 'Churches to Ring the Alarm on Climate Change' November 12th 2009.
14. Rowan Williams, 'Environment Service at Westminster Central Hall' December 5th 2009.
15. Suzanne Goldenberg and Jonathan Watts, 'Kerry's promise of support in US Congress raises hopes for deal' in the *Guardian*, December 7th 2009.
16. Amanda Debard, 'Climate-research furor might not stop US deal – Scientist decries "smear"; lawmakers want answers' in the *Washington Times*, December 5th 2009.
17. Marlowe Hood, 'Copenhagen scientists, negotiators slam "Climategate"' AFP, December 7th 2009.
18. Ben Webster and Murad Ahmed, 'Climate scientists' email was hacked by professionals' in *The Times*, December 7th 2009.
19. Jan M. Olsen, 'UN climate boss to CEOs: play a more serious role' AP, December 11th 2009.
20. Prince of Wales, 'The eyes of the world are upon you' in the *Guardian*, December 15th 2009.
21. Ben Webster, '2010 will be the warmest year on record, predicts the Met Office' in *The Times*, December 11th 2009.
22. Anna Cuenca, 'Maverick trio scoff at the West at climate summit' in AFP, December 16th 2009.
23. 'Violence breaks out at Copenhagen climate protests' in AFP, December 12th 2009.
24. Richard Ingham, 'After gruelling summit, a contested deal emerges on climate' in AFP, December 19th 2009.

32
NEVER AGAIN

1. John Vidal, 'Rich nations accused of Copenhagen "power grab"' in the *Guardian*, December 9th 2009.
2. John Vidal and Dan Milmo, 'Copenhagen: Leaked draft deal widens rift between rich and poor nations' in the *Guardian*, December 9th 2009.
3. Draft Copenhagen Agreement – the 'Danish Text' via http://www.guardian.co.uk/environment/2009/dec/08/copenhagen-climate-change
4. 'China Climate envoy criticises rich nations' AFP, December 10th 2009
5. '"Death of Kyoto would be the death of Africa": AU' AFP, December 15th 2009.
6. *Earth Negotiations Bulletin* Vol. 12 No. 456, December 16th 2009, p. 1
7. 'Battle of the texts looms at UN climate talks' AFP, December 10th 2009.
8. Brian Winter, 'China lashes out at US at climate conference' in *USA Today*, December 12th 2009.
9. '"Developed Countries Have Not Delivered": Chinese Vice Foreign Minister He Yafei on Climate Change' in the *Wall Street Journal* online, December 13th 2009.
10. ibid.
11. ibid.
12. ibid.
13. Jonathan Watts, 'Blair tells world to get moving as time runs short for deal' in the *Guardian*, December 14th 2009.
14. Barack Obama, 'Remarks by the President at the Morning Plenary Session of the United Nations Climate Change Conference' December 18th 2009.
15. Manmohan Singh, Speech to the UNFCCC Plenary, Copenhagen, December 18th 2009
16. Wen Jiabao, Speech to the UNFCCC Plenary, Copenhagen, December 18th 2009.
17. Tobias Rapp, Christian Schwägerl and Gerald Traufetter, 'The Copenhagen Protocol: How China and India Sabotaged the UN Climate Summit' Spiegelonline, May 5th 2010.
18. Mark Lynas, 'How do I know China wrecked the Copenhagen deal? I was in the room' in the *Guardian*, December 22nd 2009.
19. Stephen Collinson, 'Chaos greets new climate pact' AFP, December 18th 2009.
20. ibid.
21. ibid.
22. Charles Babington and Jenifer Loven, 'Obama raced clock, chaos and comedy for climate deal' AP, December 19th 2009.
23. ibid.

33
AFTERMATH

1. David Espo, 'Obama hails 60th Senate vote for health care' AP, December 19th 2009.
2. IISD, *Earth Negotiations Bulletin* Vol.12 No. 459, December 22nd 2009, p. 8.

3. John Vidal and Jonathan Watts, 'Copenhagen closes with weak deal that poor threaten to reject' guardian.co.uk, December 19th 2009.
4. Richard Ingham, 'After gruelling summit, a contested deal emerges on climate' AFP, December 19th 2009.
5. Vidal and Watts, 'Copenhagen closes with weak deal that poor threaten to reject'.
6. Ban Ki-moon, 'Remarks to the UNFCCC COP-15 closing plenary' December 19th 2009.
7. 'Copenhagen climate accord "essential beginning": Ban' AFP, December 19th 2009.
8. UNFCCC, 'Copenhagen Accord' (December 18th 2009), para 2.
9. Nicolas Sarkozy, 'Press conference given by Nicolas Sarkozy after the Copenhagen summit' December 18th 2009.
10. Ed Miliband, 'The road from Copenhagen' in the *Guardian*, December 21st 2009.
11. 'China hits back at Britain in escalating climate talks row' in the *Guardian*, December 21st 2009.
12. Gordon Brown, 'Transcript of the PM's podcast on Copenhagen' December 22nd 2009.
13. Benjamin Sportouch, 'France, China sign aviation, nuclear deals' AFP, December 21st 2009.
14. Simit Bhagat, 'Can't settle for less than Kyoto: PM' in the *Economic Times*, December 20th 2009.
15. John J. Tkacik Jr, 'China's imprints all over Copenhagen talks fiasco' in the *Washington Times*, January 14th 2010.
16. 'Opposition flays govt on climate "accord"' in the *Economic Times*, December 21st 2009.
17. Mary Kissel, 'Climate Change "Quagmire"' in the *Wall Street Journal*, March 10th 2010.
18. Phil Jones email to Mike Lockwood, May 7th 2009.
19. Kevin Rudd, 'The PM's address to the Lowy Institute' in *The Australian*, November 6th 2009.
20. John Howard interview with author, November 28th 2011.
21. 'Brazil points finger at US over climate failure' AFP, December 21st 2009.
22. PTI, 'BASIC countries slam Canada's withdrawal from Kyoto Protocol' February 14th 2012 http://ibnlive.in.com/generalnewsfeed/news/basic-countries-slam-canadas-withdrawal-from-kyoto-protocol/963616.html
23. 'Copenhagen climate accord "essential beginning": Ban'.

34
REFLECTIONS

1. T. Rees Shapiro, 'Stephen H Schneider, climate expert, dies at sixty-five' in the *Washington Post*, July 20th 2010.
2. Thomas S. Kuhn, *The Structure of Scientific Revolutions* (1996), p. 47.
3. The account of and quotes from Stephen H Schneider's 'Climate Change: Is the Science "Settled"' have been transcribed from http://www.youtube.com/watch?v=mmlHbt5jja4
4. Kevin Trenberth email to Michael Mann, October 12th 2009.

5. Kevin Trenberth, 'Communicating Climate Science and Thoughts on Climategate' (January 2011), p. 3.
6. Richard S. Lindzen, 'Resisting climate hysteria' in Quadrant Online (July 26th 2009).
7. Martin Rees & Ralph Cicerone, 'What's happening to the climate is unprecedented' in the *Financial Times*, April 9th 2010.
8. Theodore R. Marmor with the assistance of Jan S. Marmor, *The Politics of Medicare* (1970), p. 1.
9. Thomas McKeown, *The Role of Medicine: Dream, Mirage or Nemesis?* (1979), p. xi.
10. ibid., p. 177.
11. Yohe, G.W., R.D. Lasco, Q.K. Ahmad, N.W. Arnell, S.J. Cohen, C. Hope, A.C. Janetos and R.T. Perez, '2007: Perspectives on climate change and sustainability' in *Climate Change 2007: Impacts, Adaptation and Vulnerability. Contribution of Working Group II to the Fourth Assessment Report of the Intergovernmental Panel on Climate Change*, p. 827.
12. Stephen H. Schneider, 'Stephen Schneider vs Sceptics' goodplanet.org (December 2009).
13. Robert May, 'Under-informed, over here' in the *Guardian*, January 27th 2005.
14. David King, 'Sir David King: IPCC runs against the spirit of science' in the *Daily Telegraph*, February 6th 2010.
15. Conrad Keating, *Smoking Kills: The Revolutionary Life of Richard Doll* (2009), p. ix.
16. ibid., p. 82.
17. ibid., p. 83.
18. ibid., p. 86.
19. ibid., pp. 446–7.
20. Ernst L. Wynder, 'Tobacco and Health: a Review of the History and Suggestions for Public Policy' in *Public Health Reports*, January-February 1988, Vol. 103, No. 1, p. 9.
21. Keating, *Smoking Kills: The Revolutionary Life of Richard Doll* (2009), p. 155.
22. ibid., pp. 93–4.
23. ibid., p. ix.
24. ibid., p. 372.
25. ibid.
26. Kuhn, *The Structure of Scientific Revolutions* (1996), p. 148.
27. Popper, *Conjectures and Refutations* (2002 edition), p. 45.
28. William D. Nordhaus, 'Why the Global Warming Sceptics Are Wrong' in the *New York Review of Books*, March 22nd 2012.
29. Eric Feldman and Ronald Bayer, *Unfiltered: Conflicts over Tobacco Policy and Public Health* (2004), p. 1.
30. David Henderson, 'Governments and Climate Change Issues: The case for rethinking' in *World Economics*, Vol. 8, No. 2 (2007), pp. 183–228.
31. 'IPCC scientist dismisses furore over climate change report' in the *Daily Telegraph*, February 15th 2010.
32. Karl Popper, 'Science: Problems, Aims, Responsibilities' in *The Myth of the Framework*, (ed.) M.A. Notturno, (2006), p. 93.
33. Royal Society, *People and the Planet* (2012), p. 105.
34. ibid., p. 63.

INDEX

ACKNOWLEDGEMENTS

This book could not have been produced without the help of a large number of people. Global warming is a highly contentious subject and a history must be faithful both to evidence and to context. One sets sail in the knowledge of gales and squalls ahead. For this reason, I am grateful to those who agreed to be interviewed for it, particularly those who might not have been especially sympathetic to my point of view. All are identified in the text, so there is no need to list them again here. To each I provided a draft of wherever I quoted or cited what they had told me and gave them the opportunity to correct or alter it.

Global warming spans many disciplines. Rightly or wrongly, I took the view that while I might not know the right answer, I would know the right questions to ask and, one way or another, track down the right people to answer them. I am indebted to everyone who helped me in this – whether spending time answering my questions, providing leads, sourcing material or critically reviewing early drafts – Mark Banfield, Reinaldo Bellinello, Gordon Binder, Robert Bradley, Ian Byatt, Bob Carter, Jayant Chavda, John Constable, John Emsley, Guy Esnouf, Steve Hayward, Mike Hulme, Evelyne Joslain, William Kininmonth, Jürgen Krönig, Theodore Marmor, Julian Morris, Ross McKitrick, Neil O'Brien, Benny Peiser, Silvia Pondal Rios, Jeremy Rabkin, Bob Reinstein (especially for access to unpublished material), Andrew Riley of the Thatcher Foundation, Richard Ritchie, Ian Rowson, Jane Rundle, David Satterthwaite, Peter Snow, Tim Stone, Donald Sturrock, Max Telford, Michael Tooley, Gerald Traufetter, Andrew Turnbull, Harlan Watson and Nick Wood. Ian Tanner generously proofread the first draft.

Much time researching the book was spent in the library of the LSE, a rare institution retaining the belief in open access to knowledge. Photographs bring history alive. Two are sourced from the LSE, which holds the Kibbo Kift Foundation archives. Duke Blackwood and Steve Branch of the Ronald Reagan Presidential Library, John Keller of the William J. Clinton Presidential Library and Polly Nodine at the Jimmy Carter Library provided the photographs of their respective presidents and Patricia Powell in William Reilly's office provided the ones of him. Bridget Gillies of the University of East Anglia, Philip Coupland, Judge Smith facilitated approval of the inter-war years photos and Catherine Sibut-Pinote of UNCTAD provided those of Guevara and Prebisch.

Crispin Odey gave tremendous support and encouragement. Naim Attallah read an early draft of the first chapter – immediately committing to publishing the book. Without such an endorsement, it would not have been written. The rest of the team at Quartet Books showed similar confidence. In Gavin James Bower, I am fortunate to have had as editor a twice-published novelist, who while always respecting the writer's intent and expression, attended to the needs of the text, the result being a better book, and in Grace Pilkington tirelessly working to raise the book's profile.

I first heard David Henderson when he gave the Reith Lecture in the mid 1980s. When some two decades later I first met him, I was somewhat surprised to have him challenge me as to why I didn't believe in the triple bottom line. I soon learnt that a sense of humour is a prerequisite for the fields he was then interested in, one of which was global warming. The urge to write a book on the subject was sparked by conversations with him and the logic of its historical approach was refined and developed as a result of his criticisms and encouragement. The book profited immensely by drawing liberally from the deep well of his knowledge.

Age seems to confirm the veracity of clichés – and the one about a book being a third person in a marriage turns out to be true. This book has occupied half of my marriage to my wife Alice. Throughout she has always been supportive and tolerant of its excessive demands. To her the book is dedicated as some recompense – with love.